STUDY GUIDE

Edward J. Gac
University of Colorado, College of Business and Administration

Rhonda Carlson
Legal Writer

CONTEMPORARY BUSINESS & E-COMMERCE LAW

Fourth Edition

Henry R. Cheeseman

Pearson Education

Upper Saddle River, New Jersey 07458

Acquisitions editor: David Parker
Assistant editor: Ashley Keim
Production editor: Wanda Rockwell
Manufacturer: Courier, (Bookmart Press, Inc.)

Prentice Hall

ISBN 0-13-047385-5
10 9 8 7 6 5 4

CONTEMPORARY BUSINESS AND E-COMMERCE LAW:
Legal, Global, Digital, ad Ethical Environment 4E

DEDICATION

In loving memory of

Mary Kathleen Wilder, 1947-2001

teacher, scholar, writer, mother, and friend,
who made the world a better place

PREFACE

This study guide is a companion publication to *Contemporary Business and E-Commerce Law: Legal, Global, Digital, and Ethical Environment 4E, by Henry R. Cheeseman.* The key ingredient of a study guide is that it should *add value* for the student. By adding value, we mean that the student's learning experience should be enriched, not burdened, by the use of this supplemental material. This enrichment should include more than just opportunities to memorize information bytes for the inevitable mid-term and final exams. The unfortunate consequence of this test bank mentality is that it all too often fosters what we call bathtub learning. Students absorb a lot of material for a computer-graded exam and then flush it afterwards. No thought is given to why the answer was right or wrong. *The only thing that is worse than a wrong answer on an exam is a right answer for the wrong reason.*

It is the development of the underlying right reasoning process that the student should be learning from the study guide. The study guide should give the student guidance, first and foremost, on where to start to study these materials. *What are the right question students should be asking themselves to gain some insights into how law works and how the public policies behind those laws developed?*

This study guides gives the student a *systematic key* to learning the material at hand. The study of law is essentially the study of a special type of logic. Thinking like a lawyer is not just a mystical antidote used by law professors to make their students wonder how their minds work. It is rather a real set of logical and progressively insightful inquiries, which will lead the student to a defensible conclusion. That conclusion, hopefully, will have taken into account the best elements of both sides of the controversy as well as the general welfare of the society within which and under whose auspices the controversy takes place.

Once that process has begun, the study guide provides opportunities for *meaningful self-testing.* That is to say, the testing process should provide an opportunity for the student to interpolate the key issues into an integrated way of thinking about the materials in the chapter. Testing methods such as true/false, multiple choice, and identification are used. They are efficient and reflect the most likely way a student will be actually tested. But even if the testing is to be limited to this format, the answers to these questions should be provided as they are in this volume to give some insight into the issues being tested. *All of the answers to the objective questions are explained in a narrative format rather than just the correct letter, true/false answer, or page reference to the text.*

To gain added value, the self-testing process includes multi-issue essay questions that will give the student an opportunity to see how issues fit into the big picture of the law. Very few issues studied in modern business law today can readily stand alone. A real appreciation of the integration of those issues means just that—the problem at hand most likely has consequences in a number of related areas. For example, in the area of consumer law, if a product injures a person, we recommend that students ask a series of sequential and related questions:

1. What were the basic terms of the contract between the buyer and the seller?
2. Did this contract contain any express or implied promises of warranty?
3. Were there any attempts to impose limitations on those warranties? If so, do these limitations meet the requirements of both substantive and procedural due process?
4. What are the possible noncontractual avenues that may be available to the consumer? These would include the remedies afforded by traditional tort law and/or the more recent remedies found in the law of strict or products liability?
5. In addition to any contract and/or tort law remedies, are there any relevant statutory protections available to the consumer at either the local or national level?

6. Finally, what is your (the student's) overall view of the proper equities in this case? Remember, first and foremost, the duty of the law and its agents is to reach an equitable result whenever possible rather than just a technically correct one.

These are the objectives for which we wrote the student study guide, a philosophy that attempts to balance between having neither too little nor too much material, but rather the right materials that will give students a sense of guidance and direction towards seeing the big picture. It is not intended to be a substitute for proper advice from a licensed legal professional, but rather to give students a greater overall appreciation of how the law works. An analogy that we raise in our Business Law course outline goes along these lines—we study music appreciation, not to play the instruments but rather to appreciate the value of the harmony they create. So it is with law. As business students, we study law not necessarily to practice law, but to appreciate its importance in every business and personal decision we make.

With that extended introduction to the general objectives of the book, the individual components of the text are outlined on the next few pages:

I. The Overview:

The overview is designed to introduce the student to the big picture of what is going on in this area of the law covered by each chapter. Wherever possible, we tried to interject any historical, economic, political, or policy references that might help the student bring this big picture into focus. For example, Professor Cheeseman properly points out that the word tort is derived from the French word for wrong. We might add that the French word's Greco-Roman origins, in turn, are derived from the word *torquere* meaning to be twisted. When you think of what has happened to you if you are at the receiving end of the tort, you have been twisted in one way or another! These overviews are intended to be running essays of personalized commentary on various legal issues of the day that will help students understand the public policy undercurrents behind the laws reviewed in each chapter.

II. Hypothetical Multi-Issue Essay Questions and
VI. Answers to Essay Questions:

We confess to having a very strong prejudice in favor of essay questions. We believe we need to give students many more opportunities to practice, develop, and hone their writing skills. This is particularly true of the law teaching profession. We have a long and distinguished history of education based on effective communication skills coupled with the substantive knowledge of law. To forego the use of those precious skills is to waste one of the key patrimonies of our profession.

The questions are purposely designed to be multi-issue, fairly complex, and challenging. Yet they are not so oblique or obfuscated as to cause the student to despair in frustration. They are meant to be an integrative exercise that brings together the key objectives of the materials covered in each chapter. In addition, they are purposely designed to provide a sense of levity and humor to the entire process. Many of them are patterned after real-life cases or situations, although all the names have been changed. Any coincidental use of real persons, living or dead, is strictly unintentional.

III. Outline:

The outline serves as an outline of materials in the chapter and reflects the content of the accompanying PowerPoint slides.

IV. Objective Questions and
V. Answers to Objective Questions:

There are thirty objective questions for every chapter in the book: ten are key term identification questions, ten are true/false, and ten are multiple choice. We use narrative descriptions of the underlying reasoning and logic used for the particular answers is given for all the questions. These materials are designed to help students study the law in a new and innovative way by going beyond the official answer. They are designed to help students prepare for the sort of testing that is widely used in college classes and on professional qualification examinations. New questions reflecting the emphasis of e-commerce in this text are included.

We also strongly suggest that you regularly check out all of the excellent material on the Cheeseman website at http://www.prenhall.com/cheeseman This material is designed to help both professors and students understand the current context of business law.

Finally, the single most constant source of satisfaction for us personally in writing this text continues to be the encouragement of our students. As teachers, we must remind ourselves that we are lucky to hold such a position of trust and service. Writing a text such as this is one way to express our personal gratitude for being in such a fortunate position. We hope you find this book of value both as a student and as a matter of personal edification.

Edward J. Gac, J.D.
Associate Professor of Business Law and Taxation
University of Colorado Leeds School of Business
Visiting Associate Professor of Tax Policy
University of Colorado School of Law; Economics Institute
Edward.Gac@Colorado.edu

Rhonda Carlson, J.D., M.L.S.
Legal Writer
Blkcocker@aol.com

April 2002

ACKNOWLEDGMENTS

A work of this scope is not the product of any one person's efforts. Many people have had an indispensable hand in helping this book become a reality.

First and foremost, we are grateful to Professor Henry R. Cheeseman for his ongoing faith and encouragement of our work. His excellent text is the constant guidepost and inspiration for this entire project.

We believe we really had a dream team of editorial support for this project. David Parker, Acquisitions Editor for General Business and Business Publishing, our new editor, has been a real pleasure to work with. We also want to acknowledge the support of Ashley Keim, editorial assistant, who always went out of her way to help us quickly and efficiently. We are very grateful to both of them, and we expect them to have successful careers at Prentice Hall.

As with any such work, there may be points of disagreement or error in the text. In spite of all the efforts listed above, any such errors are ours alone.

Chapter 1
LEGAL, BUSINESS, AND E-COMMERCE ENVIRONMENT

Never doubt that a small group of thoughtful committed citizens can change the world. Indeed it's the only thing that ever has.

Margaret Mead

I. Overview

One of the first objectives of this opening chapter centers illustrates attempts by great thinkers to answer the unanswerable. How law and ethics are defined by any one individual is ultimately related to his or her personal view of the larger global issues including morality, religion, and philosophy as well as the social sciences. When painting the picture entitled "What is law?," the broadest brush possible must be used. The interweaving of all these disciplines into one area of study, generically called jurisprudence, is what makes law study so formidable, complex, and perplexing. By the same token, those difficulties are what make it so challenging and stimulating. The study of law must start with open channels for new ideas not only from the mind, but also from the heart. Law and its various mechanisms is not just a set of rules of conduct but rather the end product of how society has decided to help or hurt its members. It is no wonder that the fascination with law and law studies is at its zenith today. Students have come to appreciate that law represents the embodiment of the social environment we all live and work in. By analogy, compare the legal environment to the water in which the life of the sea takes place. Big fish may eat little fish and so forth, but if the water itself is contaminated, all its inhabitants suffer. So too must the law be constantly examined for its societal safety. Law must be clean and based on solid moral underpinnings and proper ethics. Where it fails, the legal environment may become polluted.

The categorization of schools of jurisprudence starts with a comparison of the old and the new. The traditional schools of thought find their roots in ancient history. The Natural School of Jurisprudence is not only a school of thought vis-à-vis law but also finds proponents throughout many religions of the world. It is still considered one of the foundations of legal thinking today and is espoused by many, including Justice Clarence Thomas. Friedrich Karl von Savigny, who believed law was a product of a nation's history, was the founder of the Historical School of Jurisprudence. Advocates of a more proactive role for law may find comfort in the Sociological School of jurisprudence. Members of that school see law as a tool to be used in social engineering. Compare that philosophy with the views proffered by members of the Command School of Jurisprudence. Advocates of this school are usually members of the ruling class or royalty. Absolute monarchies are an anathema to participatory democracy because such forms of government are really only dictatorships-by-birth.

In spite of the divergent views espoused by the traditional schools of jurisprudence, the schools have a common denominator. They are all rooted on some sort of morality-based foundation. The modern schools of jurisprudence have stepped away from this commonality by finding justification for their points of view in economics, political science, or sociological agendas. The efficacy of these newer theories is untested by the litmus test of time and continues to be a source of spirited and sometimes raucous debate.

One interesting aspect of all schools of jurisprudence is that their proponents try to establish norms of ethical behavior. Ethics is derived from the Latin word *ethicus*, meaning moral. In that sense all these systems represent ways of finding moral foundations upon which a system of law can be built. These moral systems have been identified by a number of notable scholars ranging from Karl Marx, John Stuart Mill, and Immanuel Kant to John Locke. In all these systems, guideposts for behavior are set forth. If these guideposts are universally accepted, the odds are very high that they will no longer be "advisory" but rather "required by law." The process by which moral-based ethical behavior is first *desired*, then *expected*, and finally *mandated* is really the evolution of law. Hopefully with each new generation, the law will evolve in a positive sense, which will best balance the legitimate interests of all parties concerned, both public and private.

This chapter also introduces students to e-commerce and Internet law. Many changes have occurred as a result of the development of the Internet. How is an Internet contract formed? What about signatures and privacy? What criminal laws apply to the Internet? This chapter and this book address the endless list of questions pertaining to Internet technology.

Your role as a student of law is to try to develop sensitivity to how complicated the administration of law is in our society today. This course is not designed to act as a substitute for professional law school training. It is designed to help you appreciate the value and importance of law and ethics in both your personal and business behavior. The law is too complex and perplexing, but imagine a world without it. Imagine a world of Cimmerian darkness—a world described by Homer in the Eighth Century B.C. as a region of perpetual gloom.

II. Hypothetical Multi-Issue Essay Question

Ms. Jane Doe was one of the first executives hired by the Mega-Widgets Corporation. She is a recent graduate of the I.O.U. College of Business and has a degree in Artificial Accounting Intelligence. Her relationship with her supervisor, Mr. Leer, started on good terms, but it soon became clear that Mr. Leer sought some kind of personal relationship with her. He began bombarding her with unwelcome invitations for drinks, lunch, dinner, breakfasts, and trysts. Ms. Doe made it clear she was not interested in any personal relationship with her married supervisor.

When Ms. Doe rejected Mr. Leer's advances, he became belligerent and decided to "get even" by downgrading Ms. Doe's job performance ratings and holding her up to scorn and obloquy in front of her coworkers. Ms. Doe suffered emotional and physical setbacks and took prolonged sick leave. She finally quit.

During the period of Ms. Doe's employment, Mega-Widgets had a published anti-sexual harassment policy. It called for a reporting of all incidents directly to one's immediate supervisor. Ms. Doe did not follow the steps spelled out in the policy, and Mega-Widget got wind of all the problems only after she quit. Should Mega-Widgets now be held liable for the actions of Mr. Leer?

III. Outline

What Is Law?

Law consists of rules that regulate the conduct of individuals, businesses, and other organizations within society. It is intended to protect persons and their property from unwanted interference from others.

Functions of Law

- Facilitating orderly change
- Facilitating planning
- Keeping the peace
- Maintaining the status quo
- Maximizing individual freedom
- Promoting social justice
- Providing a basis for compromise
- Shaping moral standards

Qualities of the Law

- Fairness
  - The American legal system is one of the most comprehensive, fair, and democratic systems of law ever developed and enforced
- Flexibility
  - The American legal system is generally responsive to cultural, technological, economic, and social changes

Schools of Jurisprudential Thought

- Analytical school—Believes law is shaped by logic
- Command school—Believes law is a set of rules developed and enforced by the ruling party
- Critical legal studies school—Believes legal rules are unnecessary and that legal disputes should be solved using rules of fairness
- Feminist school—Believes a female perspective should be considered when laws are developed, interpreted, and applied
- Historical school—Believes law is an aggregate of social traditions and customs
- Law and economics school—Believes promoting market efficiency should be the central concern of legal decision making
- Sociological school—Believes law is a means of achieving and advancing certain sociological goals
- Natural school—Believes law is based on what is "correct"

English Common Law

- Common law
  - Developed by judges who issued their opinions when deciding a case. The principles became precedent for later judges deciding similar cases.
- Law court
  - A court that developed and administered a uniform set of laws decreed by the kings and queens after William the Conqueror. Legal procedure was emphasized over merits.
- Court of Chancery (equity court)
  - Court that granted relief based on fairness
- Merchant court
  - Courts established to administer the "law of merchants"

Sources of Law
- Codified law
- Constitutions
- Executive orders
- Judicial decisions
- Regulations and administrative orders
- Treaties

Constitutions
- The U.S. Constitution established the structure of the federal government
  - Legislative
  - Executive
  - Judicial
    - The constitution and treaties take precedence over all other laws

Treaties
- A treaty is a compact made between two or more nations

Codified Law
- Statute
  - Written law enacted by the legislative branch of federal and state governments that establishes certain courses of conduct that must be adhered to by covered parties
- Ordinances
  - Laws enacted by local government bodies such as cities and municipalities, counties, school districts, and water districts

Administrative Agency Rules and Regulations
- The legislative and executive branches of federal and state governments are empowered to establish administrative agencies to enforce and interpret statutes enacted by Congress and state legislatures
- Many agencies regulate business

Executive Orders
- An executive order is an order issued by a member of the executive branch of government

Judicial Decisions
- A judicial decision is a decision about an individual lawsuit issued by federal and state courts

Doctrine of Stare Decisis
- Based on the common law tradition, past court decisions become precedent for deciding future cases
- Precedent is a rule of law established in a court decision. Lower courts must follow the precedent established by higher courts
- Stare decisis—Latin for "to stand by the decision"—means adherence to precedent

Critical Legal Thinking
- Critical legal thinking is the process of specifying the issue presented by a case, identifying the key facts in the case and the applicable law, and then applying the law to the facts to come to a conclusion that answers the issue presented

IV. Objective Questions

Terms:

1. That which must be obeyed and followed by citizens subject to its sanction or legal consequences; a body of rules of action or conduct prescribed by controlling authority, and having binding legal force is called ______________.

2. A follower of the Sociological School of jurisprudence who believes that the purpose of law is to shape social behavior is known as a(n) ______________.

3. The Natural School of Law adheres to the position that law should be based on what is correct. This is also referred to as a ______________ ______________ of law.

4. "Battered woman's syndrome," and the "reasonable woman standard" are two legal doctrines which are evolving out of a new school of jurisprudential thought known as ______________ ______________ ______________.

5. The famous case of *Brown v. Board of Education* overturned the old ______________ ______________ ______________ doctrine that condoned separate school facilities based on race.

6. The school of jurisprudential thought that believes promoting market efficiency should be the central concern of legal decision making is known as the ______________ ______________ ______________ school.

7. Because there were many unfair results due to the limited remedies allowed in the early English Law Courts, a second set of courts called the ______________ ______________ were established in England.

8. The highest source of law found in the U.S. is found in the ______________.

9. Under the U.S. Constitution, with the consent of the Senate, the President may enter into ______________ with other nations.

10. Adherence to precedent (a rule of law established in court decisions) is known by the Latin term ______________ ______________, meaning to stand by the decision.

True/False:

1. ____ The Law and Economics school professes that market efficiency should be the ultimate goal of legal decision making.

2. ____ The Court of Chancery provided for only limited law remedies in the early English court system.

3. ____ The Critical Legal Studies movement is widely accepted and very influential.

4. ____ Law includes constitutions, statutes, court decisions, and administrative regulations.

5. ____ One function of law is to limit individual freedoms.

6. ____ Our jury system requires the same results to occur in the interpretation of the same criminal statute.

7. ____ The Analytical School of Jurisprudence maintains that the law is an aggregate of social traditions and customs that have developed over the centuries.

8. ____ Rules of the Law Merchant were based on common trade practices and usage.

9. ____ Business law students are expected to be computer literate.

10. ____ Presidential executive orders are expressly provided for in the U.S. Constitution.

Multiple Choice:

1. Which of the following is not a function of law?
   A. Facilitating planning.
   B. Maintaining the status quo.
   C. Shaping moral standards.
   D. All of the above are functions of law.

2. Which of the following is not a goal or belief of the Critical Legal Studies movement?
   A. Legal rules are used by the powerful to maintain the status quo.
   B. Subjective decision-making by judges should be permitted.
   C. Conservative, not liberal, legal theorists are responsible for the unfairness of the legal system.
   D. All of the above are goals or beliefs of the Critical Legal Studies movement.

3. The famous U.S. Supreme Court decision of *Roe v. Wade* gave women the constitutional right to:
   A. Engage in military combat.
   B. Use the defense of battered woman syndrome in criminal uses.
   C. An abortion.
   D. Job security for pregnant women.

4. Documents such as the U.S. Constitution, the Magna Carta, and the United Nations Charter all reflect, at least in part, the theories espoused by proponents of the:
   A. The Historical School of jurisprudence.
   B. The Law and Economic School of jurisprudence.
   C. The Command School of jurisprudence.
   D. The Natural Law School of jurisprudence.

5. Judge Richard Posner is the unofficial head of the following school of jurisprudential thought:
   A. Critical Legal Studies School.
   B. The Law and Economics School.
   C. The Command School.
   D. The Analytical School.

6. Tilly Target's significant other beat her until she could no longer tolerate it. Last week, she proceeded to remove a key part of her boyfriend's anatomy, which enables him to sing in a boy's choir. Ms. Target's best defense would be found in which of the following schools of jurisprudential thought?
   A. The Command School.
   B. The Law and Economics School.
   C. The Feminist Legal Theory School.
   D. The Sociological School.

7. The Merchant Court was first established as a separate set of courts in:
   A. England.
   B. U.S.A.
   C. New York City.
   D. San Francisco.

8. The Hindu Law called *dharmasastra* in San Skrit, has evolved in the Anglo-Hindu system now used in:
   A. North Yemen.
   B. North Korea.
   C. Indochina.
   D. India.

9. Islamic law, also known as *Sharia*, is the only law of:
   A. Sub-Saharan Africa.
   B. Spain.
   C. Turkey.
   D. Saudi Arabia.

10. Which of the following was not a branch of the early English system of courts?
   A. Small Claims Court.
   B. Law Courts.
   C. Chancery Courts.
   D. Merchant Courts.

V. Answers to Objective Questions:

Terms:

1. *Law.* This is the end product of an evolutionary process which ultimately reflects how a society decides to both help and punish its members.
2. *Realist.* This school of jurisprudence advocates the use of law as an instrument for change and often uses social engineering. Affirmative action principles are reflective of this sort of thinking.
3. *Moral theory.* This is still one of the most important schools of jurisprudence and was espoused by Justice Clarence Thomas in his confirmation hearings to the U.S. Supreme Court.
4. *Feminist Legal Theory.* This school of thought advocates a woman's perspective on a number of key legal issues.
5. *Separate but equal.* This case illustrates the need to have the law be flexible based on changes in our nation's cultural, technological, social, and economic environment. In this case, the U.S. Supreme Court held that "separate but equal" is inherently discriminatory and unconstitutional.

6. *Law and Economics.* This school of thought was founded at the University of Chicago and has gained many high profile advocates from both the law profession, such as Judge Richard Posner, and the economics profession, such as Milton Friedman.
7. *Chancery Courts.* These were the first equity-based courts and were headed by the Lord Chancellor. The early appointees to this position were invariably clerics who were to represent the "King's conscience."
8. *Constitution.* This is the supreme law of the land and any federal, state, or local law that is found to be in conflict with it is unconstitutional and therefore unenforceable.
9. *Treaties.* Once ratified, treaties become part of the supreme law of the land. Only the federal government may enter into treaties with other nations.
10. *Stare decisis.* This is the first of many Latin terms you will be exposed to in your study of law.

True/False:

1. True. Judge Posner is considered the unofficial head of this school of jurisprudence.
2. False. It was the limitation on remedies in the law court that led to the creation of the equity-based Court of Chancery.
3. False. The Critical Legal Studies movement has not been widely adopted. Although some law schools teach and study this theory, it is not nearly as widely recognized as the Law and Economics school.
4. True. All of these are sources of law, although much debate revolves around the notion of judge-made law.
5. False. Maximizing individual freedom within the bounds of the needs of the larger society is one of the hallmarks of the U.S. Constitution.
6. False. As illustrated in the text accompanying the case of *Standefer v. United States,* different juries may reach different results under any criminal statute.
7. False. This definition applies to the Historical School of jurisprudence.
8. True. These rules developed during the Middle Ages among merchants who traveled about England and Europe.
9. True. Universities and colleges are now rated on how well they are connected to computer technology. The new generation of students studies on-line, as well as shopping and even dating on-line.
10. False. These powers are derived express delegations from the legislative branch and otherwise implied from the U.S. Constitution.

Multiple Choice:

1. D. The law is designed to foster all of these functions.
2. D. Advocates of the Critical Studies movement believe that both conservative and liberal thinkers are responsible for the unfair and inequitable state of the legal system.
3. C. *Roe v. Wade* gave women a constitutional right to an abortion. This issue remains one of the most inflammatory controversies in the land.
4. D. Elements of most of these schools could arguably be found in all these documents, but the Natural Law School has long been held as the most important contributor to the underlying philosophies of these key historical documents.
5. B. The Law and Economics School was founded at the University of Chicago where Judge Posner taught before being appointed to the Federal Court of Appeals.

6. C. Although Ms. Target may have taken "command" of the situation and may now get rich on the book and movie rights to her story, the most likely source of sympathy for her defense will be found in the Feminist Legal Theory of the "battered woman syndrome."
7. A. The rules of the Law Merchant were first administered as a separate court in England. That court was absorbed in the regular law court system in England in the early 1900s.
8. D. Most of the world's population of Hindus live in India, a country with a population of over 900 million people. Anglo-Hindu law is a combination of Hindu law and English common law.
9. D. In most Islamic countries, the *Sharia* forms the basis of family law but coexists with other laws, except for Saudi Arabia where it is the only law.
10. A. Small claims courts are a more modern development designed to afford quick inexpensive access to the courts where the issues in dispute are relatively small in value.

VI. Answers to Essay Question

There has always been a social norm both in and out of the work place that personal relationships should be just that--a matter of personal choice. Where there is any sort of coercion, duress, or physical or emotional abuse in the process of cultivating a personal relationship, many social norms are violated. Personal relationships are not barred *per se* in the workplace. To do so would create constitutional problem with regard to freedom of association. However, where the relationship sought is in violation of the norms discussed above, there is a possible violation of a combination of religious, peer group, ethical, public policy, and legislative prohibitions.

In this case, it appears that a *prima facie* harassment case on the basis of sex can be brought against Mr. Leer. Under Title 29 of the Code of Federal Regulations, Section 1604.11, harassment on the basis of sex is a violation of Sec. 703 of Title VII. Unwelcome sexual advances, and requests of a sexual nature constitute sexual harassment when:

(1) Submission to such a conduct is made either explicitly or implicitly a term or condition of an individual's employment.
(2) Submission to or rejection of such conduct by an individual is used as basis for employment decisions affecting such individuals.
(3) Such conduct has the purpose or effect of unreasonably interfering with an individual's work performance or intimidating, hostile, or offensive working environment.

All of the schools of ethics and jurisprudence would condemn this sort of behavior to one degree or another. The Natural Law School has traditionally been less protective of women's rights, but even there, no one can really defend this sort of obnoxious attitude toward women. Unfortunately, in the evolution of law, this is still a very large issue. In 1987, 42 percent of all women (and 14 percent of all men) surveyed by the U.S. Merit System Protection Board reported that they experienced some sort of uninvited and unwanted sexual attention on the job.

The critical issue here is that where there is a showing of an illegal act of sexual harassment by an employee, should the employer be held accountable in addition to the individual? Title 29 of the Code of Federal Regulations, Section 1604.11(c) goes on to say:

> Applying general Title IV principles, employers, (and) employment agents, ...are responsible for (their) acts and those of (their) agents and supervisory employees with respect to sexual harassment...

Given this standard, it appears that Mega-Widgets Corporation may not have done enough to protect Ms. Doe. Their reporting system assumed that the immediate supervisor is not a harasser. The policy gives the supervisor the opportunity not only to cut off any complaints but also to aggravate the situation. For an employer to be shielded from the wrongful acts of its employees, it must show that its anti-sexual harassment policy was *effective*. Here it was not, and it would be equitable and proper to hold Mega-Widgets liable.

Chapter 2
JUDICIAL AND ALTERNATIVE DISPUTE RESOLUTION

He that wrestles with us strengthens our nerves and sharpens our skill. Our antagonist is our helper.

Edmund Burke

I. Overview

One of the most prevalent myths in the folklore of law and lawyers is found in the dramatization of trials. In the world of pop culture, no one knows who really did it until the end when a surprise witness shows up to "finger" the bad guy. In more modern versions, the attorney first has an illicit affair with the client and then proceeds to get her acquitted. Regardless of the outcome, the process is always full of glamour and intrigue.

Welcome to Real World 101! A trial rarely resembles the goings on found in the entertainment media. Trials are long, tedious, emotionally and financially draining processes for all parties concerned. In many ways, a trial represents a failure by the parties to reach some sort of satisfactory solution of the issue beforehand. Rarely do the parties actually want to go through the machinations of being led through a labyrinth of pleadings, motions, and the like, feeling all the while totally dependent on the sometimes questionable competence of their attorneys. Unlike the make-believe world of entertainment, the job of an attorney is to keep his or her client out of court. The attorney's professional advice should anticipate and resolve potential legal problems before, rather than after, the fact.

Assume a dispute does arise which cannot be avoided. The first step taken may not be at the courthouse door. Because of all the frustrations associated with the trial process, one of the fastest growing arenas of law-related activities is found in alternative dispute resolution mechanisms. The more traditional methods are arbitration and mediation. In both cases, a disinterested third party is called in to assist the disputing parties. With arbitration, the parties agree to be bound by the arbitrator. In mediation, the third party acts to bring the parties together but cannot bind them to an agreement. Newer mechanisms include judicial referees, fact-finders, and ombudsmen. In addition, most governmental agencies have internal administrative remedies available for resolution of problems persons may have with the agency. If none of these out-of-court methods are feasible, the court trial process may be the only option.

II. Hypothetical Multi-issue Essay Question

Mr. Cav E. At-Emptor is fed up. He has suffered through the ownership of a model XXX Luxmobile auto since he bought it new from Loud and Obnoxious Motors nine months ago.. The car has spent more time in the repair shop than on the road and is now out of warranty. Mr. At-Emptor still owes over four years worth of payments on the car and has decided he is tired of paying for the car twice--once to buy it and once again to repair it. His sales contract has a clause allowing for arbitration of disputes on a voluntary basis through an automobile industry funded panel. Mr. At-Emptor decides not to take advantage of this procedure. He cites the "lemon law" and seeks a full refund instead. Assuming this law requires the consumer to use all non-court measures first, what can Mr. At-Emptor do?

III. Outline

State Court Systems
- Limited-jurisdiction trial courts
  - Courts that hear matters of specialized or limited nature
  - Small claims courts are a good example of this
- General-jurisdiction trial court
  - Courts that hear cases of a general nature that are not within the jurisdiction of limited-jurisdiction trial courts
- Intermediate appellate courts
  - Courts that hear appeals from trial courts
- Highest state court
  - The highest court in a state system

Sample State Court System

U.S. Supreme Court
- Created by Article III of the US Constitution
- Highest court in the US
- Located in Washington D.C.
- Composed of nine justices who are nominated by the President and confirmed by the Senate
- Types of decisions:
  - Unanimous
  - Majority
  - Plurality
  - Tie

Special Federal Courts
- US Tax Court
- US Claims Court
- US Court of International Trade
- US Bankruptcy Court

US Courts of Appeals
- These are the federal court system's intermediate appellate courts
- There are 13 courts of appeals

US District Courts
- These are the federal court system's trial courts of general jurisdiction
- Presently, there are 96 district courts

Federal Court System

Jurisdiction of Federal and State Courts

Anatomy of a Jury Trial
- Jury selection
- Opening statement
- Plaintiff's case
- Defendant's case
- Rebuttal and rejoinder
- Closing arguments
- Jury instructions
- Jury deliberations
- Verdict
- Entry of judgment

E-Filings in Court

The Appeal
- In a civil case, either party can appeal the trial court's decision once a final judgment is entered
- In a criminal case, only the defendant can appeal
- An appellate court will reverse a lower court decision if it finds an error of law in the record
  - It will generally not reverse a finding of fact

Types of Alternative Dispute Resolution
- Arbitration
- Conciliation
- Fact finding
- Judicial referee
- Mediation
- Minitrial

Arbitration
- Arbitration is a form of ADR in which parties choose an impartial third party to hear and decide the dispute
- Many contracts require that disputes arising out of the contract be submitted to arbitration

Mediation and Conciliation
- Mediation
  - A form of ADR in which the parties choose a neutral third party to act as the mediator of the dispute

Conciliation

A form of mediation in which the parties choose an interested third party to act as the mediator

Minitrial

A minitrial is a session, usually lasting a day or less, in which the lawyers for each side present their cases to representatives of each party who have authority to settle the dispute

Fact-Finding

Fact-finding is a process where the parties hire a neutral person to investigate the dispute

The fact-finder reports his findings to the adversaries and may recommend a basis for settlement

Judicial Referee

IV. Objective Questions

Terms:

1. The process of bringing, maintaining, and defending a lawsuit is called ________________.

2. An arrangement in which the lawyer receives a percentage of the amount recovered from the client upon winning or settling the case is called a ________________ ________________ ________________.

3. A plaintiff must have some stake in the outcome of the lawsuit to have ________________ ________________ ________________.

4. An electronic filing is called an ________________ ________________.

5. A summons served within the territorial boundaries of a state is called ________________ ________________ ________________.

6. A statute that extends a state's jurisdiction to nonresidents who were not served a summons within the state is called a ________________ ________________ ________________.

7. The party accusing the defendant of a legal wrong is called the ________________.

8. In addition to a defendant's answer, a defendant may cite why her action was justified or that the statute of limitations has run. This additional statement is called a(n) ________________ ________________.

9. The act of other parties taking an interest and becoming involved in a lawsuit is called ________________.

10. A pretrial motion that asserts that there are no factual disputes to be decided by the jury and that the judge should apply the relevant law to the undisputed facts to decide the case is called a ________________ ________________.

True/False:

1. ____ A corporation is only subject to personal jurisdiction in the state in which it is incorporated.

2. ____ State long-arm statutes require that a nonresident have at least minimum contact with the state.

3. ____ Courts will freely grant requests for change of venue if one of the parties makes such a request.

4. ____ An answer is a defendant's response to a plaintiff's complaint.

5. ____ A major purpose for pretrial hearings is to promote settlement by the parties.

6. ____ To be granted a jury trial, both parties must agree to and request a jury trial.

7. ____ In a criminal case, only the defendant, and not the prosecutor, can appeal a case after the trial court's final judgment.

8. ____ Generally, courts are inclined to uphold arbitration decisions rather than overturn them.

9. ____ In mediation and conciliation, the third party renders a decision.

10. ____ Whether an issue is subject to arbitration is a matter for the arbitrator to decide.

Multiple Choice:

1. John and Jane have a car accident with Harry. Harry is a friend of John and Jane and decides not to sue. However Harry's girlfriend, Helga, wants to sue John and Jane. Helga cannot sue because she does not have:
   A. Standing to sue.
   B. Personal jurisdiction.
   C. Venue.
   D. Subject matter jurisdiction.

2. A plaintiff, by filing a lawsuit, automatically becomes subject to this. However, the defendant must be served a summons to become subject to:
   A. Standing.
   B. Personal jurisdiction.
   C. Venue.
   D. Subject matter jurisdiction.

3. State Zeon wishes to extend its jurisdiction over Defendant Darin through use of its long arm statute. Under what circumstances may Zeon do this?
   A. If Darin committed a tort in Zeon.
   B. If Darin made a contract in Zeon and then breached it.
   C. If Darin transacted business in Zeon that caused injury to another person.
   D. All of the above.

4. Plaintiff Pam obtains a judgment in state Zeon against Defendant Debbie, who lives in state Yalut. Pam tries to collect the judgment by attaching Debbie's property in Yalut. This is known as:
   A. *In rem* jurisdiction.
   B. A long-arm statute.
   C. *Quasi in rem* jurisdiction.
   D. None of the above.

5. Plaintiff Paul decides to sue Defendant Dan. The suit involves a car accident between Paul and Dan. Paul must:
   A. File a reply.
   B. File an answer.
   C. File a complaint.
   D. None of the above.

6. Assume the same facts as in question 5. Once Paul initiates the suit by filing the first document, Dan must:
   A. File a reply.
   B. File an answer.
   C. File a complaint.
   D. None of the above.

7. Assume the same facts as in question 5. Dan now decides the car accident was really Paul's fault. Paul has already initiated his lawsuit. What must Dan do to assert his claim against Paul?
   A. File a reply.
   B. File a cross-complaint.
   C. File an answer.
   D. None of the above.

8. Assume the same facts as in question 7. After Dan asserts his claim against Paul, what must Paul do?
   A. File a reply.
   B. File a cross-complaint.
   C. File an answer.
   D. None of the above.

9. Assume the same facts as in question 8. Paul believes Dan's claims are ridiculous. Paul enters a motion stating that there are no factual disputes, but only a disagreement over the relevant traffic laws. Therefore, Paul asks that the judge apply the law and decide the case. Paul's motion is a motion for:
   A. Discovery.
   B. Intervention.
   C. Summary judgment.
   D. None of the above.

10. Assume the same facts as in question 9. Now assume that before Paul initiates the suit in court, the two decide to resolve their dispute outside of court since litigation is so costly. Paul and Dan agree to have a neutral third party decide the dispute and render a decision. Paul and Dan have elected to resolve the dispute by:
    A. Mediation.
    B. Minitrial.
    C. Judicial referee.
    D. Arbitration.

V. Answers to Objective Questions:

Terms:

1. *Litigation.* Most disputes are resolved by non-litigation processes.
2. *Contingency fee arrangement.* There continues to be much controversy over this form of fee arrangement.
3. *Standing to sue.* An interest in a lawsuit must be more than just secondary or tangential; standing to sue requires a legally recognized stake in the outcome.
4. *E-filing.* E-filings include using CD-ROMs for briefs, scanning evidence and documents into a computer for storage and retrieval, and e-mailing correspondence and documents to the court and opposing counsel.
5. *Service of process.* Process servers have a high-risk job due to the emotionally charged nature of many court proceedings.
6. *Long-arm statute.* Without such statutes, state boundaries could be used as moats to avoid legal proceedings.
7. *Plaintiff.* This party carries the initial burden of proving the alleged wrong.
8. *Affirmative defense.* An affirmative defense will, in effect, add an additional burden of proof to the opposing party's case.
9. *Intervention.* Intervention allows public interest groups to take part in court cases that have socio-economic implications for society as a whole.
10. *Motion for summary judgment.* This motion has also been disparagingly referred to as "being laughed out of court."

True/False:

1. False. A corporation may be subject to personal jurisdiction in the state where it is incorporated, in the state where it has its principal office, and in the state where it is doing business.
2. True. Minimum contacts usually includes doing business in the state.
3. False. Courts will usually grant requests for a change of venue if such a change is necessary to avoid prejudice resulting from publicity or some other influence. Courts will not grant a change of venue if the parties are merely forum shopping.
4. True. Without an answer, the defendant may lose by default.
5. True. Settlement is encouraged at all stages of the trial process.
6. False. Only one party must request a jury trial.
7. True. In a civil case, either party may appeal the trial court's decision.
8. True. Generally courts show great deference to arbitration decisions in order to promote alternative dispute resolution.
9. False. In mediation and conciliation, the third party acts as a communication channel but does not render a decision. The parties must come to settlement on their own.
10. False. Whether an issue is subject to arbitration is a matter for the courts, not the arbitrator, to decide.

Multiple Choice:

1. A. Helga cannot sue. To have standing, a party must have a stake in the outcome of a suit. Since Helga was not involved in the accident, she does not have a stake in the suit and cannot sue on behalf of Harry. B, C, and D are incorrect because these are elements that the court of the state or region must satisfy. The plaintiff must bring the suit in a court that satisfies these elements, but they are not requirements that the plaintiff herself must personally satisfy.
2. B. A defendant must be served a summons to be personally subject to the jurisdiction of a court. A is incorrect because standing is an issue involving a plaintiff's potential stake in the outcome of a suit. C and D are incorrect because, although they are both required to properly bring a suit, they are not established by the defendant being served a summons.
3. D. A, B, and C are all circumstances where a state may extend its jurisdiction through use of a long-arm statute.
4. C. *Quasi in rem* jurisdiction is a means by which a prevailing party in a lawsuit may try to satisfy a judgment. A is incorrect because *in rem* jurisdiction is jurisdiction to hear a case because of jurisdiction over the property of the lawsuit. B is incorrect because a long arm statute is a statute that extends a state's jurisdiction to nonresidents who were not served a summons within the state.
5. C. A complaint is a document filed by the plaintiff that defines the problem. A is incorrect because a reply is an answer in a cross-complaint situation which the original plaintiff files to the cross-complainant. B is incorrect because an answer is a document filed by the defendant in response the plaintiff's complaint.
6. B. Once Paul files the complaint, Dan must file an answer, which is a response to Paul's complaint that must be filed with the court and served on Paul. A is incorrect because a reply is an answer in a cross-complaint situation which the original plaintiff files to the cross-complainant. C is incorrect because a complaint is the document filed by Paul that defines the problem.
7. B. A cross-complaint is an addition to Dan's answer that says Paul injured Dan, not vice-versa. A is incorrect because a reply is an answer in a cross-complaint situation which the original plaintiff files to the cross-complainant. Here, Paul would file a reply. C is incorrect because an answer is a response to the plaintiff's complaint that must be filed with the court. Although a cross-complaint is an addition to Dan's answer, B is the correct answer since it more specifically answers the question.
8. A. A reply is an answer in a cross-complaint situation that the original plaintiff (Paul) files to the cross-complainant (Dan). B is incorrect because a cross-complaint is the addition to Dan's answer that said Paul injured Dan, not vice-versa. C is incorrect because an answer is Dan's response to Paul's complaint.
9. C. A motion for summary judgment is a pretrial motion that asserts that there are no factual disputes to be decided by the jury and that the judge should apply the relevant law to the undisputed facts to decide the case. A is incorrect because discovery is a process during which both parties engage in various activities to discover facts of the case from the other party and witnesses prior to trial. B is incorrect because intervention is the act of other parties taking an interest and becoming involved in a lawsuit.
10. D. Arbitration is a form of alternative dispute resolution in which the parties choose an impartial third party to hear and decide the dispute. A is incorrect because in mediation, the third party does not render a decision, but rather helps the parties come to a resolution themselves. B is incorrect because a minitrial is a short session in which the lawyers for each side present their cases to representatives of each party who have the authority to settle the dispute. C is incorrect because a judicial referee is a court appointee who conducts a private trial and renders a judgment.

VI. Answers to Essay Question:

The vast majority of consumer complaints are resolved outside the court system. It is in the best interests of business goodwill to cultivate the loyalty of happy customers. It is never too late in the dispute process to reach some sort of private settlement. Unfortunately, the more time and money that is spent on the issue, the greater the likelihood of both sides hardening their positions. There is an optional arbitration provision available to Mr. At-Emptor. He may not trust this mechanism based on his negative experience with his car and the dealer. Although contractually optional, he may have to use the arbitration procedure as a precondition to being able to file a complaint in court under the state lemon law. Connecticut and California have recently incorporated such provisions in their versions of lemon laws.

If the out-of-court methods do not provide an adequate resolution of the problem, Mr. At-Emptor must initiate a lawsuit in a court having a proper basis for jurisdiction. Here a state court will most likely have jurisdiction over both the subject matter and the parties. Assuming proper service of process is made on the defendants, the lawsuit will have begun.

Once the actual pretrial process has begun, both parties will have their day in court. The defendants may choose to counterclaim based on the contract amounts still due or make pretrial motions to dismiss.

The actual trial will allow both parties to present their sides of the case to the jury in accordance with proper procedures as supervised by the judge. In civil cases, a judge retains the ultimate power to enter the final judgment jury verdict notwithstanding if he or she feels that the jury decision was somehow based on biased, emotional, or inflamed factors.

Because this is a civil case, either party can appeal the judgment rendered by the trial court.

FINAL EXAM REVIEW FOR LEG565AD

(reference: Cheeseman, "Contemporary Business & E-Commerce Law 4th Edition" Prentice Hall, NJ, 2003.)

Week	Topic	Objective(s)	Ch/Page
6	1. Holders in Due Course, Liability and Defenses b. Elements of HDC Status and its Effects on Defenses to Enforcement	1,2	20/p.487-489
6	2. Checks, Wire Transfers, and the Banking System c. Rights, Duties and Liability of Parties	5	21/p.517,521
6	3. Credit, Suretyship and Secured Transactions b. Creating and Perfecting a Security Interest under UCC Article 9	5	22/p.535
6	4. Bankruptcy and Reorganization b. Chapter 7 Liquidation	1,3	23/p.555
7	1. Agency a. Overview of the Agency Relationship and Key Terms c. Contract and Tort Liability to Third Parties	2	24/pp583-584,595
7	2. Sole Proprietorships and Franchises a. Overview of Business Entity Choice b. Introduction to Sole Proprietorships c. Introduction to Franchise Law	1 5	27/p.665-666, 675
7	3. Partnerships and Limited Liability Companies a. The Creation, Operation and Dissolution of Partnerships	1	28/p.694
7	4. The Nature, Formation and Financing of Corporations a. Overview of Corporate Law	1	29/p.722
8	1. Rights, Duties, and Liability of Corporate Directors Officers, and Shareholders b. Rights and Duties	3	30/p.750
8	2. Mergers, Acquisitions, and Termination of Corporations	5	31/p.779

	b. Mergers and Acquisitions		
8	3. Securities Regulation and Investor Protection	3	32/p.801-803
8	4. Antitrust Law	4	35/p. 880,894
	a. Overview of the Sherman Act		
9	1. Consumer Protection	8,10	33/p.846
	c. Federal Consumer/Debtor Protection Laws		
9	2. Environmental Protection	1	34/p.856
	a. Overview of Environmental Law Issues		
9	3. Employment and Labor Law	7	25/p.624
	c. Employment Law		
9	4. Equal Opportunity in Employment (EEO)	7	26/p.649
	a. Overview of EEO Laws		
10	1. Personal Property and Bailments	3	36/p.913
	a. Overview of Personal Property Law		
10	2. Real Property and Landlord-Tenant Relationship	2,4,5	37/p.932-934,940,944
	a. Ownership of Real Property		
10	3. Insurance, Wills and Trusts	6	38/p.971
	b. Overview of Wills		

Chapter 3
CONSTITUTIONAL LAW FOR COMMERCE AND INFORMATION TECHNOLOGY

The United States Constitution has proven itself the most elastic compilation of rules of government ever written.

Franklin D. Roosevelt

I. Overview

The scales of justice have long been one of the key symbols of our judicial system. They represent the system's effort to show evenhandedness and give fair weight to the evidence presented by competing parties. This symbol can also be used to illustrate the key-learning objective of this chapter. Our founding fathers struggled long and hard to create a balance of power among the three branches of government in much the same way as the justice system attempts to weigh the evidence. This process of balancing among the competing sovereignties of federal, state, and local governments is a process of constant evolution and, like the scale, will constantly shift with the addition and deletion of new political, social, and demographic weights.

What makes this constitutional balancing both fascinating and yet sometimes traumatic is that business people and their entities, represented by all sorts of firms from the smallest sole proprietorship to the largest multinational corporation, are caught in the middle of these sovereign "turf wars." In today's legal environment, there is little doubt about the power of the government to create, regulate, control, and even ban business activities. The real question appears to be which flag to salute—the federal government, the state, the local government, or some combination of all of the above? Where the rules of the game seem to shift almost daily between competing referees, it becomes most difficult for businesses to play the game.

The basic ground rules for business are set out in the U.S. constitutional provisions directly addressed to business, such as the Commerce Clause, and in larger protections accorded to all persons in the Bill of Rights and other key amendments. Many business entities are classified as *juristic* persons (law created entities which are given a limited and controlled legal existence) such as corporations or trusts. Subsequent chapters will go into more detail on how and when these forms of doing business may be used. In the context of constitutional allocation of control over business activity, the most important provision is the Supremacy Clause. Under this clause, the national sovereign can preempt control over any particular form of business it chooses to. This clause, combined with very broad judicial interpretation of the Commerce Clause, has been a part of the process which attempts to create an even playing field for business, i.e., the federal sovereign has long believed that a states power to control commerce under its constitutional police power should be harnessed if it encumbers the national flow of commerce or if it somehow unduly favors local business at the cost of competing political entities such as other cities, states, or even regional interests. The balancing of the scale continues.

What makes this process of setting out the business playing rules even more difficult is that many of our basic rights, set out in the Bill of Rights such as freedom of speech, press, religion, and assembly, have been held by our courts to apply in varying degrees to business enterprises. How, when, where, and to what extent these rights should apply to business continues to be both a source for constant academic debate and a regulatory dilemma for both government and business alike. For example, what constitutional limits should be set on e-commerce, which is now a virtual "wild west"? Law works best where all the players involved have a "bright line" or clear signal to use as a guidepost for their actions. When that line is blurred, all are disserved.

II. Multi-issue Essay Question

Brew Ha Ha Brewery is located in New Salem, California, and is engaged in the business of producing specialized low alcohol products for specifically identified niche markets. It produces diet beers for the overweight crowd, young malts for the young at heart, and aged malts for the over-the-hill yuppie couch potatoes.

Brew Ha Ha's newest product seeks to cash in on the latest craze, a revival of witchcraft! Their new product is named "The Devil Made Me Do It!" The "Devil" beer has the same low alcohol content as all its other products but is designed to look like a steamy caldron of some witch's brew that will conjure up images of the rich and diverse history of occult practices throughout the world.

The Texas State Liquor Control Board took umbrage at this product's motives and felt that it would foster and encourage the practice of witchery in its fair state. It thus decided to ban the sale of "The Devil Made Me Do It!" within the state on the grounds that this product would undermine the morals of its citizens through the practice of witchcraft. The Board claims its actions are allowable under both the U.S. and local constitutions. Brew Ha Ha claims its constitutional rights are being violated. What result?

III. Outline

Major Functions of the Constitution
- Creates the three branches of government
  - Executive
  - Legislative
  - Judicial
- Protects individual rights by limiting the government's ability to restrict those rights

Separation of Powers
- Article I
  - Establishes the legislative branch of government
- Article II
  - Establishes the executive branch of government
- Article III
  - Establishes the judicial branch of government

Supremacy Clause
- Establishes that the federal Constitution, treaties, federal laws, and federal regulations are the supreme law of the land

Commerce Clause
- Grants Congress the power "to regulate commerce with foreign nations, and among the several states, and with Indian tribes"
- Gives the federal government authority to regulate interstate commerce

State "Police Power"
- States can enact laws to protect or promote public health, safety, morals, and general welfare
- State and local laws cannot unduly burden interstate commerce

Incorporation Doctrine
- A doctrine that states that most fundamental guarantees contained in the Bill of Rights are applicable to state and local government action

Federal Telecommunications Act
- Preempts state and local laws dealing with the telecommunications industry

Freedom of Speech
- Fully protected speech
  - Speech the government cannot prohibit or regulate
- Limited protected speech
  - Commercial speech
  - Offensive speech
- Unprotected speech
  - Dangerous speech
  - Fighting words
  - Speech that incites the violent or revolutionary overthrow of the government
  - Defamatory language
  - Child pornography
  - Obscene speech

Freedom of Religion
- Establishment Clause
  - Prohibits government from establishing a government sponsored religion or promoting one religion over others
- Free Exercise Clause
  - Prohibits government from enacting laws that prohibit or inhibit people from participating in or practicing their chosen religion

Equal Protection Clause
- Strict scrutiny test
  - Applied to classifications based on race
- Intermediate scrutiny test
  - Applied to classifications based on protected classes other than race (e.g. sex or age)
- Rational basis test
  - Applied to classifications not involving a suspect or protected class

Due Process Clause
  Substantive due process
    Requires governmental laws to be clear and not overly broad; the test is whether a reasonable person can understand the law
  Procedural due process
    Requires government to give people proper notice and hearing before depriving them of life, liberty, or property

Privileges and Immunities Clause
  Prohibits states from enacting laws that unduly discriminate in favor of their residents

IV. Objective Questions:

Terms:

1. Under the U.S. Constitution, the states delegated certain powers to the federal government. These powers are known as ________________ powers.

2. Under the U.S. Constitution, the three branches of government are given different powers. This is known as the doctrine of ________________ ________________ ________________.

3. When a state law conflicts with a federal law, the federal law takes precedent. This concept is known as the ________________ ________________.

4. The constitutional provision that addresses the issues of a national market and free trade among the states is known as the ________________ ________________.

5. Two provisions of the Telecommunications Act restricted the ability of cyberspace operators from transmitting certain images. They were the ________________ ________________ ________________ and the ________________ ________________ ________________.

6. Dangerous speech, defamatory language, and obscene speech are examples of ________________ ________________.

7. Examples of speech that are not prohibited under the First Amendment, but may be restricted are ________________ speech and ________________ speech.

8. The First Amendment contains two religion clauses. The ________________ ________________ prevents a state from establishing or promoting a state religion. The ________________ ________________ ________________ prohibits the government from prohibiting or inhibiting people from participating in their religion.

9. Although the freedom is not absolute, the ________________ is the only business given explicit constitutional protection.

10. The Fourteenth Amendment's Due Process Clause has two categories of due process; ________________ due process and ________________ due process.

True/False:

1. ____ The Bill of Rights was part of the U.S. Constitution as originally adopted in 1787.

2. ____ The federal government has the exclusive right to regulate foreign commerce.

3. ____ If a state's law is in its best interest, it can preempt any relevant federal law.

4. ____ All provisions of the Bill of Rights are applicable only to the federal government.

5. ____ The Freedom of Speech Clause extends to protect both speech and conduct.

6. ____ Offensive speech, like obscene speech, can be forbidden under the First Amendment.

7. ____ Commercial speech can never be restricted under the First Amendment.

8. ____ Government action that "impacts" on religion, but does not "promote" it is valid.

9. ____ The requirements of the Due Process Clause apply to both state and federal government, but never to private individuals.

10. ____ The U.S. Supreme Court uses different standards when reviewing equal protection cases, depending on the classification affected.

Multiple Choice:

1. The federal government is given the power to set international trade policies in the U.S. Constitution under which doctrine or clause?
   A. Doctrine of Separation of Powers.
   B. Commerce Clause.
   C. Preemption Doctrine.
   D. Supremacy Clause.

2. Under the *Miller v. California* standard for determining whether speech is obscene, there is a three-part test that is analyzed from the viewpoint of:
   A. An objective standard.
   B. The person who produced the allegedly obscene speech.
   C. The person who is offended by the obscene material.
   D. Contemporary community standards.

3. If the government believes certain offensive language is inappropriate to air on a television program, the government:
   A. Can do nothing; even though it is offensive, it is completely protected by the First Amendment.
   B. Can completely ban the speech since offensive language is like obscene language and can be entirely forbidden.
   C. Can regulate the language, restricting it to certain times the language can be aired.
   D. Can use its discretion to do any of the above.

4. If the government believes certain commercial speech should be regulated, the government must:
   A. Prove a substantial interest in restricting the speech.
   B. Prove the restriction advances the government's interest.
   C. Prove the restrictions are not more extensive than necessary.
   D. All of the above.

5. Ian Informant calls Janet Journalism at the Daily Reporter to tip her off about a scandalous politician. Janet meets with Ian and writes a very controversial story on the politician. Assume a lawsuit arises under a state law and this state has a "shield law."
   A. The First Amendment does not protect Janet and the Daily Reporter because a journalist can never write detrimental articles about politicians.
   B. Janet and the Daily Reporter are protected by the First Amendment and cannot be forced to reveal that Ian was the source because of the shield law.
   C. Janet and the Daily Reporter are protected by the First Amendment but can be forced to reveal that Ian was the source.
   D. None of the above.

6. Assume the same facts as in question 5 except that the suit is brought under a law allowing federal prosecution.
   A. The First Amendment does not protect Janet and the Daily Reporter because a journalist can never write detrimental articles about politicians.
   B. Janet and the Daily Reporter are protected by the First Amendment and cannot be forced to reveal that Ian was the source because of the shield law.
   C. Janet and the Daily Reporter are protected by the First Amendment but can be forced to reveal that Ian was the source.
   D. None of the above.

7. State X passes a law requiring all Caucasian citizens to pay an additional fee of $20 when applying for a driver's license. In examining the constitutionality of this requirement, the Supreme Court is likely to use which standard?
   A. Strict scrutiny.
   B. Intermediate scrutiny.
   C. Rational basis.
   D. Any of the above.

8. State X passes a law requiring that all lifeguards hired for public swimming pools must be men. In examining the constitutionality of this law, the Supreme Court is likely to use which standard?
   A. Strict scrutiny.
   B. Intermediate scrutiny.
   C. Rational basis.
   D. Any of the above.

9. The police throw Patron Paul in jail for thirty days without a hearing for allegedly punching Customer Carl at the Big Bear Bar. Have the police violated any constitutional provisions?
   A. No, because Carl told the police what happened, so the procedural due process requirement has been met.
   B. No, because a reasonable person could understand it is illegal to injure someone else, so the substantive due process requirement has been met.
   C. Yes, because procedural due process requires a notice and a hearing.
   D. Yes, because a law against punching someone in a bar violates substantive due process.

10. Competitive Car Company lobbies several Congress members to pass a law giving U.S. car companies a tax break. Is this action protected under the Constitution?
    A. Yes, it is protected by the freedom to petition.
    B. No, the freedom to petition only protects individuals, not companies.
    C. No, this issue is governed by the individual states.
    D. None of the above.

V. Answers to Objective Questions:

Terms:

1. *Enumerated.* These powers are expressly named or granted to the federal government by the Constitution.
2. *Separation of powers.* Under this doctrine, one branch is not permitted to encroach upon the domains of the other two branches.
3. *Preemption doctrine.* If certain matters are considered of national importance and the federal government has addressed the issue, states may not pass laws that are inconsistent with federal law. Federal law takes precedence over state law.
4. *Commerce Clause.* This provision is found in Article I, Section 8, Clause 3 of the U.S. Constitution and is the cornerstone of an enormous amount of exclusive legislative and regulatory power over interstate commerce by the federal government.
5. *Computer Decency Act; Signal bleed provision.* The Supreme Court overturned the Computer Decency Act and the signal bleed provision claiming that the Internet must be given the highest possible level of First Amendment free speech protection.
6. *Unprotected speech.* These are all forms of expression that have been found in varying degrees to be unprotected by the freedom of speech rights guaranteed by the First Amendment to the U.S. Constitution.
7. *Offensive, commercial.* Both of these forms of speech enjoy limited protection under the 1st Amendment. As such, they may not be totally barred, but are subject to restrictions on a case-by-case basis.
8. *Establishment Clause, Free Exercise Clause.* While not seeking to establish an "official" state religion, our system seeks to strongly protect the individual's right to practice his or her own religious beliefs.
9. *Press.* Our founders appreciated the necessity for a free and open press. The press is sometimes called the "Fourth Estate."
10. *Substantive, procedural.* Substantive goes to content, and procedural goes to application. In both cases, constitutional guarantees are sought to be meaningful in both context and application. They are based on the Fifth and Fourteenth Amendments.

True/False:

1. False. The Bill of Rights was not part of the original Constitution and was not adopted until 1791.
2. True. The Commerce Clause of the U.S. Constitution grants the federal government the exclusive right to regulate commerce with foreign countries.
3. False. Under the Supremacy Clause of the U.S. Constitution, federal laws, treaties, and regulations are the supreme laws and may preempt any conflicting state laws.
4. False. Through the incorporation doctrine, the Supreme Court has held that most provisions of the Bill of Rights apply to the states as well as to the federal government.
5. False. The Freedom of Speech Clause protects oral, written, and symbolic speech, but not conduct.
6. False. Offensive speech can be restricted by the government, but cannot be completely forbidden.

7. False. Commercial speech, like offensive speech, can be subject to certain time, place, and manner restrictions.
8. True. Under the Establishment Clause, the government can impact on religion with laws that have a secular, as opposed to a religious, purpose.
9. True. The Due Process Clause imposes no duty on private individuals.
10. True. Depending on the classification that a given law may impact, the Court may apply a strict scrutiny, intermediate scrutiny, or rational basis test.

Multiple Choice:

1. B. The Commerce Clause gives the federal government the exclusive power to establish and regulate international trade.
2. D. The Court in *Miller* set out a three-part test that requires application of "contemporary community standards" in determining whether certain speech appeals to prurient interest, is patently offensive, or has any social value.
3. C. Offensive speech has limited protection under the 1st Amendment. The government cannot forbid such speech, but can restrict it in certain ways.
4. D. For the government to be allowed to restrict commercial speech, it must prove A, B, and C. This is because commercial speech has limited protection under the 1st Amendment. Therefore, the government bears some burden in showing restriction of such speech is justified.
5. B. The press is the only business granted express protection under the First Amendment. Therefore, as long as Janet and the Daily Reporter act within the law, this speech is protected. The state's shield law protects Janet from having to divulge that Ian was the informant. However, this protection is not derived from the 1st Amendment. A is incorrect because the First Amendment does protect such speech provided it does not violate another law. C is incorrect because the shield law protects Janet's informants.
6. C. State shield laws only protect the press under state law. If a suit is brought under a federal law, the First Amendment protects the speech, but not Ian's identity. A is incorrect because the 1st Amendment does generally protect such speech. B is incorrect because the shield law is not effective in a federal court.
7. A. This law classifies persons based on race, which is a suspect class. Therefore, strict scrutiny would be the appropriate test. B is incorrect because intermediate scrutiny is used for other protected classes such as sex, age, or national heritage. C is incorrect because rational basis applies when the classification is neither suspect or protected.
8. B. Because this classification is based on gender, the appropriate test is intermediate scrutiny. A is incorrect because strict scrutiny is generally used for the suspect class of race. C is incorrect because rational basis applies when the classification is neither suspect or protected.
9. C. Regardless of the facts of this punching incident, Paul is entitled to notice and a hearing before he is jailed. A is incorrect because Carl's recitation of the facts does not satisfy the procedural due process requirement of a notice and hearing. B is incorrect because even if substantive due process is satisfied, this alone is not enough to put Paul in jail. D is incorrect because such a law does not violate substantive due process. However, procedural due process must also be satisfied.
10. A. The freedom to petition applies to both individuals and businesses. C is incorrect because the states must provide their citizens this freedom under the 1st Amendment.

VI. Answers to Essay Question:

The issue of the regulation and control of the sale of alcoholic products has had a long and controversial history in the U.S. In 1919, Congress passed the Eighteenth Amendment to the U.S.

Constitution that banned the manufacture, sale, or transportation of intoxicating liquors in the U.S. After years of futile efforts to enforce this ban, the Twenty-First Amendment to the U.S. Constitution repealed the Eighteenth Amendment in 1933.

As part of the repeal of the Eighteenth Amendment, Section 2 of the Twenty-First Amendment allowed state and local jurisdictions the option to regulate the transportation, importation, or possession of intoxicating liquors within their jurisdictions. Because the Twenty-First Amendment specifically gave states local options over the control and regulation of the sale of intoxicating liquors, it appears that the Texas Liquor Board does have the presumptive power to act on this issue. In addition, traditional police powers to protect the health, safety, and morals of its citizens are also entrusted to state and local authorities.

States have very broad powers regarding the control of intoxicating products, but this power is not absolute. There is an expectation by the federal government that any exercise of the local option power of liquor control will be suspect and preempted by the national goals of free commerce required by the Commerce Clause. Thus, even though the state may be the first to act in these types of cases, it may not always have the last word in these matters.

In addition to concerns over power-sharing between national and state governments in cases involving the sale of intoxicating commercial products, we have a possible "larger" issue—that of free speech. Any attempt to limit free speech, albeit by a commercial entity rather than a private person, is always subject to great scrutiny by the courts.

This question was based on a real incident that occurred in Texas in 1991. The Dixie Brewing Co. of New Orleans sought to market a beer in Texas called "Blackened Voodoo Lager." The Texas liquor authorities sought to ban the product because it believed the lager would promote occult practices. It later rescinded the proposed ban for fear of legal challenges.

Those challenges would most likely have been made on the grounds of commercial free speech and interference with the free flow of goods in interstate commerce. At a minimum, the state of Texas would have the burden of proof in showing that the sale of this product would damage the health, safety, morals, and welfare of its citizens as required under the use of its state police powers.

Chapter 4
INTENTIONAL TORTS, NEGLIGENCE, AND STRICT LIABILITY

We are all servants of the laws to the end that it may be possible for us to be free.
Cicero

I. Overview

The word tort is not readily familiar to students who have not yet studied law. Like so many words of art found in the law, it has a special meaning derived from an essentially simple principle. The word is literally derived from the French word for wrong. That French term, in turn, is derived from the Latin phrase *torguere* meaning to twist, or *tortus* meaning to be twisted or wrested aside. When we consider that the odds are high that sooner or later we may be twisted by an auto accident, job-related injury, or some similar event, the term starts to make sense. Injuries of all sorts are an unfortunate fact of life. What makes those injuries actionable, i.e., where a court will recognize legal grounds for bringing a lawsuit, is when someone may have committed a noncontractual wrong for which a court, after appropriate proceedings, may grant a remedy. In other words, sometimes when we get twisted, the law provides an opportunity to correct the situation if the twist is found to be wrongful.

Tort law structure is divided into three main classifications:

1. Intentional torts.
2. Unintentional torts (negligence).
3. Strict liability.

Of these three, the first categories, intentional torts, are the most readily identifiable because they are very often concurrently classified as crimes. A crime is generally defined as an act in violation of a local or national penal statute. Governmental authorities, on behalf of the victim, seek the remedy for crimes for all of us because the act is classified as an act against the common welfare of society. The civil remedy for the same act is an action for intentional tort. The victim brings his lawsuit against the perpetrator of the intentional tort. Damages are paid directly to the victim after proper proceedings are held. The major classifications of intentional torts are acts against the person, and acts against property. Acts against persons include assault, battery, false imprisonment, defamation of character, invasion of privacy, slander, libel, and abuse of the legal process. One recent example of the abuse of legal process was found in Colorado where a group of tax protesters filed liens on the personal assets of government employees as a way to object to government tax collection procedures. An example of an act against property would be intentional trespass upon someone else's land.

The second category, unintentional or negligence torts, is both far more prevalent and more difficult to clearly identify. The first thing to remember is that responsibility can be imposed regardless of intent. This is a consequence-oriented result in the law as compared to the intent-oriented result discussed in the law of intentional torts.

The second major point is that the term negligence is in fact a legal conclusion arrived at only when a special set of sequential circumstances have taken place. These circumstances are generally called elements. Just as the makeup of all physical substances can be broken down into key components, so may many legal doctrines. The doctrine of negligent tort is made up of five key elements. These elements are:

1. A showing of a duty of care owed by the defendant to the plaintiff.
2. A showing of a breach of that duty.
3. A showing of injury to the plaintiff.
4. A showing of causal nexus or connection between the breach of duty and the injury (actual cause).
5. A tying of that actual cause to the legal responsibility for the consequences of the act or omission. This connection between actual and legal cause is referred to as proximate cause.

If all five circumstances are found, and there are no recognized defenses, excuses, or statutory immunities, a legal conclusion of negligence can be reached. Immunity is an exception from liability recognized in the law, usually based on an overriding public policy. Thus the key to the study of negligent torts is the proper identification and use of the elements of negligence. Think of the use of elements as essential building blocks in the construction of a foundation on which a conclusion of negligence can be built. If any of the blocks are missing, the building will not stand.

The third major area of tort law is found in what has generally come to be classified as strict liability torts, i.e., liability without fault. This is not to say that it is automatic liability, but rather that a different set of elements have evolved in certain cases involving special kinds of activities which are considered ultra-hazardous and with regard to the use of certain products. This doctrine is fully discussed in Chapter 21 of this text. At this juncture, it is important to remember that this doctrine may represent one of the options available to an injured tort victim.

Under all three classifications of tort, the burden of proof remains with the plaintiff to show that he or she has been wronged and that he or she is entitled to some sort of court-ordered remedy from the defendant.

II. Hypothetical Multi-issue Essay Question:

One day while Gabby and Crabby were driving through the intersection of College and 28th in Boulder, Mario struck their car. Mario ran the light at high speed but said, in his defense, that he thought he had the green light. In fact, Mario is color blind, but relied on the expectation that the green light was at the bottom of the signal. The Boulder City light repair crew had just reversed them in order to make more drivers think twice before going for the yellow. Thus, the red was at the bottom.

In the meantime, Gabby was unconscious with grave head injuries and was rushed to a nearby hospital. Crabby, uninjured but somewhat hysterical, accompanied him to the hospital. When they arrived, the only surgeon on duty was an exhausted Dr. Franklin Pierce. He had just completed a complicated lobotomy on a former Hell's Angel, now an outstanding business consultant. Dr. Pierce told Crabby that her husband Gabby needed brain surgery but that he was too worn out to undertake it.

Crabby, quite frantic, seized Dr. Pierce by the shoulders and said intensely: "Doctor, you must! Money means nothing to me when my dear Gabby's life is concerned. I will see that you are paid $50,000. Oh, doctor, I promise!" Dr. Pierce agreed to do the operation and completed it in a professional manner. Gabby died four days later, and Crabby now refuses to pay. Discuss the legal consequences of this case to all parties involved.

III. Outline

Three Kinds of Torts
- A tort is a wrong
- There are three kinds of torts
  - Intentional torts
  - Unintentional torts (negligence)
  - Strict liability

Intentional Torts against Persons
- Assault
- Battery
  - False intent doctrine
- False imprisonment
- Defamation of character
  - Poses a particular problem for Internet service providers
- Invasion of the right to privacy
- Misappropriation of right to publicity (appropriation)
- Intentional infliction of emotional distress
- Malicious prosecution

Intentional Torts against Property
- Trespass to land
- Trespass to and conversion of personal property

Elements of Unintentional Torts (Negligence)
- To be successful in a negligence lawsuit, the plaintiff must prove:
  - The defendant owed a duty of care to the plaintiff
  - The defendant breached this duty of care
  - The plaintiff suffered injury
  - The defendant's negligent act caused the plaintiff's injury

Causation
- Causation in fact
  - The defendant's negligent act must be the causation in fact or actual cause of the plaintiff's injuries
- Proximate cause
  - A point along the chain of events caused by a negligent party after which this party is no longer legally responsible for the consequences of his or her actions

Professional Malpractice
- Professionals, such as doctors, lawyers, architects, accountants, and others owe a duty of ordinary care, called the reasonable professional standard, in providing their services
- A professional who breaches this duty is liable for professional malpractice

Negligent Infliction of Emotional Distress
- Negligent infliction of emotional distress is a tort that permits a person to recover for emotional distress caused by the defendant's negligent conduct
- A plaintiff must prove that
  - A relative was killed or injured by the defendant
  - The plaintiff suffered severe emotional distress
  - The plaintiff's mental distress resulted from a sensory and contemporaneous observance of the accident

Special Negligence Doctrines
- Negligence per se
- Res ipsa loquitur
- Dram shop acts
- Social host liability
- Guest statutes
- Good Samaritan laws
- Fireman's rule
- "Danger invites rescue" doctrine
- Liability of common carriers and innkeepers
- Liability of landowners

Defenses Against Negligence
- Assumption of the risk
- Contributory negligence
- Comparative negligence
- Superseding or intervening event

Business Torts
- Unfair compentition
- Disparagement
  - False advertising
  - Intentional interference with contractual relations
  - Breach of the implied covenant of good faith and fair dealing
  - Entering certain businesses and professions without a license
  - Intentional misrepresentation (fraud)
  - Civil RICO

Strict Liability
- Strict liability is liability without fault
- Strict liability doctrine
  - There are certain activities that place the public at risk of injury even if reasonable care is taken
  - The public should have some means of compensation if such injury occurs
- Strict liability was first imposed for abnormally dangerous activities

IV. Objective Questions

Terms:

1. A tort is a civil wrong where an injured individual sues an offending party seeking _______________ or monetary compensation.

2. An unintentional tort where the offending party commits an act or omission that a reasonable man would not commit is categorized as a _______________ tort.

3. The intentional tort of _______________ requires only a reasonable apprehension of immediate harm, whereas the tort of _______________ involves harmful or offensive contact with another person.

4. A false statement about a person, which is subsequently intentionally or accidentally published to a third person, is called the tort of _______________ _______________ _______________.

5. A false statement which is spoken to a third person is _______________; the name for a false statement which is written in a letter, newspaper, book, etc., or broadcast via movie, video, television show, or radio is _______________.

6. If a defendant acts with _______________ _______________, he makes a false statement knowingly or with reckless disregard of its falsity.

7. The law against _______________ _______________ _______________ protects a person's right to live without unwarranted or undesired publicity.

8. Proving causation for an unintentional tort case requires two elements. The act of the person charged must be both the actual cause or _______________ _______________ _______________, as well as the legal or _______________ cause.

9. Under the _______________ _______________ doctrine, damages are computed according to the fault of the plaintiff as well as the fault of the defendant so that the defendant only pays for the percentage of damage that he caused.

10. The obligation not to cause anyone harm or risk of harm is known as the _______________ _______________ _______________ element in the law of negligence torts.

True/False:

1. ____ All torts are also crimes.

2. ____ To be liable for an intentional tort, the perpetrator must have intended the result of his act as well as the act itself.

3. ____ A person is liable for false imprisonment if he restrains a person with threats of immediate or future physical harm.

4. ____ "Publication" to a third party in defamation of character means that a statement must appear in written form in a newspaper, book, magazine, etc.

5. _____ Public figures and officials have the same degree of protection under defamation laws as common citizens do.

6. _____ In the tort of invasion of privacy, unlike defamation of character, truth is not a defense, because a statement need not be false to be a violation.

7. _____ In determining a person's duty of care under negligence, whether the person acted within the reasonable person standard is determined by that person's subjective intent at the time.

8. _____ Usually, if certain harm was a foreseeable result of a person's negligent act, the act is said to be the proximate cause of the damages.

9. _____ Under *res ipsa loquitur*, there is an inference that the defendant was negligent and must therefore bear the burden to prove he was not negligent.

10. _____ Under the traditional common law doctrine of contributory negligence, a plaintiff cannot recover damages from a defendant even if the plaintiff was only at fault for a very small part of his own injury.

Multiple Choice:

1. Mr. Temper, a 5-foot, 4-inch, 100-pound hothead, tells Mr. Big, a professional wrestler, that he is going to "make him regret he set foot in this bar." At the same time, Temper clenches and raises his fists. Big looks at Temper from head to toe and responds, “Yeah, right.”
   A. Big can sue Temper only for assault.
   B. Big can sue Temper for assault and battery.
   C. Big can sue Temper for intentional infliction of emotional distress.
   D. Big has no cause of action.

2. Thug approaches Joe, who is standing in line to buy a movie ticket, from behind. Thug plans to hit Joe over the head with a baseball. Completely unaware, Joe bends over to tie his shoe just as Thug swings. Consequently, Thug strikes Ann who was standing in front of Joe, fracturing her skull.
   A. Joe can sue Thug for assault; Ann can sue Thug for battery.
   B. Joe has no cause of action against Thug; Ann can sue only for battery.
   C. Ann can sue Joe for negligence; Joe can sue Thug for contributory negligence.
   D. Joe has no cause of action against Thug; Ann can sue for both assault and battery.

3. To be liable for false imprisonment, the perpetrator may intentionally confine a person with:
   A. Physical force or barriers.
   B. Threats of physical harm.
   C. False arrest.
   D. All of the above.

4. Mr. Gossip tells Ms. Nosy that: “Flashy is a drug dealer. That's how she can afford to drive that Ferrari.” Is Gossip liable for defamation of character against Flashy?
   A. Yes, because it is a false statement that was told to a third party.
   B. Yes, because all that is necessary is that the statement is false.
   C. No, because telling only Nosy does not really hurt Flashy.
   D. No, because Gossip only said this, but did not write it.

5. Speedy approaches an intersection in his car, as the light turns yellow. Instead of slowing down, Speedy speeds up and runs a red light. Mr. Walker, a pedestrian waiting to cross the street, yells and waves his fist at Speedy as he drives by. Unfortunately, Walker startles Ms. Precious who is also waiting to cross, and Precious drops the $500 vase she just bought. Is Speedy liable for the damage to the vase?
   A. No, because Speedy had no duty to drive carefully.
   B. No, because running the light is not a breach of the duty.
   C. No, because Speedy's running of the light was not the proximate cause of the broken vase.
   D. Yes, because the plaintiff, Precious, suffered injuries.

6. Assume the same facts as in question 5 except that Precious does not notice Walker. After the walk signal illuminates, Precious looks both ways, does not see Speedy, and begins to cross the street. Speedy, however, hits Precious, breaking her leg, as he runs the red light.
   A. Speedy is liable for negligence, and Precious can recover damages.
   B. Speedy is liable for an intentional tort, and Precious can recover damages.
   C. Speedy is not liable for negligence because his acts were not the proximate cause of Precious' injuries.
   D Speedy is not liable for an intentional tort because intentional torts are wrongs against property only.

7. Ms. Rich's house is broken into about once a month when she leaves town. Usually, the burglar steals jewelry, watches, and other personal items from her house. One time when going out of town, Rich rigged a gun to go off when her bedroom door was opened. Slimy subsequently broke into her house and was shot. Rich's state distinguishes between types of visitors.
   A. Rich is liable because she owes Slimy and all other visitors a duty of ordinary care.
   B. Rich was perfectly justified in doing this since she knew someone would break in; she is not liable for protecting her property like this.
   C. Rich is not liable because she was not at the house when this happened.
   D. Rich is liable because she cannot willfully, or wantonly, injure Slimy.

8. Assume the same facts as in question 8 except that Rich does not set the rigged gun. Slimy enters the house through a window in the kitchen. The pipes under the kitchen sink have a leak. Slimy steps down onto the floor, slipping in a puddle of water, and falls and breaks his tailbone.
   A. Slimy has a good case against Rich since having leaky pipes is clearly negligent and therefore a breach of the duty of ordinary care owed Slimy.
   B. Slimy has a good case against Rich because she assumed the risk of a break-in when she left town.
   C. Slimy has no case against Rich because stepping in the puddle of water was an intervening event.
   D. Slimy has no case against Rich because Rich does not owe him a duty of ordinary care.

9. Couch Potato decides shoveling his walk after a huge snowstorm is just too much work. Three days later his walk is a sheet of ice. Social Visit, on the way to see her friend Couch Potato, notices the ice but proceeds up the walk anyway. Social Visit slips and cracks her kneecap. Assume Couch Potato was 60 percent at fault and Social Visit was 40 percent at fault.
   A. If this is a contributory negligence state, Social Visit can recover 60 percent of the damages.
   B. If this is a partial comparative negligence state, Social Visit can get no recovery.
   C. If this is a pure comparative negligence state, Social Visit can recover 60 percent of the damages.
   D. If this is a contributory negligence state, Social Visit can recover 40 percent of the damages.

10. Assume the same facts as in question 9 except that Couch Potato is only 40 percent responsible and Social Visit is 60 percent responsible.
   A. If this is a traditional common law contributory negligence state, Social Visit can get no recovery.
   B. If this is a partial comparative negligence state, Social visit can get no recovery.
   C. If this is a pure comparative negligence state, Social Visit can recover 40 percent of the damages.
   D. All of the above.

V. Answers to Objective Questions

Terms:

1. *Damages*. The tort must fit into one of the three recognized categories.
2. *Negligence*. This is the second of the three categories of tort law.
3. *Assault, battery*. These acts are also often classified as crimes.
4. *Defamation of character*. Note that publication to a third person is required.
5. *Slander, libel*. Oral and written statements are separately classified.
6. *Actual malice*. Malice is very difficult to prove.
7. *Invasion of privacy*. Privacy is not specifically mentioned in the U.S. Constitution, but has been recognized as a limited right by the courts.
8. *Causation in fact, proximate*. Both elements are required.
9. *Comparative negligenc*e. This is a relatively new method as opposed to the old common law rule of contributory negligence.
10. *Duty of care*. This is the first element of unintentional (negligence) tort law.

True/False:

1. False. Torts are causes of action founded in common law. In torts, one private party sues another. Crimes are statutory wrongs that are prosecuted by the government. While some torts are also crimes, not all torts fall in this category. For example, negligently denting another car's bumper is a tort, but not a crime, in most jurisdictions.
2. False. The perpetrator need only have intended his original act, not its result, to be liable.
3. False. A threat of immediate harm may be considered false imprisonment. However, a threat of future harm is not.
4. False. Publication in this context means that a third person heard or saw the untrue statement, not that it was written in a formal publication.

5. False. Public figures and officials can only recover for defamation if they prove the defendant made a false statement knowingly, or with reckless disregard for its falsity, i.e., actual malice.
6. True. The tort of invasion of privacy protects one's right to live without unwarranted or undesired publicity. Whether a fact is private is the relevant issue, not whether it is truthful.
7. False. Under a reasonable person standard, courts determine how an objective, careful person would have acted in a certain situation. The defendant's subjective thoughts or intentions are not considered.
8. True. In determining proximate cause, if a certain event was a foreseeable result of a negligent act, that will generally satisfy this part of the causation requirement.
9. True. Because the defendant has superior knowledge and because the evidence suggests there was a negligent act, *res ipsa loquitur* shifts the burden from the plaintiff to the defendant.
10. True. Under contributory negligence, if the plaintiff contributed in any way to his own injuries, he cannot recover from the defendant under the traditional common law. The more modern view is to use comparative negligence doctrines.

Multiple Choice:

1. D. Since Big is obviously not frightened by this threat, he cannot sue for assault. Because Temper never hit Big, there is no harmful or offensive touching, and therefore, no battery. C is incorrect since Big does not suffer any emotional damage from this threat.
2. B. Since Joe was never in apprehension of harm, nor actually harmed by Thug's actions, he has no cause of action. Ann can sue for battery since she was harmed, but not for assault since she was not in apprehension of harm. A is incorrect since Joe has no cause of action. C is incorrect because Ann cannot make a case that Joe was negligent for tying his shoe. Also, there is no contributory negligence here at all. D is incorrect because Ann cannot sue for assault.
3. D. All three of these types of restraint are considered false imprisonment.
4. A. The elements of defamation of character require both (1) a false statement and (2) the telling or writing of the statement to a third party. B is incorrect because the second element is missing. C is incorrect because while the degree of harm may affect the amount of damages, the elements do not require a certain level of harm. D is incorrect because defamation of character may be in either slander or libel form.
5. C. While Speedy's act was certainly negligent, the chain of events from speeding through the light to breaking the vase is very long and unforeseeable. Consequently, since foreseeability is generally the test for proximate cause, Speedy is not liable for the vase. A and B are incorrect because Speedy did have a duty and subsequently breached that duty. Therefore, even though Speedy is not liable for the vase, these are not the correct reasons. D is incorrect because although Precious suffered damages, that alone is not enough to recover damages.
6. A. Since Speedy owed Precious a duty of care, breached that duty, and caused Precious injury, he was negligent and is liable. B is incorrect because nothing suggests Speedy intended to hit Precious. C is incorrect because hitting a person in a crosswalk is certainly a foreseeable result of running a red light. Therefore, there is proximate cause. D is incorrect because this is not an intentional tort. Also, intentional torts are wrongs against both persons and property.
7. D. Although Slimy is a trespasser, Rich cannot willfully or wantonly injure him. A, B, and C are all incorrect because a trespasser is owed a certain, lower duty by a property owner.
8. D. Rich owes Slimy a duty not to willfully or wantonly injure him. However, she does not owe a duty of ordinary care. A and B are incorrect because although having the puddle may be negligent, Rich does not owe Slimy a duty higher than the willful or wanton standard. C is incorrect because the theory of intervening events is irrelevant here.
9. C. In a pure comparative negligence state, Social Visit can recover the amount Couch Potato is responsible for, regardless of whether she is more than 50 percent responsible. A is incorrect because, in a contributory negligence state, Social Visit could not get anything if she is responsible

for even a minor part of her injuries. B is incorrect because, even in a partial comparative negligence state, Social Visit could recover since she is less than 50 percent responsible. D is incorrect because, in a contributory negligence state, Social Visit would get no recovery.

10. D. A is correct because under traditional common law contributory negligence, Social Visit cannot recover because she is partially responsible. B is correct because, under partial comparative negligence, Social Visit cannot recover because she is more than 50 percent at fault for the injury. C is correct because, under pure comparative negligence, Social Visit can recover any amount for which Couch Potato is responsible.

VI. Answers to Essay Question

We have two basic sets of possible contract-tort based actions against Mario and the City of Boulder, and a contract-based dispute between the Estate of Gabby and Franklin Pierce and/or his employer, the hospital. For purposes of this chapter, we will focus on the tort issues.

In both cases, neither Mario nor the City of Boulder employees intended any harm, so an intentional tort was not likely here. Nor is there any particularly ultra hazardous activity or dangerous defect in any product apparent in the facts as presented. Thus, strict liability does not seem likely here. There is a distinct possibility of negligence on the part of both Mario and/or the City of Boulder. The following set of key questions might be appropriate for the student to use in answering the tort law aspects of the problem:

1. Classification of the tort:
   - (a) Was Mario's act a(n):
     - (1) Intentional tort?
     - (2) Unintentional or negligent tort?
     - (3) Strict liability tort?
   - (b) Was the City of Boulder's act a(n):
     - (1) Intentional tort?
     - (2) Unintentional or negligent tort?
     - (3) Strict liability tort?

Assuming you decide that both acts being examined are classified as negligent torts:

1. What was the basic duty owed by Mario to his fellow users of the road? Would this duty be in any way modified by his physical condition, i.e., his colorblindness? Mario's basic duty was to operate his car in a safe manner as expected of a reasonable person under all the circumstances. This duty may, in fact, be modified by his physical limitation of colorblindness because he must be even more careful than the average driver not to rely solely on the position of the lights to see if it is safe to enter the crossing. In other words, he may have to compensate for his physical limitations by taking extra precautions.
2. What was the basic duty owed by the City of Boulder to the users of the road? As for the City, it too had a duty to make sure that all drivers could rely on the signals. Were either of these two duties not lived up to by either Mario and/or the City of Boulder? If so, would you classify these acts and/or omissions a breech of the duties owed? In both cases, it appears that both parties acted below the standard of care and were in breach of that duty.
3. Assuming you decided that both Mario and the City of Boulder did act below the standard of care set out in the duty of care, was there an injury suffered by the plaintiff such that it would allow him or her to bring a tort action before the court? Was the injury the physical or causal result of the defendant's act or omission? Was it the proximate and legal cause of the harm? The injuries to the plaintiffs were a direct result of Mario's presence in the intersection at the wrong time. His presence could, at least in part, be attributed to Mario's reliance on switched signals.
4. If you assumed that Gabby and Crabby were harmed by Mario's and the City of Boulder's acts, how would you allocate the percentage of harm among the defendant(s)? In spite of Mario's attempt

to shift a portion of the responsibility to the City, the primary responsibility for the accident will rest with him as the driver who had the opportunity to avoid the accident in the first place.

Would you excuse either Mario and/or the City of Boulder based on:

1. Any of the recognized defenses such as assumption of risk, or last clear chance, or contributory negligence, or comparative negligence? From the facts as given, it appears that none of the defenses would apply vis-à-vis Mario.
2. Do any of the recognized immunities apply in this case such as governmental/immunity, or intrafamily immunity, or charitable immunity? As for immunities, governmental immunity may be applicable to the City of Boulder.

Chapter 5
BUSINESS AND ON-LINE CRIMES

The history of the great events of the world is scarcely more than the history of crimes.

Voltaire

I. Overview

One of the most frequently posed questions to members of the legal profession is: "How can you defend someone you know is guilty of a crime?" The answer lies in a simple truth: we are defending not only the accused but also the criminal justice system. Because of this dual role, no other area of law is more ripe with controversy than criminal law and criminal procedure.

The American system of criminal law has been at the center of constant debate since the inception of the republic. On one side, there is a clear duty on the part of the sovereign to protect its populace from the ravages of miscreants whose behavior simply cannot be tolerated. On the other hand, the easier it is for the sovereign to punish, incarcerate, or even condemn its members to death, the less free the entire society is in the end. One of the badges of a totalitarian government is the ease with which it may act against its citizens. Consider the plight of the Holocaust victims; Nazi leaders did not worry about due process.

Our criminal justice system is designed to afford maximum protection for society's members. It is the government's burden of proof to show guilt beyond a reasonable doubt under the law of criminal procedure. The accused is entitled to an entire set of legal protections generically called due process. That phrase is derived from two provisions in the U.S. Constitution: one in the Fifth Amendment pertaining to actions by the federal government and one in the Fourteenth Amendment pertaining to actions by state governments. Under these provisions and the numerous federal, state, and local criminal codes, our system tells the sovereign we are presumed innocent until properly proven guilty according to the rules of the game. What upsets so many people in our society today is a concern that the rules may have been bent too far in favor of the accused.

As students of the legal system and as potential leaders in business, your job is to know what those rules of the game are and to learn how to comply with them as responsible members of society. Failure to do so is not only bad for business; it is bad for everyone.

II. Hypothetical Multi-issue Essay Question

Mergatroyd T. Wheeler has really come up with something. He has found a way in which his newly patented automobile engine could run on alternative energy. That alternative source of energy is found in a secret process that can turn cow pies into horsepower. His tests showed he could get twenty miles per pie (mpp) behind the wheel of his prototype named B.S. Bukaroo.

Mergatroyd took his idea to a large and prestigious investment banking firm, Wee, Cheatum, and Howe, to promote and sell shares in his new company named Go Pies. The financial records of the company were certified by I, Know, and Nothing CPAs. Shares in the new company were publicly sold to investors from all fifty states.

The company failed because the engine would only run on extremely odiferous pies, and now a number of disgruntled investors are pressing to bring RICO actions, at both the criminal and civil levels, against all parties involved.

III. Outline

What Is a Crime?
- A crime is a violation of a statute for which the government imposes a punishment

Parties to a Criminal Action
- Plaintiff
  - In a criminal lawsuit, the government (not a private party) is the plaintiff
- Defendant
  - The accused is the defendant

Classification of Crimes
- Felonies
- Misdemeanors
- Violations

Essential Elements of a Crime
- Criminal act
  - Defendant must have actually performed the prohibited act
  - Actual performance is called actus reus (guilty act)
- Criminal intent
  - Defendant must be found to have possessed the requisite state of mind when the act was performed
  - The state of mind is called mens rea (evil intent)
- Strict liability (liability without fault)
  - Some states impose criminal liability without a finding of mens rea (intent)

Stages of Pretrial Criminal Procedure
- Arrest
- Indictment or information
- Arraignment
- Plea-bargaining

Criminal Trial
- All jurors most unanimously agree before the accused is found guilty of a crime
- If even one juror disagrees, the accused is not guilty
- If all jurors agree the accused did not commit the crime, the accused is innocent

Crimes Affecting Business
- Robbery
- Burglary
- Larceny
- Theft
- Receiving stolen property
- Arson
- Forgery
- Extortion
- Credit card crimes
- Bad checks

White-Collar Crimes
- Embezzlement
- Criminal fraud
  - Mail fraud
  - Wire fraud
- Bribery
- Racketeer Influenced and Corrupt Organizations Act (RICO)

Inchoate Crimes
- Inchoate crimes include incomplete crimes and crimes committed by nonparticipants
- Examples include
  - Criminal conspiracy
  - Attempt to commit a crime
  - Aiding and abetting the commission of a crime

Computer Crimes
- Federal and state governments have enacted laws that address computer crimes
  - Counterfeit Access Device and Computer Fraud and Abuse Act of 1984
  - Electronic Funds Transfer Act

Corporate Criminal Liability

Constitutional Safeguards

Fifth Amendment Privilege Against Self-Incrimination

Attorney/Client Privilege

Accountant/Client Privilege

5th Amendment Protection Against Double Jeopardy

6th Amendment Right to a Public Jury Trial

8th Amendment Protection Against Cruel and Unusual Punishment

IV. Objective Questions:

Terms:

1. In a criminal context, the ______________ ______________ refers to the guilty act, while the ______________ ______________ refers to the accused's guilty intent.

2. Crimes can be divided into three separate classifications: __________________, ______________, and __________________.

3. To be found guilty of a crime, the accused must be found to have had the required state of mind. Where the accused purposely commits an act, this is ______________ intent. Where the accused acts recklessly, this is _______________ intent.

4. To arrest someone, the police must usually obtain a warrant. The warrant must be based on a showing of ______________ ______________.

5. The Act making it a criminal offense for anyone to intentionally access and obtain information from a protected computer without authorization is called the ______________ ______________ ______________ ______________.

6. At a criminal trial, the government's burden of proof requires that it prove ______________ ______________ ______________ ______________ that the accused is guilty.

7. Assault, battery, and murder are examples of crimes against the _______________ while robbery, burglary, and thefts are examples of crimes against ______________.

8. Criminal conspiracy, attempt to commit a crime, and aiding and abetting are all examples of _____________ crimes.

9. The Supreme Court created the ______________ ______________ ______________ to allow use of otherwise illegal evidence from a search if the police reasonably believed they were acting lawfully.

10. The Fifth Amendment's privilege ______________ ______________ provides that no person will be forced to be a witness against himself.

True/False:

1. ____ In a criminal action, the parties are the victim and the defendant.

2. ____ If a statute imposes absolute liability for a crime, *mens rea* does not have to be proven.

3. ____ Even today, corporations are never held criminally liable because it is impossible for a corporation to have a criminal mind.

4. ____ If police do not have time to get an arrest warrant against a fleeing person, they can arrest the person without probable cause.

5. ____ A grand jury determines a person's guilt in serious crimes such as murder.

6. ____ If the accused enters a plea of *nolo contendere*, the government has the option of accepting the plea or requiring the defendant to plead either guilty or innocent.

7. ____ In a criminal trial, all jurors must unanimously find the accused guilty beyond a reasonable doubt for a conviction.

8. ____ Under RICO, to prove a pattern of racketeering, no less than two predicate acts must be committed within a ten-year period.

9. ____ To be liable for criminal conspiracy, the accused must have actually committed a crime.

10. ____ The exclusionary rule prevents illegally obtained evidence from being used against the person searched.

Multiple Choice:

1. Vice President Vance works for Capital Corp. Vance forged several official documents to get money from Bigtown Bank. The bank ultimately loaned money to Capital Corp. for various business operations.
   A. Only Vance can be liable for these actions since an employee of a corporation is always liable for his illegal action, and the corporation can never be held liable because it cannot form *mens rea*.
   B. Only the corporation can be liable for these actions because the forgeries benefited the corporation. Vance is not liable because he did not personally benefit.
   C. Neither the corporation nor Vance can be liable because the forgeries benefited only the corporation. Since the corporation cannot form *mens rea*, it cannot be guilty of a crime.
   D. Both Vance and the corporation can be liable. Vance is liable for forgery regardless of whether he or the corporation benefits. The corporation may be liable for the crimes of its employees.

2. Accused Adam is tried by a jury. Only one of the jurors is not convinced beyond a reasonable doubt that Adam is guilty. All other jurors are sure Adam is guilty.
   A. Adam is convicted because a majority of the jurors believe he is guilty.
   B. Adam is acquitted because in a criminal trial, the jury must be unanimous to convict.
   C. Adam is convicted because although one juror is not convinced beyond a reasonable doubt, the government only has to prove Adam's guilt by a preponderance of the evidence.
   D. None of the above.

3. Andy Accountant manages the financial accounts of Multimill Co. Andy decides Multimill will not notice a $10,000 withdrawal from a few of its accounts. Ultimately, Andy takes about $100,000 from Multimill Co. Andy has committed:
   A. Theft.
   B. Extortion.
   C. Embezzlement.
   D. Larceny.

4. Hardcore Cop enters a house with a valid search warrant that states he can search for and seize any firearms such as guns, pistols, etc. However, Hardcore sees a large bag of cocaine lying on the coffee table and seizes that as well.
   A. Hardcore legally seized the cocaine because it was sitting out in plain view.
   B. Hardcore legally seized the cocaine because, once he had a valid search warrant, he could take anything he thought was suspicious.
   C. Hardcore illegally seized the cocaine because, if it is not listed on the warrant, it can never be seized.
   D. None of the above.

5. Hardcore Cop arrests Running Randy for alleged bank robbery. Before he says anything else, Hardcore says, "You might as well admit it now and save us both a lot of time." Randy looks at him and replies, "All right, all right, I did it--Are you happy?"
   A. Randy's confession is admissible evidence since Hardcore did not coerce him.
   B. Randy's confession is inadmissible because an accused person must have a lawyer for a confession to be used as evidence.
   C. Randy's confession is inadmissible because Hardcore did not inform him of his Miranda rights.
   D. None of the above.

6. Defendant Dan is acquitted on murder charges. Under such circumstances, the government can:
   A. Conduct a new trial.
   B. Appeal the decision to a higher court.
   C. Do nothing, this decision stands.
   D. None of the above.

7. Defendant Dan is convicted on murder charges. Under such circumstances, Dan can:
   A. Conduct a new trial.
   B. Appeal the decision to a higher court.
   C. Do nothing, this decision stands.
   D. None of the above.

8. Convicted Chris is found guilty of criminal RICO violations. Chris had bought several expensive cars and stock in a well known, legitimate computer company with money from his racketeering activity. What is the appropriate punishment for this crime under these circumstances?

A. Imprisonment.
B. Forfeiture of the car.
C. Forfeiture of the stock.
D. All of the above.

9. Shady Sharon and Shifty Sherman agree they are going to rob Wealthy Wendy. Sharon and Sherman drive to Wendy's house. At this point, Sharon and Sherman would be liable for:
   A. Criminal conspiracy.
   B. Attempt to commit a crime.
   C. Aiding and abetting the commission of a crime.
   D. None of the above.

10. Assume the same facts as in question 9 except that Sharon and Sherman have only agreed to do this, but are not yet in the car. At this point, Sharon and Sherman would be liable for:
    A. Criminal conspiracy.
    B. Attempt to commit a crime.
    C. Aiding and abetting the commission of a crime.
    D. None of the above.

V. Answers to Objective Questions

Terms:

1. *Actus reus, mens rea*. Both elements are necessary to the definition of a crime.
2. *Felonies, misdemeanors, violations*. These classifications are set out in statutes.
3. *Specific, general*. The statutes describe the type of intent required.
4. *Probable cause*. Case law interpretation of this standard is controversial.
5. *Information Infrastructure Protection Act.* The IIP Act provides protection for any computer attached to the Internet.
6. *Beyond a reasonable doubt*. This does not mean absolute certainty.
7. *Person, property*. These classifications are similar to those used in the law of intentional torts.
8. *Inchoate*. This means partial, unfinished, or incomplete.
9. *Good faith exception*. This is a very controversial issue and is subject to ongoing revision by the U.S. Supreme Court.
10. *Against self-incrimination*. This is one of our basic projections against governmental abuse of power.

True/False:

1. False. In a criminal trial, the parties are the government and the defendant. The victim does not bring criminal charges, but may have a separate civil action.
2. True. A statute imposes absolute liability on a specific type of crime where proof of *mens rea* is not a necessary element.
3. False. Corporations can be held criminally liable even though the corporation itself does not have *mens rea.* The corporation is liable for the criminal acts of it managers, agents, and employees.
4. False. Even if police do not have time to get an arrest warrant, any arrest must be based on probable cause.
5. False. A grand jury does not determine guilt or innocence. The grand jury decides if there is sufficient evidence to hold the accused for a trial.

6. True. The government can choose to accept this plea or require the accused to plead either guilty or innocent.
7. True. All jurors must unanimously agree that the government has convinced them beyond a reasonable doubt that the accused is guilty. If even one juror is not so convinced, the jury is considered a hung jury.
8. True. Under RICO, a pattern of racketeering activity can be established by any two predicate acts committed within a ten-year period.
9. False. To be liable for criminal conspiracy, the accused must only have overtly acted to further the crime. Actual commission of the crime is not required.
10. True. Where the search is unreasonable, evidence obtained will not be admissible against the person searched.

Multiple Choice:

1. D. A corporate officer or director is always individually liable for the crimes he/she commits whether to benefit himself/herself or the corporation. In addition, the corporation is liable for the criminal acts of its officers or directors. A is incorrect because the corporation can be held liable even though it does not have *mens rea*. B is incorrect because Vance is liable for this illegal act even if he did not personally benefit. C is incorrect for reasons explained above.
2. B. In a criminal trial, all jurors must unanimously agree beyond a reasonable doubt that the accused is guilty. A is incorrect because a majority is not sufficient for conviction. C is incorrect because the preponderance of evidence standard only applies to civil trials, not criminal trials.
3. C. Because Andy has stolen money that his employer entrusted to him, this is embezzlement. A and D are incorrect because Andy was entrusted with the money. B is incorrect because extortion is a threat to expose something about another person.
4. A. Under the plain view doctrine, Hardcore can seize evidence that is in plain view. This evidence is admissible even though it is beyond the specific scope of the warrant. B is incorrect because a warrant does not generally allow seizure of items not on the warrant unless they fit into some exception. C is incorrect because certain exceptions do allow such seizures.
5. C. A confession that is obtained before a person is advised of his Miranda rights is generally not admissible as evidence. A is incorrect because although police coercion is illegal, the critical issue in this question is whether Randy was advised of his rights. B is incorrect because Randy need only be aware of his rights before a confession can be used. The presence of a lawyer is an option for Randy, but not required if he wishes to confess.
6. C. Once a defendant is acquitted, the government cannot appeal or ask for a new trial.. Such actions would violate the double jeopardy clause of the Fifth Amendment.
7. B. If Dan is convicted, he can appeal this decision to a higher court.
8. D. Under criminal RICO, the guilty defendant can be imprisoned or fined. In addition, forfeiture of any other property gained through RICO violations is permitted.
9. A. Since Sharon and Sherman actually drove to Wendy's house, this would be considered an overt act. An overt act is all that is required to prove criminal conspiracy. Actual commission of the crime is not necessary. B is incorrect because they have not actually tried to commit the crime yet. C is incorrect because the crime has not yet been committed.
10. D. Until Sharon and Sherman make an overt act, they are not liable for anything. There must be something beyond a mere agreement.

VI. Answers to Essay Question:

The RICO (Racketeer Influenced and Corrupt Organization) statute is a federal statute designed to go after white-collar criminals associated with racketeering activities. Even though it can preempt state and local laws covering similar activities, it can be prosecuted in tandem with those statutes. We will assume only the federal law applies here. The elements called for under the statute are:

(1) A prohibited pattern of acts, which in this case may involve securities and or mail fraud.

(2) The specific intent to defraud investors.

If it can be shown that the accused parties made fraudulent misrepresentations about the performance capabilities of the "cow pie" engine to investors, then this behavior may come under the purview of the RICO statute. There is great controversy over the proper fit between the acts defined in the RICO statute as opposed to the normal business risks that come with investing in new ventures. Many commentators have argued that RICO should not be used as a wedge to recover losses from simple bad judgment.

Assuming that the inventor, the investment bankers, and the CPA firm were all properly warned of their rights and all business records were properly searched, it appears likely that they may be criminally liable for securities fraud.

If any of the procedures were not used properly, the criminal case against the accused would most likely fail based on the exclusionary rule. That rule would keep out tainted evidence. It is interesting to note, however, that even though the criminal case may fail, civil damages may still be imposed in a separate trial. The lower burden of proof coupled with the possibility of nonexclusion of the tainted evidence in that forum only exacerbates the problem for the accused under this broad-reaching statute.

As a matter of public policy, government does need statutes like RICO to go after white-collar crime and racketeering-related activities. As mentioned, the concern of many critics of the law is that it is casting too large a net and dragging in innocent, legitimate business entities.

Chapter 6
INTERNATIONAL AND COMPARATIVE LAW

The world must be made safe for democracy.
Woodrow Wilson

I. Overview

The word oxymoron is defined as a combination of contradictory or incongruous words or phrases. To many, the term international law represents one such example. At best, defining international law along the traditional domestic lines of a body of rules of behavior and mechanisms designed to enforce those rules cannot work as well when the protagonists are sovereign nations. By definition, sovereignty incorporates the notion of freedom from external controls and supreme power over one's own affairs. Perhaps that might be a starting point to resolve this oxymoronic dilemma. A nation does need to have control over its internal affairs, but its national interests and the welfare of its citizens do not end at its borders. Nations, just like individuals, can only find protection for their own rights when they are willing to honor the rights of others. Law eventually works its way through to a system of cooperative behavior for the larger mutual good. All law calls for some sacrifice of individual freedoms for the betterment of the corporate body. This is the fundamental reality upon which all law is ultimately based. The onsequences of failures to honor that reality at the global level are readily apparent to all of us. If the role of law is to act as a mechanism for civilization that makes violence the last resort, then international law is a goal worth striving for by all nations.

The second interesting aspect of any comparison between international law of nations and international commerce is the sheer breadth and scope of commerce. International trade has become the lifeblood of nations, and international business entities are its circulatory system. In quite blunt terms, the scale and power attained by many large multinational banks, manufacturers, and diversified business organizations have simply surpassed the ability of any one country to control them. That is not to say that these are renegade free agents with no accountability. It is to say, however, that the worldwide staging area of economic competition is such that these players are big enough and strong enough to have the rules written for in their favor. Witness the worldwide and sometimes fawning competition by countries for new plants and financial investments created by these entities. Also, witness the impact of e-commerce on international law.

This chapter seeks to outline some of the basic terminology used in international law. It also describes some of the key players on that stage. Like any good play, this stage is filled with intrigue, pathos, and sometimes-epic Greek tragedy. The failures of whole nations only magnify our own fallibilities, because international law is the most fragile of all legal systems. Yet in an age of possible nuclear self-destruction, do we have any choice? Like it or not, as William James said: "The things we cherish most are at the mercy of the things we cherish least."

II. Hypothetical Multi-issue Essay Question

Mario LaFast is a U.S. resident who has finally made it to the top of the auto-racing world. He just signed a three-year contract to drive for Enzo Frank and his Formula One Supercharged Turbo Racing Team in Italy. Mario's retainer is over $10 million annually plus a percentage of the winning purse if he finishes third or higher. The Formula One race calendar starts in South America in late spring and finishes in late fall in Australia. The races are held in a total of sixteen different countries.

All of the host countries have their own tax systems, and they all seek to tax, as income, winnings from auto races. In addition, many of these tax laws measure income on a worldwide basis. Thus income earned in a race in Italy may be reportable on Mario's tax returns filed in New York. If all the possible taxes are independently and fully imposed by every nation in which Mario won, his tax bill would exceed his income because of "double-taxation" on the same income by more than one taxing entity. What can Mario do?

III. Outline

United States and Foreign Affairs
- Commerce Clause
  - Vests Congress with the power "to regulate commerce with foreign nations"
- Treaty Clause
  - States that the President "shall have power, by and with the advice and consent of the Senate, to make treaties, provided two-thirds of the senators present concur"

Sources of International Law
- Custom
  - Consistent, recurring practices between two or more nations over a period of time that have become recognized as binding
- General principles of law
  - Principles of law recognized by civilized persons
- Judicial decisions and teachings
  - Judicial decisions and writings of the most qualified legal scholars of the various nations involved in a dispute
- Principle of comity
  - Courtesies between countries based on respect, goodwill, and civility, rather than law
- Treaties and conventions
  - Agreements or contracts between two or more nations that are formally signed by an authorized representative and ratified by the supreme power of each nation

United Nations
- An international organization created by a multilateral treaty in 1945
- Is governed by:
  - The Security Council
- Composed of 15 member nations
- The General Assembly

Composed of all member nations
The Secretariat
Administers day-to-day operations

International Regional Organizations
European Union
Latin, Central, and South American Economic Communities
African Economic Communities
Middle Eastern Economic Communities
World Trade Organization

International Courts
International Court of Justice
European Court of Justice
National courts
The majority of cases involving international law disputes are heard by the national courts of individual nations

International Contract Clauses affecting Jurisdiction and Choice of Law
Forum selection
Designates the judicial or arbitral forum that will hear and decide the case
Choice of law
Designates the law to be applied by the court or arbitrator in deciding the case

Act of State Doctrine and Sovereign Immunity Compared
Act of state
An act of a government in its own country that is not subject to suit in another country's courts
Sovereign immunity
Act of a government in a foreign country that is not subject to suit in the foreign country

International Arbitration
Nonjudicial method of dispute resolution where a neutral third party decides the case
Parties agree to be bound by the arbitrator's decision

Extradition
Extradition
Sending a person back to a country for criminal prosecution
The United States has entered into extradition treaties with many countries

IV. Objective Questions:

Terms:

1. Canada, Mexico, and the United States entered into an international trade agreement that created the largest free trade area in the world. This treaty is called the ________________ ________________ ________________ ________________ ________________.

2. The most important of all international organizations is the ________________ ________________.

3. The power of Congress to regulate commerce with foreign nations is granted by the _______________ _______________.

4. An agreement or contract between two or more nations that is formally signed by an authorized representative and ratified by the supreme power of each nation is called a _______________.

5. The largest and most important economic community of nations in Africa is the _______________ _______________ _______________ _______________.

6. The influence of this economic group is inexorably tied to the worldwide supply and price of oil. It is the _______________ _______________ _______________ _______________ _______________.

7. The extension of courtesies to other countries that are not obligations of law, but rather based on respect, goodwill, and civility is called _______________.

8. This international court is located in the Hague and is the judicial branch of the United Nations. It is called the _______________ _______________ _______________ _______________ or the _______________ _______________.

9. The doctrine which provides that judges of one country cannot question the validity of an act committed by another country within that other country's own borders is known as the _______________ _______________ _______________ _______________.

10. A nonjudicial method of dispute resolution whereby a neutral third party decides the case is called _______________.

True/False:

1. ____ Under the Treaty Clause of the U.S. Constitution, states may enter into treaties with foreign nations.

2. ____ The President is the agent of the United States in dealing with foreign nations.

3. ____ International courts are not bound by the doctrine of *stare decisis*.

4. ____ In 1996, the World Trade Organization issued three new conventions.

5. ____ Individuals or businesses may have cases decided by the International Court of Justice.

6. ____ The European Court of Justice is a lower court branch of the World Court.

7. ____ The United States grants absolute immunity to foreign governments from suit in U.S. courts.

8. ____ An action based on a commercial activity carried on in the United States will not make a foreign nation immune from suit in U.S. courts.

9. _____ The U.S. is a signatory to the United Nations Convention on Recognition and Enforcement of Foreign Arbitral Awards.

10. _____ A nation has the authority to criminally prosecute its citizens that commit crimes in other countries.

Multiple Choice:

1. Which of the following statements is true of international law?
   A. There is no single source of international law.
   B. No single world court is responsible for interpreting international law.
   C. There is no world executive branch that can enforce international law.
   D. All of the above.

2. Which of the following is not a source of international law?
   A. Custom.
   B. State law.
   C. Treaties.
   D. Comity.

3. To establish that a practice has become a custom, you must show:
   A. Consistent and recurring action by two or more nations over a considerable period of time.
   B. Recognition that the custom is binding.
   C. A and B.
   D. Neither A nor B.

4. Which of the following is not a significant international organization?
   A. Supreme Court of the United States.
   B. United Nations.
   C. International Court of Justice.
   D. International Monetary Fund.

5. Which of the following is not a branch of the United Nations?
   A. Security Council.
   B. General Assembly.
   C. Secretariat.
   D. Common Market.

6. Which of the following are regional development banks?
   A. Inter-American Development Bank.
   B. Asian Development Bank.
   C. Caribbean Development Bank.
   D. All of the above.

7. Which of the following is not an Africa-based economic organization?
   A. CARICOM.
   B. EFTA.
   C. LAIA.
   D. All of the above.

8. The North American Free Trade Agreement:
   A. Created the world's largest free trade area.
   B. Eliminates all tariffs on nonagricultural trade by 1995.
   C. Establishes tariffs for many agricultural products.
   D. B and C.

9. Organizations that conduct international arbitration include:
   A. American Arbitration Association.
   B. International Chamber of Commerce.
   C. United Nations Commission on International Trade Law.
   D. All of the above.

10. To avoid the problems of judicial resolution of international disputes, the parties to an international transaction should include the following in their contracts:
    A. Choice of law clause.
    B. Exclusionary clause.
    C. Choice of forum clause.
    D. A and C, but not B.

V. Answers to Objective Questions:

Terms:

1. *North American Free Trade Agreement.* This treaty creates the largest free trade area in the world. It was ratified after much heated debate over jobs, environmental issues, and immigration issues.
2. *United Nations.* This organization, created after World War II, is the spiritual successor to the League of Nations, which was dissolved in April 1946.
3. *Commerce Clause.* Only the federal government may enter into treaties with other nations, and these treaties become part of the law of the land.
4. *Treaty.* Treaties can be classified as bilateral, multilateral, and/or conventions.
5. *Organization of African Unity.* This is a relatively new organization formed in 1991, but because of its large membership and wide ranging treaty provisions, its importance is expected to grow dramatically.
6. *Organization of Petroleum Exporting Countries.* The influence of this group will never be insignificant, but it has been reduced by the discovery of oil in areas like the North Sea and Alaska.
7. *Comity.* This term is sometimes referred to as "comity of nations" in the international setting.
8. *International Court of Justice or World Court.* Remember the decisions of this court are only binding if a nation agrees to be bound.
9. *Act of State Doctrine.* The underlying reasoning for this doctrine is restraint on the part of the judiciary that permits the executive branch to arrange affairs with foreign governments.
10. *Arbitration.* Arbitration is almost always faster, cheaper, and more efficient than going to a judicial tribunal.

True/False:

1. False. Only the federal government can enter into treaties.
2. True. Treaty negotiations are often long and tortuous processes where much goes on behind the scenes. Witness the hard-fought give and take between the U.S. and Mexico on issues relating to employment and environmental law enforcement mechanisms relating to the NAFTA treaty.

3. True. As a matter of practicality, they do follow their own past decisions for guidance in making decisions, even though they are not bound by them.
4. False. It issued two conventions: the Copyright Treaty and the Phonogram Treaty.
5. False. Only nations can have cases decided by the International Court of Justice, but nations may bring a case on behalf of an individual or business.
6. True. This court, located in Luxembourg, represents the judicial branch of the European Community or Common Market.
7. False. The U.S. grants qualified or restricted immunity to foreign governments, as outlined in the Foreign Sovereign Immunities Act of 1976.
8. True. One of the exceptions to absolute foreign immunity provided by the Foreign Sovereign Immunities Act of 1979 allowing for a suit to be brought on the basis of commercial activity.
9. True. The U.S. signed this convention in 1970 and amended the preexisting Federal Arbitration Act to reflect this international law.
10 True. Enforcement of these criminal laws is carried out with the help of extradition treaties. Unfortunately, these treaties are not fully in place throughout the world, and many safe havens are available to rich miscreants fleeing from justice.

Multiple Choice:

1. D. This loose knit structure of international law is what has led some critics to call international law an oxymoron.
2. B. State laws are generally not considered a source of international law although state statutes can be applied to international entities doing business in those states. If, however, state laws are in conflict with a treaty, the Preemption Clause of the Constitution renders the conflicting state or local law void.
3. C. The nation's leaders may find custom in any number of sources such as government policy statements, diplomatic correspondence, and speeches.
4. A. Even though the impact of the court's decisions may have an impact around the world, it is not technically classified as an international organization in the same sense that the International Court of Justice is.
5. D. The organization of the Common Market is separate from the United Nations even though both organizations have member nations that belong to both groups.
6. D. The main objective of these organizations is to assist in the economic development of its member nations.
7. D. All of the above represent regional organizations based outside of Africa. The Organization of African Unity, made up of 51 African countries, represents the largest organization of countries from the African continent.
8. A. The NAFTA free trade area encompasses not only the world's largest geographic trade area. It is also is the richest with a $6.5 trillion market.
9. D. All of these organizations provide arbitration services to international clients. More and more international business contracts have arbitration clauses included at the outset.
10. D. The use of these clauses is a practical, if not legal, necessity in today's international legal environment.

VI. Answers to Essay Question:

Mario certainly has a mixed blessing in having made it to the top of the racing world. On the plus side, he is a world celebrity who really does live out the fantasy of "life in the fast lane!" The down side of all this is that life has gotten far more complicated since the days when he was drag racing behind the high school parking lot.

Under our taxing structure, both the federal and state governments can impose income taxes. Some taxes paid to the states can be taken as either a business expense above the line for arriving at adjusted gross income or as a personal itemized deduction after having arrived at adjusted gross income. This would all be well and good if Mario's racing activities were limited to the confines of the U.S.

But his income source will not only be the U.S. Because our tax laws seek to impose a tax on world-based income, Mario must seek ways to avoid double taxation. The federal government will be the source of any protection that may be afforded to him on this issue in the U.S. He must look at the national governments of other nations seeking to impose their taxes. Those national policies will preempt any state laws that may be in conflict with the national policies enumerated in the tax laws.

The best source of protection available to Mario will be tax treaties. These are either bilateral or multilateral agreements entered into by various nations that are designed to allow each other to tax a "fair share" of Mario's overall worldwide income without creating confiscatory double taxation. Under these treaties, each nation decides how to defer taxation to other nations where that income is already earned and taxed in the other nation. By use of certain agreed upon parameters of exclusions from income or credits for taxes paid to other countries, the net intended effect is to allow taxation of the same income only once in each country where it is earned.

All this is easier said than done. The world of international taxation is one of the most difficult areas of law, accounting, and business practice. It takes years just to learn the ropes and is extremely technical in nature. Mario should seek advice and counsel from professionals trained in these matters.

In case two or more of the countries seeking to tax Mario reach an impasse on how his income should be taxed, they may choose to voluntarily submit their disagreement to the one of the international tribunals such as the U.N. International Court of Justice. As an alternative, the countries may choose to seek a voluntary form of arbitration. If arbitration is chosen, enforcement will still be an issue if they are not signatories to an arbitration enforcement convention.

In any event, Mario's best route on this issue is to have his contract set out, at the beginning, a choice of law and choice of jurisdiction. Even there, remember, each country as a sovereign can still ultimately control the taxation of income earned within its own borders.

Mario has decided to move to Monaco because of the favorable tax laws there. Even if he is still taxed elsewhere, he will do just fine, thank you, at the casinos in Monte Carlo.

Chapter 7
ETHICS AND SOCIAL RESPONISBILITY OF BUSINESS

There are no shortcuts to any place worth going.
Beverly Sills

I. Overview

The novel *Spartacus* was written by the French author Bernard Joseph Saurin in 1760. The novel was named after the Thracian slave who died in 71 B.C. after leading a revolt against the Roman authorities. In this novel, Saurin wrote: "The law often permits what honor forbids." That quote truly encapsulates what the study of ethics vis-à-vis law is all about. While it may have been legal to trade in human misery and subjugation, could it ever be morally acceptable under any of the systems of ethics discussed in this chapter?

The study of ethics revolves around the examination of rules, conduct, and character through a morally tinted microscope. That law should be grounded in some sort of morality-based foundation is self-evident. Yet, painful history has shown us over and over the prices paid when there is an abyss between law and ethics. Sometimes these gaps can only be closed by conflagrations and wars so costly that we are reminded of an old anonymous proverb: "Adam ate the apple, and our teeth still ache."

The goals of all the ethical schools of thought are to seek some sort of morally based rationale for human behavior. This rationale may be found in outside sources as seen in schools of ethical fundamentalism or in the rule that provides the greatest good to society as illustrated by utilitarianism. Others such as Kant and Rawls have sought to devise formulas of behavior based on universal rules or social contract, respectively. In all these systems, a morally based methodology is sought as a guidepost for behavior. If these guideposts are universally accepted, the odds are very high that they will no longer be advisory, but rather required by law. The process by which morally based ethical behavior is first desired, then expected, and finally mandated is really the evolution of law. Note that law is the last stage in this often-painful process.

Because so much of our legal and economic activities are conducted in the corporate format, juristic (law-made) business entities cannot ignore this constant and dynamic tug and pull between ethics and law. We have either come a long way or become hopelessly bogged down in political correctness depending on your personal point of view. This dynamic is well illustrated by the differences between the thinking of the court in *Dodge v. Ford Motor Company* in 1919 and today's social audits of corporate citizenship The bottom line in the study of ethics is ultimately personal. Imagine a world without a concern for law and ethics. As French author Stephanie Félicité Genlis said in 1793, "The man whose probity consists in merely obeying laws cannot be truly virtuous or estimable; for he will find many opportunities of doing contemptible and even dishonest acts, which laws cannot punish."

II. Hypothetical Multi-Issue Essay Question

Taxes are a cost of not only doing business but also an expense of living in a free society. As Justice Oliver Wendell Holmes said in 1927, "Taxes are what we pay for civilized society." But must we pay even in death? The federal government has a long established set of wealth transfer taxes designed to pay back the government for the protection it provided to a decedent during his or her accumulation of wealth. One such tax is the federal estate tax that is imposed on decedents whose estates are valued at over $600,000 at the time of death.

You are a newly graduated associate of the Ras, Ma, and Tas law firm that represents the estate of Dealing Don, the late, great Aspen rental real estate mogul. The fair market value of his rental real estate holdings is over $6 million. Your senior partner finds an Internal Revenue Code section that allows a "use" value for family-owned businesses. This use value is intended to preserve family farm ownership. Should you convert these city rental units to agricultural "use" in order to reduce their reported value to under $600,000?

III. Outline

Law and Ethics

Theories of Ethics
- Ethical fundamentalism—People look to an outside source or central figure for ethical guidelines
- Utilitarianism—People choose the alternative that would provide the greatest good to society
- Kantian ethics—A set of universal rules based on reasoning establishes ethical duties
- Rawls's social justice theory—Moral duties are based on an implied social contract
- Ethical relativism—Individuals decide what is ethical based on their beliefs about right and wrong

Social Responsibility of Business
- Social responsibility of business is a duty owed by businesses to act socially responsible in producing and selling goods and services

Theories of Social Responsibility
- Maximize profits
  - To maximize profits for stockholders
- Moral minimum
  - To avoid causing harm and to compensate for harm caused
- Stakeholder interest
  - To consider the interests of all stakeholders, including stockholders, employees, customers, suppliers, creditors, and local community
- Corporate citizenship
  - To do good and solve social problems

Examples of Ethical Questions
- Is it ethical to pay bribes in a foreign country where that is the common business practice?
- Is it ethical to use cartoon characters to advertise cigarettes?

IV. Objective Questions:

Terms:

1. A theory that says law should be based on morality and ethics is called _______________ _______________ _______________ _______________.

2. When a person looks to an outside source for ethical rules or commands, he or she might be categorized as a _______________ _______________.

3. A moral theory that dictates that people must choose the action or follow the rule that provides the greatest good to society is called _______________ _______________.

4. A moral theory that says that people owe moral duties that are based on universal rules is called _______________ _______________.

5. Basing one's behavior on the adage: "Do unto others as you would have them do unto you" is called the _______________ _______________.

6. A moral theory that says each person is presumed to have entered into a social contract with all others in society to obey moral rules that are necessary for people to live in peace and harmony is called a ___________ ___________.

7. A theory of social responsibility that says a corporation must consider the effects its actions have on persons other than its stockholders is called _______________ _______________.

8. A theory of social responsibility that says a business has a responsibility to do good is called _______________ _______________.

9. Under the traditional business judgment rule, directors of a corporation owe a _______________ _______________ to act on an informed basis, with reasonable care, and in good faith.

10. This act generally makes it a crime for U.S. companies to bribe a foreign official for purposes of conducting business. It is called the _______________ _______________ _______________ Act of 1977.

True/False:

1. ____ Under Kantian ethics, morality is judged by a person's act or motives.

2. ____ Moral Minimum is a theory of social responsibility that says that business is responsible for helping to solve social problems that it did little if anything to cause.

3. ____ Law and ethical standards never conflict with one another. If something is a law, it is, by definition, ethical.

4. ____ Company codes of ethics are a form of Ethical Fundamentalism.

5. ____ Under Ethical Fundamentalism, a person looks solely within himself or herself for ethical codes or commands.

6. ____ Immanuel Kant would support the Golden Rule.

7. ____ John Locke and Jean Jacques Rousseau would support the Social Contract theory of morality.

8. ____ Ethical Relativism has been widely adopted as a moral theory because of its strict guidelines.

9. ____ A corporation which moves jobs overseas to take advantage of lower pay scales is adhering to the Stakeholder Interest school of corporate ethics.

10. ____ Courts expect due diligence to include searching the Internet when searching for public information.

Multiple Choice:

1. Literal Larry believes that all ethical standards and rules are set out in a book called *Ethics For All* that was written by a prominent ethical theorist. Larry refers to this book for guidance and rules by which to lead his life. Which of the following moral theories most accurately describes Larry's conduct?
   A. Kantian Ethics.
   B. Utilitarianism.
   C. Ethical Relativism.
   D. Ethical Fundamentalism.

2. Societal Sarah believes people should take actions that benefit society as a whole the most. Which of the following moral theories most accurately describes Sarah's beliefs?
   A. Kantian Ethics.
   B. Utilitarianism.
   C. Rawl's Social Justice.
   D. Ethical Fundamentalism.

3. Individual Ingrid firmly believes that each person must look at her own feelings of what is right and wrong to decide what is ethical. Which of the following moral theories most accurately describes Ingrid's beliefs?
   A. Utilitarianism.
   B. Ethical Fundamentalism.
   C. Rawl's Social Justice.
   D. Ethical Relativism.

4. Dutiful Daisy always considers how she would feel in another person's place before she takes any action. Then, Daisy only acts in accord with how she would want others to behave in the same circumstances. Which of the following moral theories most accurately describes Daisy's conduct?
   A. Kantian Ethics.
   B. Ethical Relativism.
   C. Ethical Fundamentalism.
   D. Utilitarianism.

5. Fair Fran believes that fairness is the essence of justice. In addition, Fran believes all people in a society have a contract with one another to obey moral rules. Which of the following moral theories most accurately describes Fran's beliefs?
   A. Utilitarianism.
   B. Ethical Fundamentalism.
   C. Rawl's Social Justice.
   D. Ethical Relativism.

6. C.E.O. Ceo believes her corporation must consider what effects it may have on its employees, suppliers, customers, etc. Which of the following social responsibility theories does Ceo most likely subscribe to?
   A. Corporate Citizenship.
   B. Stakeholder Interest.
   C. Moral Minimum.
   D. Maximizing Profits.

7. C.E.O. Crusader believes his corporation can be influential and should be responsible for helping solve many social problems. Crusader believes his corporation should address problems above and beyond those actually caused by the corporation. Which of the following social responsibility theories does Crusader most likely subscribe to?
   A. Corporate Citizenship.
   B. Stakeholder Interest.
   C. Moral Minimum.
   D. Maximizing Profits.

8. Advocate Andrew believes corporations should not only be held accountable for their financial states, but also for some level of social responsibility. Andrew's beliefs would most likely be defined as:
   A. A code of ethics.
   B. A social contract.
   C. A corporate social audit.
   D. A distributive justice theory.

9. Corporate codes of ethics can be derived from principles educated by:
   A. Ethical Fundamentalism.
   B. Utilitarianism.
   C. The Categorical Imperative.
   D. All of the above.

10. Nobel Prize winner Professor Emeritus Milton Friedman would favor a corporate code of conduct that would:
    A. Maximize profits at all costs.
    B. Maximize profits based on open and free competition.
    C. Maximize profits as long as the corporation plays within the rules without fraud or deception.
    D. B and C only.

V. Answers to Objective Questions:

Terms:

1. *Moral theory of law.* Without it, law has no foundation.
2. *Ethical fundamentalism.* The Natural Law School of jurisprudence is considered to be one of the oldest, and it is still quite important in the study of law and ethics.
3. *Utilitarianism.* Many people have difficulty with this theory because of its lack of absolute or bottom line standards.
4. *Duty ethics.* This is traceable to religious doctrines.
5. *Categorical imperative.* The key advocate was Kant.
6. *Social contract.* Many of the newer schools of jurisprudence are variations on this basic theme.
7. *Stakeholder interest.* The difficulty lies in defining who is a stakeholder.
8. *Corporate citizenship.* This debate has gone on since the inception of the corporate form.
9. *Fiduciary duty.* It is the highest standard of duty implied by law.
10. *Foreign corrupt practices.* Note, the payment may be legal if it was lawful under local written laws of the foreign country where the payment was made.

True/False:

1. True. The test is based on the categorical imperative.
2. False. Moral Minimum is a theory of social responsibility which says that corporate social responsibility is to make a profit while avoiding harm to others. Corporate Citizenship is a theory of social responsibility that says business is responsible for helping to solve social problems that it did little if anything to cause.
3. False. The law may be contrary to a person's ethical standards.
4. True. The real question is whether they are being enforced or whether this is just the latest form of public relations.
5. False. Just the opposite is true. Under ethical fundamentalism, one looks to an outside source. This source may be from religious or political teachings and the like.
6. True. The Golden Rule is generally accepted as being based on the categorical imperative preferred by Kant.
7. True. Under this theory, there is an implied contract which states: "I will follow the rules if everyone else does."
8. False. Even though ethics remain an extremely personal decision, few people deny that some universal ethical rules should apply to everyone.
9. False. Most proponents of this school of thought would argue that such a move is based on the shareholder's interest alone.
10. True. More and more cases are ruling that searching the Internet is a necessary part of legal research.

Multiple Choice:

1. D. One of the key criticisms of any form of fundamentalism is that it tends to be too simplistic by trying to answer all questions at all times.
2. B. This approach tends to balance the various interests involved and weigh in favor of the most benefit possible for the least social cost or harm.
3. D. Many people have difficulty with this theory because of the lack of outside controls on Ingrid's decisions.
4. A. This theory is also often referred to as the Golden Rule.

5. C. This theory represents one of the newer schools of jurisprudence.
6. B. The debate goes on as to how big the stakeholder class should be.
7. A. Most corporations have only bought into this theory on a limited basis.
8. C. A corporate social audit, in theory, would hold a corporation accountable for its social responsibility as well as its financial status.
9. D. All of these theories can and do provide rational bases for guidelines which can be used in a corporate code of ethics.
10. D. Professor Friedman holds the traditional view that the social responsibility of business is to serve its shareholder first through profits lawfully gained.

VI. Answers to Essay Question:

A tax law practitioner has a duty to represent his or her client in an ethical manner as prescribed by the professional canons of ethics. For example, candidates for admission to the bar swear to "not counsel or maintain any suit or proceeding which shall appear to be unjust, nor any defense except ... to be honestly debatable under the law of the land."

The behavior expected here would be to use all *legal* arguments to favor your client's position. This would include any reasonable interpretations of Internal Revenue Code sections that may be applicable in the case at hand.

The law firm should affirm maximum advocacy for its client, but guard against unethical behavior that would pervert the true intention of this code section, i.e., it cannot make white black or vice versa only to save taxes. If there is no plausible way these parcels could be used as agricultural properties, you and the firm should ethically advise your client against using such an IRC code section in this case.

Perhaps, some earlier estate planning should have been done to prevent this problem. In any event, the present tax problem must be ethically dealt with by legitimately arguable positions regarding the code sections involved.

The bottom line rests within ourselves. We all have a duty to pay taxes. These taxes can be avoided legally and ethically. They cannot, however, be evaded by specious claims not based in any sort of reality. In the end, we must develop our best personal instinct to draw the bottom line, but also to do so ethically.

Chapter 8
TRADITIONAL AND INTERNET CONTRACTS

We have too many high-sounding words, and too few actions that correspond with them.

Abigail Adams

I. Overview

A contract is an agreement between two or more parties, which provides not only for performance of duties called for under the agreement, but also for legal remedies in case of breach. Because of the possibility of having to ask a court of law for a remedy, think of a contract as a private arrangement on the surface. But like an iceberg, the bulk of its real existence is found below the surface. The difference between an agreement at the surface and a contract below the surface is the law's recognition that certain requirements, called elements, are in place. If these elements are not properly arrived at between private parties, the larger society, represented by its courts, will enter the fray and enforce the rights and duties called for under the contract.

The first objective of this chapter is to introduce students to the four basic elements of contract and some of the terminology that they will be using throughout the remainder of the course. The elements of contract are analogous to the legs on a chair or table. They not only provide support but are also its essence, i.e., a table will not only fall if a leg is missing, it is not even really a table. An agreement with all four elements is elevated in the eyes of the law to the status of contract. Without these legs, the agreement fails legally and will not stand. Thus, even though all contracts are agreements, not all agreements are contracts. To be a contract the agreement must show:

1. Mutual assent.
2. Consideration.
3. Capacity.
4. Legality.

The second major objective of this chapter is to introduce students to key terms and phrases used in the study of contract law. One of the hallmarks of the technical information explosion is techno-babble. Walk into a room of specialists from any given area of endeavor, and you will hear them talking in code, using language which is often unintelligible to the layperson. Yet that specialized language provides the key to the body of knowledge. The study of law is no different than any other academic specialty in that sense. Lawyers have been referred to as "word merchants," and the profession's use of words of art has long predated the hi-tech world of today. In spite of the movement to use clearer and more straightforward language in legal documentation, the need to know some basic terms remains essential.

The key terms used here tend to be dichotomous, and you can use that dichotomy as a learning tool. Take, for example, the number of parties to a contract. At least two parties are required in all contracts. One of those two parties has to initiate the contract formation process. The person starting the mutual assent process with a promise is the *offeror*, the other person is the *offeree*. Next, look at the dichotomy of the promises being used: is it a promise for a promise (*bilateral*), or is it a promise for an act (*unilateral*)? Have these promises been *expressly* made or can they somehow be *implied* from the circumstances? Does the form that this agreement is taking require certain *formalities* (such as a negotiable instrument), or can it be in any manner chosen by the parties (*informal*) as long as the elements of contract are met?

Once the parties have formed an agreement, are the performance obligations already fully met (*executed*), or are there still remaining performance obligations on the part of one or more of the parties (*executory*)? In addition, you may have to examine issues of enforceability. If all the elements are in place, the agreement is now considered a *valid contract*. If one or more of the essential elements is missing, the agreement is not raised to the status of contract and may be legally *void*. There are also certain situations where a contract is created, yet will not be enforced. If a legal defense is found to be in place, such as a writing requirement, the contract may be an *unenforceable contract*. Sometimes, certain persons are given a legally recognized power to avoid the contract after it has been entered into. These contracts are *voidable*, and examples of this sort of situation will be found in cases involving young people or people with limited mental capacity.

II. Hypothetical Multi-issue Essay Question

T. Bear is a champion purebred American Cocker Spaniel. He is big and strong and has a heart of gold. T. Bear's only bad habit is that he loves to explore and has a tendency to run off if not kept on a lead. Last week, his owner made the mistake of letting T. Bear run loose near Boulder Creek. T. Bear ran off into the woods and got lost. He has been missing for a week, and his frantic owner has posted signs all over Boulder and in the newspapers as follows:

"Reward of $1,000.00 to anyone
who finds and returns my dog, T. Bear."
Signed, Owner

Last night, after reading the poster, Ms. Margaux d'Vin goes into the woods in search of T. Bear. She finds him sleeping with wolves in the woods. Ms. d'Vin picks up T. Bear and proceeds to take him towards the owner's home. Half way back, T. Bear recognizes the neighborhood, jumps out of Ms. d'Vin's arms, and runs home to his happy owner. Ms. D'Vin becomes less than ecstatic when the owner decides not to pay the reward, claiming that T. Bear came home on his own. Does Ms. D'Vin have any contract rights against the owner?

Note: The T. Bear in this question is our beloved first black American cocker spaniel champion. He died on April 12, 2000. Ms. D'Vin was his mate, Ch. Tamra's Crystal Pistol TT, who died on March 7, 1999. They had one litter of beautiful puppies, Julie, Hellen, Sabrina, and Bordeaux. They are sadly missed.

III. Outline

What Is a Contract?

A contract is an agreement that is enforceable by a court of law or equity

Parties to a Contract

Offeror and offeree

Elements of an Enforceable Contract
- Agreement
- Consideration
- Contractual capacity
- Lawful object

Sources of Contract Law
- Common law of contracts
- Uniform Commercial Code
- Restatement of the Law of Contracts

Classifications of Contracts
- Bilateral and unilateral contracts
- Express and implied contracts
- Formal and informal contracts
- Executed and executory contracts
- Valid, void, voidable, and unenforceable contracts

Equity and Contracts
- Equity
  - A doctrine that permits judges to made decisions based on fairness, equality, moral rights, and natural law
- Quasi- or implied-in-law contract
  - An equitable doctrine whereby a court may award monetary damages to a plaintiff for providing work or services to a defendant even though no actual contract existed. The doctrine is intended to prevent unjust enrichment or unjust detriment

IV. Objective Questions

Terms:

1. A contract entered into by way of exchange of promises of the parties, a "promise for a promise," is called a ________________ ________________.

2. A contract that has been fully performed on both sides, a completed contract, is called a ________________ ________________.

3. The National Conference of Commissioners on Uniform State Laws drafted the ________________ ________________ ________________ ________________ ________________ , which establishes uniform legal rules for the formation and enforcement of electronic contracts and licenses.

4. A contract that requires a special form or method of creation is called a ________________ ________________.

5. A contract where agreement between parties has been inferred because of their conduct is a(n) ________________ ________________.

6. A contract that is not formal, yet is fully enforceable and may be sued upon if breached is a ________________ ________________.

7. An agreement by the issuer of the letter to pay a sum of money upon the receipt of an invoice and bill of lading is a ________________ ________________ ________________.

8. The theory that says the intent to contract is judged by the reasonable person standard is known as the ________________ ________________ ________________ ________________.

9. A contract that has no legal effect because one of the essential elements of a contract is missing is called a ________________ contract.

10. A contract where a party has the option to avoid his contractual obligations, and if a contract is avoided, both parties are released from their contractual obligations, is called a ________________ contract.

True/False:

1. ____ When parties voluntarily enter into a contract, the terms of the contract become private law between the parties.

2. ____ Form contracts, which offer goods on a take-it-or-leave-it basis, are the common type of contracts today.

3. ____ The Restatement of the Law of Contracts is binding law.

4. ____ For an agreement to be formed, there must first be an offer and acceptance.

5. ____ Click-wrap licenses are not enforceable.

6. ____ Under a bilateral contract, there is no contract until the offeree performs the requested act.

7. ____ If a contract contains ambiguous language as to whether it is unilateral or bilateral, it is assumed to be unilateral.

8. ____ If an offeree in a unilateral contract has already partially completed performance, the offer cannot be revoked.

9. ____ Actual contracts are always express contracts, and never implied-in-fact contracts.

10. ____ Parties to a contract may voluntarily perform a contract that is unenforceable.

Multiple Choice:

1. Joe is a house painter who offers to paint Jennifer's house for $2000 by May 1. Jennifer promises to pay Joe $2000 for the paint job if he completes it by May 1. This is an:
   A. Bilateral contract.
   B. Unilateral contract.
   C. Voidable contract.
   D. Executed contract.

2. Hellen wants to order a cup of cafe lattè, but she cannot decide on a single or a double. She is thinking about a single, but holds two fingers up to the clerk when she orders without actually stating which strength she wants. The clerk prepares a double, and Hellen now refuses it, saying she only wanted a single. In deciding the terms of Hellen's order, a court will look at:
   A. The objective intent of the offeree.
   B. The subjective intent of the offeree.
   C. The objective intent of the offeror.
   D. The subjective intent of the offeror.

3. Which of the following is not a source of law or guidance in contract law?
   A. The common law of contracts.
   B. The Uniform Commercial Code.
   C. The Restatement of the Law of Contracts.
   D. All of the above are sources of contract law.

4. Regarding contract law, which of the following is incorrect?
   A. The offeror is the party who makes an offer.
   B. The offeree is the party to whom the offer is made.
   C. The parties to a contract can form an enforceable contract for any goods or services they choose.
   D. All of the above are correct.

5. On September 1, Max makes an offer to sell his car to Nelson. Nelson tells Max he "needs a few days." On September 5, Nelson tells Max he will buy the car. Which of the following is correct?
   A. Max is an offeree.
   B. A contract was formed on September 5.
   C. A contract was formed on September 1.
   D. Nelson may void the contract once it is formed since he accepted the offer and, therefore, may subsequently reject it.

6. Which of the following is not a basic element needed to form a contract?
   A. The contract must be written.
   B. There must be an agreement.
   C. Consideration must be given for performance.
   D. All of the above are necessary elements to form a contract.

7. Homeowner Helen tells Neighborhood Kid, "If you mow my lawn, I'll pay you $5." Which of the following is correct?
   A. Helen and Kid have a bilateral contract.
   B. Helen and Kid have a formal contract.
   C. Helen and Kid have a unilateral contract.
   D. Helen and Kid have a quasi-contract.

8. Assume the same facts as in question 7. Also assume Kid begins mowing the lawn and has only part of it left when Helen tells him, "I've decided to revoke my offer. I don't want my lawn mowed, and I'm not going to pay you." Which of the following is correct?
   A. Helen cannot revoke the offer since Kid has substantially performed.
   B. Helen can revoke the offer any time prior to the completion of mowing the lawn.
   C. Kid can recover payment for the portion of the lawn he has mowed, but not the whole $5.
   D. None of the above.

9. Which of the following is *not* required to establish an implied-in-fact contract?
   A. The plaintiff provided property or services to the defendant.
   B. The plaintiff expected to be paid and did not provide the property or services gratuitously.
   C. The contract was voidable, but not void.
   D. The defendant was given an opportunity to reject the property or services, but failed to do so.

10. On November 12, Debbie offers to sell her china cabinet to Ester for $500. Ester accepts the offer and pays Debbie $500 on November 13. However, Debbie is very busy and unable to deliver the china cabinet to Ester until November 19. Which of the following is correct?
    A. On November 13, the contract was executory.
    B. On November 20, the contract is executed.
    C. This contract is an informal contract.
    D. All of the above are correct.

V. Answers to Objective Questions

Terms:

1. *Bilateral contract.* Most contracts are bilateral rather than unilateral.
2. *Executed contract.* Full performance discharges the parties from any further obligations.
3. *Uniform Computer Information Transactions Act.* The UCITA is a model act.
4. *Formal contract.* In spite of appearances to the contrary, most contracts are informal.
5. *Implied-in-fact contract.* Often courts are asked to write in the terms based on the behavior of the parties.
6. *Informal contract.* The contract may be extensive, yet still informal, so long as the elements are still in place.
7. *Letter of credit.* This is a good example of a formal contract that calls for specific terminology, etc.
8. *Objective theory of contracts.* This theory is designed to give courts a measure by which to ascertain the intent of the parties.
9. *Void contract.* No contract is created in the first place.
10. *Voidable contract.* A contract is created but may be avoided by one or both of the parties.

True/False:

1. True. But, that agreement is subject to court review.
2. True. This is sometimes known as the "battle of the forms."
3. False. The Restatement of the Law of Contracts is a compilation of law principles as agreed upon by the drafters. However, it is not law.
4. True. This is the first of the four elements of a contract.
5. False. Courts have disagreed with this theory and recognized click-wrap contracts as enforceable.
6. False. A bilateral contract is formed upon the exchange of promises. No act of performance is necessary to create a bilateral contract. However, under a unilateral contract, no contract is formed until the offeree performs the requested act.
7. False. If a contract contains ambiguous language as to whether it is unilateral or bilateral, it is assumed to be bilateral, not unilateral.
8. True. Equity calls for giving the offeree a fair opportunity to complete the act.
9. False. Actual contracts may be express contracts or implied-in-fact contracts.

10. True. Courts look favorably upon such actions if they are based on good ethics and moral obligations.

Multiple Choice:

1. A This is a bilateral contract because both parties have engaged in making a promise in exchange for the other's promise. It is not unilateral because a unilateral contract calls for one party's promise in exchange for the other party's act. It is not voidable, under the facts as given, because there is no showing of any problem with legality, capacity, and the like. At this stage, the contract is executory rather than executed because duties of performance still exist by both parties.
2. C. The objective intent of the offeror, as reasonably interpreted by the offeree, will prevail. A court will take into account the words, conduct, and surrounding circumstances involved. Hellen's silence, coupled with her two-finger signal, shows her objective intent to order a double strength cafe lattè.
3. D. A, B, and C are all sources of contract law.
4. C. The parties to a contract cannot form an enforceable contract for any illegal purpose or result. A is a correct statement because an offeror is the party who makes an offer. B is a correct statement because the offeree is the party to whom the offer is made. D is incorrect because C is an incorrect statement.
5. B. A contract is formed when an offeree accepts the offeror's offer. When Nelson agreed to buy the car, there was an agreement. A is incorrect because Max is an offeror. C is incorrect because a contract was not formed until September 5, when Nelson accepted Max's offer. D is incorrect because, once Nelson accepts the offer, there is an agreement and an enforceable contract. Nelson cannot revoke his acceptance once he has entered the agreement.
6. A. An enforceable contract may be written or oral. B and C are necessary elements to form a contract. D is incorrect because A is not a necessary element to form a contract.
7. C. Helen has made a promise that is contingent on Kid's performing the act of mowing her lawn. Because the contract does not involve a return promise from Kid, but rather his performance, this is a unilateral contract. A is incorrect for the reason discussed above. B is incorrect because a formal contract requires a special form. Such a form was neither necessary nor used in this case. D is incorrect because here a quasi-contract gave Helen an express promise to pay Kid if he performed the act of mowing her lawn. A quasi-contract is an equitable remedy for situations where no contract existed.
8. A. Once Kid has substantially performed the act of mowing the lawn, Helen cannot revoke her offer. B is incorrect because Helen is prohibited from revoking her offer as soon as Kid begins performance or completes substantial performance of the act of mowing the lawn. C is incorrect because Kid can recover the entire $5.
9. C. The plaintiff need not prove the contract was voidable or void to prove an implied-in-fact contract. A, B, and D are all necessary elements to prove implied-in-fact contracts.
10. D. A is correct because on November 13, only Ester has performed her half of the contract. Debbie has not delivered the china cabinet and therefore her half of the contract. Because only one party has performed the contract, it is still executory. B is correct because on November 20, both parties have performed the contract. Therefore, the contract is executed. C is correct because an informal contract is any contract that does not require a special form or method of creation.

VI. Answer to Essay Question

If there were any contract in existence in this case, it would be covered by the common law of contracts. The UCC would apply only if the subject matter of the contract is specifically covered by one its subsections such as sales of goods or negotiable instruments. Here the subject matter of the alleged contract is the providing of a service by Ms. d'Vin to T. Bear's owner. Such services are not generally covered by the UCC unless they are somehow connected to a sale or lease of goods or the use of negotiable instruments.

The offeror here is T. Bear's owner, and the offeree is anyone who finds and returns T. Bear. The offer is unilateral in nature because it makes a promise (to pay $1,000) in return for an act (finding and returning T. Bear). The contract is not formed until the act is initiated and/or completed. Had the owner asked for a promise to find his dog, the contract would be bilateral and mutual assent would be arrived at upon acceptance of his promise with a second promise by the offeree. Because the promise called for an act, once the act is initiated, equity would dictate that the owners promise should not now be withdrawn. The underlying theory is that the offeree must be given a fair opportunity to complete the act he has begun at the behest of the offeror. This is the implied-in-fact aspect of this contract.

There appears to be an affirmative detriment undertaken by Ms. d'Vin here. She did something she did not have to do (go searching in the woods for T. Bear) in reliance on the owner's promise of a reward. Although this was not bargained for in the traditional bilateral promise setting, the consideration requirement would still be met because the bargain aspect of the consideration element is met in performing the act called for as acceptance.

There is no evidence of any lack of capacity here in the legal sense. T. Bear's owner was not so smart to let T. Bear run loose knowing that he had a tendency to run off. He may be foolish, but foolishness alone is not a legal defense to lack of capacity.

There is no illegality on the face or in the underlying intent of the contract. Rewards are legal if they call for a legal act. Conversely, if T. Bear's owner offered a reward for finding dogs for illegal dogfights, than the legality element would not be met.

This contract was informal in that no specific formality was required as long as the basic elements were in place. The contract is executory in that all the performance duties have not yet been completed. T. Bear's owner should pay the reward. T. Bear may have run up to the house, but Ms. d'Vin acted in good faith and should be paid. The owner cannot use T. Bear's actions as a defense to his duty to pay.

Chapter 9

AGREEMENT AND CONSIDERATION

A lean agreement is better than a fat judgment.
Proverb

I. Overview

The first element of contract is a finding of mutual assent between the contracting parties. Mutual assent is defined as a reciprocal agreement based on a meeting of minds of all the parties to a contract. The steps leading to mutual assent start with the offer and acceptance process. These steps can be broken down into subparts, and a familiarization of those subparts is essential to the study of contract law.

By analogy, think of the game of chess. The offer is the first move made by a player in a chess game. How, when, and where that opening gambit (the offer) is made will control how, when, and where the other player responds by way of acceptance. The response must be made in a timely and effective manner in order to qualify as an acceptance. Thus one of the first questions to be asked by students is: whose move is it? Second, what moves can be made? Third, what are the legal consequences of having made or not made certain moves? Remember, in spite of all the civility involved in the contract negotiating process, it is still an adverse or game-like relationship where each side is acting in its own self-interest.

The offer is broken down into three main subcomponents: intent, certainty, and communication. As an alternate memorization device, students may consider using an anagram called the QQC test. The first Q represents quality of the offer. In the eyes of the offeree, does this offer sincerely represent an objective intent to be bound? The second Q stands for quantity of the offer. If necessary, can a court, looking at this offer, find a basis upon which it could be measured, i.e., is the quantity of the offer readily determinable? The C represents communication. The offer must be communicated to the offeree in order to be effective.

Once a good offer is made, the other player must make his or her opening response. Remember, that response is dictated in many ways by the terms of the offer. Under the traditional common law *mirror image rule*, the acceptance must reflect the terms of offer. If it fails to do so, it may be deemed to be a rejection of the offer. And if it brings new terms to the table, it may be deemed a counteroffer. A counteroffer is, in fact, a new offer and sets the whole cycle of play into motion again from the reverse angle. The original offeror is now the new offeree.

Once we have a good offer, coupled with a good acceptance, the first element of contract, agreement or mutual assent, is arrived at. There are many variations on this basic theme as illustrated by the common law rules on advertising, auctions, and implied contracts based on the actions of the parties.

They all have one common denominator; sooner or later some sort of basis for mutual assent must be found before a court will go forward with enforcement of the agreement.
Consideration is the second of the four elements of contract. Spotting consideration in a contractual setting sounds easier than it really is, and when consideration is missing, the result is not necessarily an absence of contract. It is one of those elements of contract law that seems unduly complex. Perhaps a historical footnote on the issue may be of value.

Consideration is that element of a contract that is designed to show the underlying inducement to enter into the contract. It is meant to place value on the contract in order to assure evidence of a bond between the parties. In the Middle Ages, contracts between members of the privileged classes did not require consideration because a man's word was his bond. Evidence of that bond was his family seal. The seal was affixed to the contract by way of pressing the signatorius annulus (signet ring) against melted wax onto the contract. Not everyone was a member of nobility, but all wanted their contracts to be binding. Thus the concept of looking to bargained-for value in place of the seal evolved into the modern day law of contracts. Today only a few states still accept seals as a substitute for consideration. Waxed seals have become an anachronism best used for love letters, and the only common modern use of seals is the embossed seal of a notary public, whose job it is to act as a legal witness to an administration of an oath.

Consideration today is made up of two subcomponents: detriment and bargain theory. Detriment represents the value of the contract, i.e., the glue that brought the parties to the table in the first place. It is actually a very practical requirement. Without value being put at issue, a court looking at the situation may simply say the matter is moot (nothing in controversy). Detriment is usually divided into two main categories: affirmative detriment where a person promises to do something he or she has no obligation to do but for the contract and negative detriment where a person abstains from doing something he or she has a legal right to do but for the contract.

The second subcomponent of consideration is found in bargain theory. Bargain theory is designed to isolate and identify the value used as consideration. Think again of a contract as a table supported by four legs, those legs being the four elements of contract. Just as a table can only stand on its own legs, so each contract needs to identify its own consideration element. Consideration bargained for in one contract cannot be used to support another.

What if you have examined an agreement for consideration and found it lacking? You may still have a contractual result based on either the equitable doctrine of promissory estoppel or on specific statutory grounds that allow for a consideration waiver or substitute. Promissory estoppel literally means that a promise made, even though not supported by consideration, will not be allowed to be withdrawn. It is estoppel prevented by one's own act because of the harm that would befall the other party if it were allowed to be withdrawn.

The second major category of consideration exceptions from contracts is found in statutory provisions based on public policy. For example, the U.S. Bankruptcy Code allows for court-approved reaffirmation of debts that have already been discharged. Another example is found in many state statutes that provide protection to eleemosynary (charitable) organizations making pledges as enforceable contracts even though the donors may not be getting any consideration in return for their gifts.

II. Hypothetical Multi-Issue Essay Question

Ed and Bob were long-time friends, having known each other since high school. Over all these years, Ed was known as a car nut. His most prized possession is a 1923 Model P2 Alfa Romeo last raced in the 1925 Grand Prix of France by the legendary race driver, Antonio Ascari. This car is considered a classic from the golden era of sports car racing.

Unfortunately, Ed's wife, Rosanne, did not look at his car the same way. She was always complaining to Ed how he would spend money on the car rather than on more mundane matters such as food, clothing, and shelter for their children, Enzo, Mario, and the Duetto twins, Guiseppe and Emilio.

One late Friday afternoon, after a round of golf, Ed and his friend Bob are comparing notes on their respective domestic situations over a beer at Putter's Tap. Ed says to Bob, "You know, I'm really tired of Rosanne being on my case over the Alfa. For $5,000 I would sell it to you just to shut her up!" Bob did not respond at the time. But on the following Monday, Bob showed up at Ed's office with a certified check for $5,000 and said, "Sold!" Ed does not want to see his precious Alfa go, and both parties are now before your court. What results?

III. Outline

Agreement
- Agreement
  - The manifestation by two or more persons of the substance of a contract
- Parties
  - Offeror—person who makes an offer
  - Offeree—person to whom an offer has been made

Requirements of an Offer
- For an offer to be effective:
  - The offeror must objectively intend to be bound by the offer
  - The terms of the offer must be definite or reasonably certain
  - The offer must be communicated to the offeree

Objective Theory of Contracts
- A theory that says the intent to contract is judged by the reasonable person standard and not by the subjective intent of the parties

Definiteness of Terms
- An offer (and contract) must contain the following terms
  - Identification of the parties
  - Identification of the subject matter and quantity
  - Consideration to be paid
  - Time of performance
  - Sometimes a term in a contract is an implied term that can reasonably be supplied by the courts

Advertisements
- A general advertisement is an invitation to make an offer
- A specific advertisement is an offer

Rewards
- To collect a reward, the offeree must
  - Have knowledge of the reward offer prior to completing the requested act
  - Perform the requested act

Auctions
- Auction with reserve
  - Unless expressly stated otherwise, an auction is an auction with reserve, i.e., the seller retains the right to refuse the highest bid and withdraw the goods from auction

Auction without reserve
An auction in which the seller expressly gives up his or her right to withdraw the goods from sale and must accept the highest bid

Termination of an Offer by Action of the Parties
Revocation of the offer by the offeror
Rejection of the offer by the offeree
Counteroffer by the offeree
Promissory estoppel

Termination of the Offer by Operation of Law
Destruction of the subject matter
Death or incompetence of the offeror or the offeree
Supervening illegality
Lapse of time

Acceptance
Only the offeree can legally accept an offer and create a contract
The offeree's acceptance must be unequivocal
Mirror image rule
Silence is not considered acceptance even if the offeror states that it is

Time of Acceptance
Acceptance upon dispatch rule
Acceptance of a bilateral contract occurs when the offeree dispatches the acceptance by an authorized means of communication
Mail box rule—an acceptance is effective when it is dispatched even if it is lost in transmission
Proper dispatch rule—an acceptance must be properly addressed, packaged, and posted

Mode of Acceptance
An offeree must accept an offer by an authorized means of communication
Express authorization
Acceptance must be by a specified means of communication
Implied authorization
Mode of acceptance is implied from what is customary in similar transactions, usage of trade, or prior dealings between the parties

Elements of Consideration
Something of legal value must be given, e.g. either a legal benefit must be received or a legal detriment suffered
There must be a bargained-for exchange
Gift promises (gratuitous promises) are unenforceable because they lack consideration
Nominal consideration
Nominal consideration is not considered legally sufficient in many states

Business Contracts
Option-to-cancel clauses
Allow one or both parties to cancel the contract
Best efforts contracts
Requires one or both of the parties to use their "best efforts" to achieve the objective of the contract

Contracts Lacking Consideration
- Illegal consideration
  - Promise to refrain from doing an illegal act
- Illusory promises
  - Promise where one or both parties can choose not to perform their obligation
- Moral obligations
  - Promise made out of a sense of moral obligation or honor or love or affection; some states enforce these types of contracts
- Preexisting duty
  - Promise based on the preexisting duty of the promisee to perform; the promise is enforceable if the parties rescind the contract and enter into a new contract or there are unforeseen difficulties
- Past consideration
  - Promise based on the past performance of the promisee

Settlement of Claims
- Law promotes the voluntary settlement of disputed claims
  - Accord
    - An agreement where the parties agree to accept something different in satisfaction of the original contract
  - Satisfaction
- The performance of an accord

Promissory Estoppel
- For the doctrine of promissory estoppel to be applied, the following elements must be shown:
  - The promisor made a promise
  - The promisor should have reasonably expected to induce the promisee to rely on the promise
  - The promisee actually relied on the promise and engaged in an action or forbearance of a right of a definite and substantial nature
  - Injustice would be caused if the promise was not enforced

IV. Objective Questions

Terms:

1. Under contract law, the person who makes the offer is called the _______________ while the person to whom the offer is made is called the _______________.

2. Auctions that are merely invitations to make an offer where the seller may refuse the highest bid are called _______________ _______________ _______________.

3. Generally, for an acceptance to create a contract, the acceptance must be _______________, i.e., without any conditions.

4. An offer may be terminated by the parties in three ways: _______________ by the offeror, _______________ by the offeree, or _______________ by the offeree.

5. Under the _______________ _______________, acceptance of a bilateral contract occurs when the offeree dispatches the acceptance by an authorized means of communication.

6. Several types of promises are enforceable without consideration. These exceptions to the general rule are usually based on ________________ ________________.

7. In examining the adequacy of consideration, the courts use the ________________ ________________ ________________ standard to determine whether the consideration is legally sufficient.

8. A contract where a buyer agrees to purchase all requirements of an item from one seller is called a(n) ________________ ________________.

9. A contract where a seller agrees to sell all production of an item to one buyer is called a(n) ________________ ________________.

10. If a party with a valid legal claim agrees not to sue the wrongdoer in exchange for a payment of consideration, this is called a ________________ ________________.

True/False:

1. ____ To be legally bound to a contract, the parties must have had subjective intent to be bound when the contract was made.

2. ____ Advertisements are generally considered offers and are binding on a company if subsequently accepted.

3. ____ Silence is not usually considered acceptance, but an offeree may indicate that silence means acceptance.

4. ____ If an offeree inquires into the offer regarding certain terms or information, this is considered rejection of the offer.

5. ____ Death of either party terminates an offer only if the other party is given notice.

6. ____ Option to cancel clauses render a contract illusory in all cases.

7. ____ Most states hold that moral obligation is treated as legal consideration.

8. ____ Generally, if an existing contract is modified, new consideration must be given to avoid the preexisting duty rule.

9. ____ If the parties to an existing contract rescind the contract and form a new contract, no new consideration is required.

10. ____ The doctrine of promissory estoppel is also sometimes referred to as the equitable doctrine of detrimental reliance.

Multiple Choice:

1. Fast Food Inc. mailed a letter to Ambitious Allen on May 1, 2003, offering to sell a fast food franchise. The offer stated it would not be withdrawn prior to June 5, 2003. Which of the following is correct?
   A. The offer cannot be assigned to another party by Allen if Allen chooses not to accept.
   B. A letter of acceptance from Allen to Fast Food sent on June 5, 2003, and which is received by Fast Food on June 6, 1993 does not create a valid contract.
   C. The offer is an irrevocable option that can be withdrawn prior to June 5, 2003.
   D. The Statute of Frauds does not apply to the proposed contract.

2. Owner Owen offered to sell a parcel of land to Purchasing Paul for $90,000. The offer was made by Owen in a signed writing and provided that it would not be revoked for five months if Paul promised to pay Mix $250 within two days. Paul agreed to do so. Which of the following is correct?
   A. Paul's agreement to pay Mix $250 is insufficient consideration to form an option contract.
   B. Owen may withdraw the offer any time prior to Paul's payment of the $250.
   C. An option contract is formed.
   D. Although an option contract is formed, the duration of such a contract is limited to three months.

3. SuperStore published advertisements containing price quotes and a description of products that it would like to sell. Customer Chris demands the right to purchase one of the products at the quoted price. Which of the following statements is correct under general contract law?
   A. SuperStore must sell the product that Chris demands at the quoted price.
   B. Chris has accepted SuperStore's firm offer to sell.
   C. SuperStore has made an offer.
   D. Chris has made an offer.

4. One morning, Helpful Harry reads in the local newspaper that Mrs. Widow is advertising a $100 reward for her lost poodle. While walking to the grocery that afternoon, Harry sees a small white poodle. As he gets closer, he realizes it is Mrs. Widow's poodle. He picks up the dog and returns it to Mrs. Widow. Which of the following is correct?
   A. Harry is entitled to the reward because he knew about it and performed the act.
   B. Harry is entitled to the reward regardless of whether he knew about it before or not.
   C. Harry is not entitled to the reward because he would have done the act anyway.
   D. Harry is not entitled to the reward because he has not met the necessary requirements to form a unilateral contract.

5. Assume the same facts as in question 4 except Harry read nothing in the paper about the reward. Which of the following is correct?
   A. Harry is entitled to the reward because he knew about it and performed the act.
   B. Harry is entitled to the reward regardless of whether he knew about it before or not.
   C. Harry is not entitled to the reward because he would have done the act anyway.
   D. Harry is not entitled to the reward because he has not met the necessary requirements to form a unilateral contract.

6. Computer Corp. contracts with Nuts-N-Bolts to buy all the screws Computer Corp. needs to assemble its computers. Which of the following is correct?
   A. The contract is unenforceable because it is so vague it is an illusory promise.
   B. The contract is unenforceable because it does not involve a bargained-for exchange.
   C. The contract is enforceable as a requirements contract.
   D. The contract is enforceable as an output contract.

7. Luxury Hotel contracts with Lush Lawn to have Lush Lawn mow the grass and keep up the gardens on the Hotel grounds for 10 years. However, there is a clause in the contract that states that if the hotel chooses, the contract may be terminated provided Luxury Hotel pays Lush Lawn a sum of $2,000 on termination. Which of the following is correct?
   A. The contract is unenforceable because an option to cancel clause makes the contract an illusory promise.
   B. The contract is unenforceable because the $2,000 payment is past consideration.
   C. The contract is enforceable because only one party has the option to cancel.
   D. The contract is enforceable because the option to cancel clause is supported by consideration.

8. Certain promises are enforceable without consideration. Which of the following is such an exception?
   A. Promises to pay debts barred by the Statute of Limitations.
   B. Promises to pay debts barred by discharge in bankruptcy.
   C. Promissory estoppel.
   D. All of the above.

9. Super-Chain Store agrees to purchase all its film needs from Photo-Perfect and no other film suppliers. There is nothing in the contract regarding 'best efforts." Which of the following is correct?
   A. This is an exclusive dealing contract and will most likely be found illusory and unenforceable.
   B. This is an exclusive dealing contract and will most likely be found enforceable after the best efforts duty is implied.
   C. This is not an exclusive dealing contract and will most likely be found illusory and unenforceable.
   D. This is not an exclusive dealing contract and will most likely be found enforceable even without the best efforts duty implied.

10. Sunshine Orange Juice contracts with Manny's Market to deliver orange juice three times per week on Mondays, Wednesdays, and Fridays. Sunshine soon realizes this is a big commitment since Manny's is a little ways out of town. After a couple of weeks, Sunshine delivers only on Monday and Wednesday. Manny calls Sunshine when the Friday shipment does not come. Sunshine says Manny will have to pay an extra $50 per week for Friday deliveries. Manny agrees because he has to have an orange juice shipment for the weekend. Which of the following is correct?
    A. The contract is unenforceable because Sunshine had a preexisting duty.
    B. The contract is unenforceable because Sunshine rescinded the old contract.
    C. The contract is enforceable because Manny agreed to it regardless of a preexisting duty.
    D. The contract is unenforceable because there was new consideration given.

V. Answers to Objective Questions

Terms:

1. *Offeror, offeree.* You must learn to readily identify the key players in the study of contract law.
2. *Auction with reserve.* With a reserve, the proposed seller keeps the back door open in case he or she considers bids to be too low. Eventually, market realities are forced upon such sellers if they want to sell their goods.
3. *Unequivocal.* If conditions are attached, it may be a rejection coupled with a counteroffer. Note the UCC has modified this rule somewhat in sales between merchants.
4. *Revocation, rejection, counteroffer.* All three of these methods may be used to kill an offer.
5. *The mailbox rule.* Remember this rule applies only to acceptances. In spite of the complications created by the rule, it has many practical benefits. Consider for example what happens every April 15th vis-à-vis constructive delivery of tax returns to Uncle Sam.
6. *Public policy.* Public policy exceptions to general rules are usually equity-based doctrines that have evolved out of these laws and/or legislative enactments.
7. *Shock the conscience.* Normally, courts do not become involved with second-guessing the fair market value of an exchange unless the amount of consideration is so grossly inadequate that it "shocks the conscience."
8. *Requirements contract.* These terms are very commonly used in commercial UCC contracts between merchants.
9. *Output contract.* The UCC allows for far more after-contract latitude in the interpretation of these contracts.
10. *Release agreement.* The giving up of a legal right to a cause of action is considered a form of valid consideration in the eyes of the law.

True/False:

1. False. Only objective intent is required. Subjective intent is immaterial.
2. False. Advertisements are generally considered invitations to make an offer.
3. True. The offeree may determine that his own silence will mean assent. However, the offeror is generally not allowed to do this.
4. False. Mere inquiries or questions about the offer do not amount to a counteroffer.
5. False. Notice is not necessary. Death or incompetence of either party terminates the offer even without notice.
6. False. These clauses do not render a contract illusory if a party to the contract has given legal consideration for the right to cancel.
7. False. A minority of states hold that moral obligations are enforceable.
8. True. This new consideration may consist of doing something new, making payment at an earlier time than originally required, or accepting consideration different from that required in the contract.
9. True. No new consideration is needed if the entire contract is new. For this exception to the general rule to apply, the parties must voluntarily agree to the changes in contract terms.
10. True. The doctrine of promissory estoppel is based on the premise of justifiable good faith reliance and the enforcement of promises that gave rise to that reliance.

Multiple Choice:

1. A. An offer cannot be assigned to another party. A valid acceptance may only come from the person to whom the offer was directed. B is incorrect because the letter of acceptance, mailed on June 5, creates a valid contract. C is incorrect because the offer is not an irrevocable option because there was no consideration given. D is incorrect because the Statute of Frauds applies to contracts that cannot be performed within one year of entering into the agreement.
2. C. An option contract is an irrevocable offer that is actually a contract if supported by sufficient consideration. Here, Owen's offer was an offer to form an option contract. Therefore, A is incorrect. B is incorrect because the formation of an option contract results in an irrevocable offer to enter into another contract for the term agreed upon. D is incorrect because an option contract may be formed for any agreed upon period.
3. D. Published price quotes and descriptions of products are merely invitations for interested parties to make an offer. Chris's demand to purchase one of the products at the quoted price is an offer by Chris to buy the product from Kraft. A is incorrect because SuperStore has the option to either accept or reject Rice's offer. B and C are incorrect because Chris does not have an offer to accept.
4. A. To be entitled to receive the award, Marty must have known about the reward and performed the required act. B is incorrect because Marty must know about the reward. C is incorrect because Marty is entitled to the reward under these facts. D is incorrect because Marty has met the requirements.
5. D. Since Harry did not know about the reward, he is not entitled to receive it.
6. C. A is incorrect because requirements contracts are not too vague to be enforced. B is incorrect because this was bargained for.
7. D. The $2,000 to be paid on termination is consideration for the right to cancel. A is incorrect because if supported by consideration, such clauses do not make the contract illusory. B is incorrect because this is not past consideration. C is incorrect because either or both parties may have the option to cancel.
8. D. A, B, and C are all examples of public policy-based exceptions to the general rule.
9. B. If the contract does not address the best efforts rule, the court will likely imply such a term. When such a term is implied, the contract is not illusory. A is incorrect because exclusive dealing contracts are not automatically illusory and unenforceable. C is incorrect because this is an exclusive dealing contract. D is incorrect because with the best efforts rule implied, it will be enforceable.
10. A. B is incorrect because the old contract was not rescinded and doing so would not render the new contract unenforceable. C is incorrect because the preexisting duty rule applies in such situations. D is incorrect because there was no new consideration given.

VI. Answers to Essay Question

The first move here would be Ed's. If he was really interested in selling the car, he would either make an offer or solicit offers for the car. In order for the statement made by Ed to be construed as an offer, the three elements of a good offer would have to be found. The quantity and communication elements are readily in place. The quantity of $5,000 is stated and the statement was directly made to Bob. The real issue is the quality of the offer. Can this statement be seen through the eyes of the offeree as evidence of an objective intent to enter into an agreement? As will be seen in subsequent chapters, the value exchange does not always have to be *quid pro quo* as long as the legal requirements of consideration are met. Sometimes people sell off valuables at less than fair market value for personal reasons. Could Bob seriously see this as a real offer knowing as much as he does about Ed's personal situation?

Assume that Ed's statement was an offer to Bob, did Bob respond in a timely manner? It appears that timing of his response would be considered reasonable in that he acted on the next regular business day. The attempted acceptance complied with the mirror image rule in that there were no material changes to the offer as presented by Ed. In addition, this response was directly communicated to Ed.

Chapter 10
CAPACITY AND LEGALITY

He who derives the advantage ought to sustain the burden.
Legal Maxim

I. Overview

Contracts are important. Personal and business success is measured in large part by how well a person can provide for his or her own best interests in contracts. A disadvantageous contract can lead to personal and business disasters such as bankruptcies, loss of credit worthiness, and criminal violations. Because of the possible severe consequences, the fabric of contract law is thoroughly interwoven with a concern that parties entering into these agreements are doing so with free will, full knowledge of all the material circumstances, and a recognized capacity to contract. Capacity is the legal qualification, competency, and fitness to enter into contracts. It is the third of the four essential ingredients to a contract. As with the other three elements, if the capacity element is missing, a contract cannot result from the agreement in question. Capacity to contract is presumed between adults. It is protected legally in both the federal and all state constitutions. Before this legally sanctioned right is limited, such limitations must be justified on the basis of sound public policy.

An important area of concern in the law of contract capacity limitation is found in the public policies that seek to protect certain individuals from the sometimes-harsh results of contracts. Even though protection can sometimes be seen as a euphemism for punishment on the part of the protected party, the simple reality is that sometimes people do need to be protected from themselves and others.

The first category of persons needing protection is the young, i.e., the legal minor. All states have established certain minimum age requirements as a threshold that must be crossed before a person is allowed to contract on his or her own (now age 18 in most jurisdictions). Prior to that threshold age, attempts to contract by such young persons may have limited success. In the area of necessities and certain statutorily recognized situations such as insurance and military enlistment, these contracts may be enforceable against a minor. But as a matter of public policy, the general rule is that the minor is allowed, for his or her own protection, to legally avoid the consequences of contractual entanglement. This avoidance is not always a free ride. If there is misrepresentation by the minor or equity concerns for an adult who dealt with the minor in good faith, some reimbursement may be due from the minor.

The second major area of public policy-based protection from contract capacity is found in the area of persons with varying degrees of limited mental capacity. Just as we do not want very young people signing mortgages, we do not want documents enforced against persons who had no real ability to comprehend the material elements of those documents. This is not a matter of excusing someone from the binding effects of a contract based on simple ignorance. If that were the case, courts would surely be even more inundated than they already are. What this area of the law is concerned about is a real lack of capacity to protect oneself. Because contract results can be harmful, a lack of a mental ability to protect one's self-interest is considered fatal to the contracting process.

A person who has been judged incompetent simply has lost the power to contract for himself or herself and needs to have someone else act in his best interest in a fiduciary (trust) role. Attempts to contract with persons legally declared incompetent are void. Persons who are not legally declared incompetent can seek to avoid contracts based on some sort of lesser capacity. Be it based on mental limitations, drunkenness, drug use, or other affliction, they must prove to a court that at the time they entered into the disputed agreement, they lacked the cognitive ability to understand the material elements of the transaction. The burden of proof is on the person raising the issue of limited capacity, and courts will look at those issues on a case-by-case basis. Simple ignorance, emotional weakness, or predilections for doing stupid things are not sufficient grounds for avoidance of a contract. As with minors, contracts with persons of limited capacity can later be ratified. If the contract is disaffirmed, restoration or restitution may be appropriate depending on the individual circumstances of each case.

The second objective of this chapter is to introduce you to the element of legality and its role in the formation and enforcement of contracts. Most people think of a contract as a two party situation: A contracts with B, each legally capable of providing recognized legal value. Note how the phrase legal keeps cropping up through the language of contracts. That is because a contract is not only a two party situation. All contracts have a hidden third party which is the larger society represented by the courts. If A and B have a dispute about what the contract terms really provided, where must they ultimately turn to get an answer? Courts are that third party, C, whose job it is to provide a resolution of the problem between A and B.

In the process of resolving a contract problem, can anyone realistically expect courts to use any other ground rules other than their own? The legality element of contract simply refers to those ground rules. If A and B want their agreement honored in a court of law as a valid and binding contract, that agreement must be recognized first and foremost as a legal agreement. To do otherwise would be asking a court to become a party to an agreement that contravenes the basis for its existence, its *raison d'etre*.

When a court examines an agreement for its legality component, it asks three key questions:

1. Is this agreement illegal *per se*?
2. If this agreement is legal on its face, may it have an intent or consequence that is against public policy?
3. If there is an element of illegality or violation of public policy, is it a material aspect of this contract, or is the problem incidental or collateral to the agreement?

The first question is really the easiest to answer. If the agreement calls for an illegal act on its face, the agreement is void and cannot be raised to the status of a legal contract. Thus, a contract for a murder by an underworld figure cannot and will not be enforced by a court of law. The real world result might be otherwise, but that result is certainly not within the purview of legality.

The second question is much more difficult to answer in that public policy is a moving target. Courts look at the reality of these agreements, even if they appear legal on their faces. For example, a simple buy/sell agreement looks perfectly legal. But what if the agreement is designed to act as a restraint of trade by discriminating as to who can sell to whom at fixed prices? Or what if we are looking at a prenuptial agreement that looks fair on its face but gives all the property to the party against whom a divorce action is brought? Is this really just a liquidation of property or an illegal punitive measure against a party who is exercising a legal prerogative? What about an exculpatory clause that goes too far?

The third area of concern goes to the issue of enforceability once the problem area is identified. Not all illegalities are fatal to the entire contract. If a court finds that the illegality or public policy concern is not at the heart of the contract, it may still enforce the agreement, because the legality is considered collateral or incidental.

II. Hypothetical Multi-Issue Essay Question

Bunco Bullhead (age 45) has a multitude of problems. He has long been addicted to alcohol, gambling, and questionable personal habits. These proclivities are all readily available to him at his favorite watering hole, Joe's Bar and Gambol, in his hometown, Tim Buck II. Bunco particularly enjoys Joe's place because Joe provides free drinks as long as players continue to gamble. In addition, Bunco's credit limit on his Happy Card was extended to $5,000. Happy Card's motto is "Why Worry About Paying—We Don't!"

Last night Bunco got drunk and lost his newly credited $5,000 at Joe's poker table. In addition, Bunco signed a note to Joe for an additional $5,000 that he also lost. Bunco now refuses to pay both his credit card bill and his note to Joe alleging a lack of capacity and legality at the time. How do you decide?

III. Outline

Persons Who May Lack the Capacity to Contract
- Minors
  - Persons who have not reached the age of majority
- Mentally incompetent persons
- Intoxicated persons

Minors
- Infancy doctrine
  - Allows minors to disaffirm (cancel) most contracts they have entered into with adults
- Duties
  - Restitution
- Minor's
- Competent party's
  - Restoration

Ratification

Special Types of Contracts for which Minors May be Liable
- Medical, surgical, and pregnancy care
- Psychological counseling
- Health and life insurance
- Performance of duties relating to stock and bond transfers, bank accounts, etc.
- Educational loan expenses
- Contracts to support children
- Contracts to enlist in the military
- Artistic, sports, and entertainment contracts

Disaffirmance of Contracts Based on Legal Insanity
- If a person is adjudged insane, the contract is void; neither party can enforce the contract
- If the person is insane, but not adjudged insane, the contract is voidable by the insane person; the competent party cannot void the contract

Disaffirmance of Contracts Based on Intoxication
- Most states provide that contracts entered into by intoxicated persons are voidable by that person
- A person who disaffirms a contract based on intoxication generally must be returned to the status quo

Illegality—Contracts Contrary to Statutes
- Usury laws
- Gambling statutes
- Sabbath laws or blue laws
- Contracts to commit a crime
- Licensing statutes

Illegality—Contracts Contrary to Public Policy
- Immoral contracts
- Contracts in restraint of trade
- Covenants not to compete
- Exculpatory clauses

Effect of Illegality
- Illegal contracts are void; therefore, the parties cannot sue for nonperformance
- Exceptions
  - Innocent persons who were justifiably ignorant of the law or fact that made the contract illegal
  - Persons who were induced to enter into the contract by fraud, duress, or undue influence
  - Persons who entered into an illegal contract withdraw before the illegal act is performed
  - Persons who were less at fault than the other party for entering into an illegal contract
- In pari delicto
  - When both parties are equally at fault in an illegal contract

Unconscionable Contracts
- Elements of an unconscionable contract:
  - The parties possessed severely unequal bargaining power
  - The dominant party unreasonably used its unequal bargaining power to obtain oppressive or manifestly unfair contract terms
  - The adhering party had no reasonable alternative

Remedies for an unconscionable contract:
- Court can refuse to enforce the contract
- Court can refuse to enforce the unconscionable clause but enforce the remainder of the contract
- Court can limit the applicability of any unconscionable clause to avoid any unconscionable result

IV. Objective Questions:

Terms:

1. Generally, in determining contractual capacity, the party asserting ________________ of the person in question bears the burden of proof.

2. Minors are allowed to disaffirm contracts they have entered into with adults under the ________________ ________________.

3. A minor who enters into a contract and subsequently reaches the age of majority may ________________ the contract, and in so doing, become bound by the contract.

4. When a minor chooses to leave home and live apart from her parents, this is known as ________________.

5. If a person has been adjudged insane, any contract entered into by that person is _______________.

6. Illegal contracts are contracts contrary to _______________ or _______________.

7. Licensing statutes enacted to protect the public from unlicensed persons trying to recover payment are called _______________ statutes.

8. In an effort to protect goodwill, the seller of a business may enter into an agreement with the buyer not to engage in a similar business in a certain area for a certain time. Such contract provisions are _______________ _______________ _______________ _______________.

9. A contractual provision that relieves one or both parties from tort liability for ordinary negligence is called a(n) _______________ _______________.

10. To prevent enforcement of oppressive, unfair, or unjust contracts, courts developed an equitable doctrine called the _______________ _______________ _______________.

True/False:

1. ____ A minor may disaffirm one part of the contract and enforce another part, binding the competent part.

2. ____ The infancy doctrine is a subjective standard, allowing courts to examine the minor's maturity, experience, knowledge, etc.

3. ____ If a minor ratifies a contract before reaching the age of majority, she is bound by the contract and can no longer disaffirm it.

4. ____ Parents are never liable for any contracts entered into by their children.

5. ____ Contracts by persons who are alcoholics or drug addicts are automatically voidable.

6. ____ Banks and financial institutions may never be exempted from usury laws.

7. ____ An exculpatory clause can relieve the parties from tort liability arising from willful conduct, recklessness, and gross negligence as well as ordinary negligence.

8. ____ A person who is protected by a regulatory statute but enters into a contract that violates the statute may enforce the contract against the other party.

9. ____ A contract can never be divided into legal and enforceable parts and illegal and unenforceable parts. That is, the whole contract must be enforced or rejected.

10. ____ The definition of unconscionability is an objective, standard definition that is applied uniformly in all cases.

Multiple Choice:

1. Drinker Doug entered into a written agreement to sell a parcel of land to Sober Shawn for $50,000. At the time the agreement was executed, Doug had consumed a large amount of alcoholic beverages that significantly impaired his ability to understand the nature and terms of the contracts. Shawn knew Doug was very intoxicated and that the land was valued at $125,000. Doug, now completely sober, wishes to avoid the contract. The contract is:
   A. Void.
   B. Legally binding on both parties in the absence of fraud or undue influence.
   C. Voidable at Doug's option.
   D. Voidable at Doug's option only if the intoxication was involuntary.

2. On May 1, 2003, Mint, a 16-year-old, purchased a sail boat from Sly Boats. Mint used the boat for six months at which time he advertised it for sale. Which of the following statements is correct?
   A. The sale of the boat to Mint was void, thereby requiring Mint to return the boat and Sly to return the money received.
   B. The sale of the boat to Mint may be voided by Sly at his option.
   C. Mint's use of the boat for six months after the sale on May 1 constituted a ratification of that contract.
   D. Mint may disaffirm the May 1 contract at any time prior to reaching majority.

3. Minor Matt, a 16-year-old, contracts with City Cycles to buy an $8,000 motorcycle. Which of the following is correct?
   A. The contract is voidable by Matt.
   B. The contract is void as soon as it is made.
   C. The contract is voidable by City Cycles.
   D. The contract is voidable by either Matt or City Cycles.

4. Assume the same facts as in question 3 except that it is two years later and Matt has reached the age of majority. Matt has not disaffirmed the contract and has made payments on the motorcycle while he has been 18. Which of the following is correct?
   A. The contract is void and Matt has no obligation to pay.
   B. The contract is now voidable by City Cycles.
   C. Matt has ratified the contract and is now bound by it.
   D. Matt has ratified the contract but is not responsible for payments from when he was a minor.

5. Insane Ida has been adjudged insane. Ida enters a contract to sell her large family estate to Swindler Sandy for $500. Which of the following is correct?
   A. The contract is voidable by Ida.
   B. The contract is voidable by Ida's legal guardian.
   C. The contract is not voidable and can be enforced.
   D. The contract is void.

6. Assume the same facts as in question 5 except Ida has not been adjudged insane, but has an impairment that makes her legally insane. Which of the following is correct?
   A. The contract is voidable by Ida.
   B. The contract is voidable by Sandy.
   C. The contract is not voidable and can be enforced.
   D. The contract is void.

7. Cleanhands Cathy hires Hitman Harry to get rid of her wealthy brother, Millionaire Mike, so that she can inherit his estate. Cathy pays Harry $5,000. Which of the following is correct?
   A. The contract is void as contrary to statute.
   B. The contract is void as contrary to public policy.
   C. The contract is voidable if Cathy disaffirms it.
   D. None of the above.

8. WallStreet Walter is being prosecuted for a white-collar crime. During jury recess one day, Walter approaches one of the jurors and offers her $50,000 if she will vote for his acquittal. Which of the following is correct?
   A. The contract is void as contrary to statute.
   B. The contract is void as contrary to public policy.
   C. The contract is voidable if the juror disaffirms it.
   D. None of the above.

9. Oppressive Orville owns several commercial buildings in a large city. Startup Sam wants to open a business making handmade yo-yos. Sam has been looking all over town but cannot find a space to rent. Orville knows rental spaces are tight right now and agrees to rent to Sam. However, Orville includes an exculpatory clause that releases him from liability resulting from his own negligence in maintaining the plumbing or wiring. Which of the following is correct?
   A. The exculpatory clause is enforceable since courts generally favor exculpatory clauses.
   B. The exculpatory clause is enforceable since renting property is an essential service.
   C. The exculpatory clause is unenforceable since there was obviously unequal bargaining power.
   D. The exculpatory clause is unenforceable since courts never enforce exculpatory clauses.

10. Wilma Widow is destroyed and heartbroken when her husband of forty-five years dies. Shyster Sherman convinces Wilma that she should employ him so that she is sure to inherit all her husband's property. Anxious and worried about the horrible possibilities Sherman described, Wilma agrees to hire him for $10,000. Which of the following is correct?
    A. Wilma can sue to recover any amount paid since this contract is void as contrary to a statute.
    B. Wilma can sue to recover any amount paid since she was ignorant of the facts.
    C. Wilma can sue to recover any amount paid since she was induced to enter the contract under duress and undue influence.
    D. None of the above.

V. Answers to Objective Questions

Terms:

1. *Incapacity.* Note that this burden of proof can vary in difficulty depending on the basis for the claimed incapacity. Diminished capacity tends to be relative and circumstantial. Thus it is often the most difficult to prove.
2. *Infancy doctrine.* The basic theory behind the doctrine is to protect the young from contract abuses of adults.
3. *Ratify.* Ratification relates back to the inception of the contract and can be evidenced by words, acts, or even silence.
4. *Emancipation.* Even though the person may still be technically a legal minor, his or her lifestyle may show otherwise. Many young persons live on their own, get married, and even have children.

5. *Void.* Remember, there is a critical difference between void and voidable. In a void situation, no contract ever came into being in the first place. In a voidable situation, someone has a back door that allows him to get out from under the obligations of an otherwise enforceable contract.
6. *Statutes, public policy.* Both of these are sources for possible declarations of illegality in contract law.
7. *Regulatory.* Most professions are covered by such statutes ranging from lawyers and doctors to most trades people.
8. *Covenants not to compete.* Courts impose definite time and space limitations regarding where these covenants are allowed. Remember, these provisions are an exception to the general rule that makes anticompetition clauses illegal.
9. *Exculpatory clause.* There are limitations on when and where these clauses can be used based on equity and the court's views on procedural due process.
10. *Doctrine of unconscionability.* Even though there is no one single definition of this doctrine, look to the overall fairness of the contract process, i.e., would it be equitable to both parties to enforce the contract? The statement quoted from Chief Justice Fuller at the beginning of this chapter provides an excellent capsulation of this concept.

True/False:

1. False. A minor must disaffirm the whole contract, or none of the contract, but cannot pick and choose certain portions.
2. False. The infancy doctrine is based on an objective standard. The standard is determined by the minor's age alone.
3. False. A person is only bound by a ratified contract if she ratifies the contract after she reaches the age of majority.
4. False. Parents are liable for their children's contracts for necessaries of life that they have not adequately provided and surety contracts they sign on behalf of their children.
5. False. A person's degree of intoxication at the time a contract is made is the relevant issue. Also, if the person's mental capacity is permanently impaired, that person may be considered legally insane.
6. False. Banks and financial institutions are sometimes allowed exceptions to usury laws for loans above a certain dollar amount.
7. False. Exculpatory clauses only release the party from the liability arising from ordinary negligence.
8. True. This is an exception to the general rule of unenforcability of illegal contracts.
9. False. A court may divide the contract and enforce part while not enforcing another part.
10. False. The definition of unconscionability is flexible and varies from case to case.

Multiple Choice:

1. C. Doug lacked legal capacity when the contract was made and therefore has the power to void the contract. A is incorrect because the contract is voidable, not void. B is incorrect because Doug is not bound by the contract since he did not have legal capacity. D is incorrect because the intoxicated person may void the contract even if intoxication was voluntary.
2. D. Tom may disaffirm the contract at any time prior to reaching the age of majority. A is incorrect because the contract is voidable, not void. B is incorrect because the adult cannot void the contract, only the minor can void it. C is incorrect because a minor cannot ratify the contract until reaching the age of majority.
3. A. Matt may void the contract if he chooses. He may also perform the contract. City Cycles is bound by Matt's decision. B is incorrect because the contract is voidable, not void. C is incorrect because City Cycles does not have any options here. D is incorrect for the same reason as C.

4. C. Matt can ratify the contact by his conduct. Since he did not disaffirm the contract and has continued to pay, he has ratified and is now bound by the contract. A is incorrect because the contract is not void. B is incorrect because once ratified, the contract is not voidable. D is incorrect because ratification relates back to the date of the contract.
5. D. Once a person is judged insane, any contracts entered into are void. A through C are incorrect because the contract is void, not voidable.
6. A If a person is legally insane, but not adjudged insane, contracts entered into by that person are voidable, but not void.
7. A. This is a contract to commit a crime that is void as against a criminal statute. B is incorrect because, although this is against public policy, the specific issue is the statute. C is incorrect because the contract is already void regardless of what Cathy does.
8. B. Contracts that obstruct the legal process are void as against public policy. A is incorrect because this is a public policy issue, not a statutory issue. C is incorrect because the contract is void regardless of what the juror does.
9. C. Exculpatory clauses that result from unequal bargaining power are generally unlawful. A is incorrect because the clause will not be enforced, and courts do not generally favor exculpatory clauses. B is incorrect because it is not applicable in this situation. D is incorrect because courts do, sometimes, enforce exculpatory clauses.
10. C. Wilma entered into this contract when she was under extreme duress from her husband's death and Sherman's pressure. Such contracts are illegal. A is incorrect because this is not a statutory issue. B is incorrect because she was not ignorant of the facts.

VI. Answers to the Essay Question

The public policy under which Bunco is seeking exemption from the consequences of his agreements with Joe and Happy Card is protection based on diminished capacity. Even though Bunco may be a certified member of both Alcoholics and Gamblers Anonymous, that fact alone will not give him a basis for avoiding contract. Because he has not been declared legally incompetent, the presumption of legal competency will stand until he can prove diminished capacity. If he can show that Joe knew Bunco had these weaknesses and took advantage of him by weakening his resistance with more alcohol, he may be able to show undue influence.

If Bunco was able to convince a court that his capacity was, in fact, lost due to undue influence, he would be able to avoid the legal consequences of his note with Joe. He would probably still have to pay Happy Card because it did not exert any undue influence over him. Bunco does have an uphill battle here. His intoxication was voluntarily induced, and if Bunco was snookered in more ways than one, he really has no one to blame but himself.

Assuming that Joe's Bar and Gambol is duly licensed to do business as a gaming and drinking establishment, there appears to be no illegality on its face. The contract does not appear to violate public policy.

On the issue of good conscience, it can be argued by Mr. Bullhead that unconscionable methods were used to reduce his mental capacities, and that behavior should now render the contracts unenforceable. As was mentioned earlier in this answer, this argument will most likely fail because his intoxication was voluntarily induced. In addition, he was fully aware of the practice before he even set foot in good old Joe's.

Chapter 11
WRITING AND E-COMMERCE SIGNATURE LAW

He is not deemed to give consent who is under a mistake.
Latin legal phrase

I. Overview

Enforcement of Contracts

The first set of defenses to the enforcement of contracts in this chapter revolves around the issue of free will. Where free will is compromised, mutual assent is also compromised, and the agreement may not stand as a contract. What makes this area of the law difficult it that courts and juries are asked to exercise 20/20 hindsight when looking back on how the parties were thinking as they were embarking on the road to contract formation. The subjectivity of measuring intent has always been a troublesome puzzle to unravel; yet without it, the objective facts placed before a court may not show the reality of consent. Because of the potential harshness of a bad contract, courts want to be very sure that the assent element of contracts is just that--free and real consent to the agreement.

Factors that mitigate or diminish the genuineness of assent can be tracked on a scale of incremental culpability. At the bottom of the scale is an innocent mistake which can be either unilateral of bilateral. In a contract mistake, one or both of the parties is acting under an erroneous belief about some aspect of the contract. Normally, if only one (unilateral) of the parties is mistaken, there will be no grounds for rescission unless that mistake is coupled with some sort of bad faith or abuse on the part of the nonmistaken party, i.e., one step further up the culpability ladder. Where the mistake is mutual (bilateral), either party may seek rescission if the mistake is considered material, i.e., so important that no real meeting of the minds ever occurred.

The next increment up the slope of culpability is found in the area of misrepresentation or concealment. The shade of gray turns darker when a person is actively seeking to mark the cards or pass them under the table. Here we can see that freedom of assent is even further compromised than in mistake alone. Now the element of *scienter* (guilty mind) enters the picture, and the grounds for rescission are greatly increased. If the misrepresentation is material, known to be so by the maker, made with the intent to deceive, and is justifiably relied upon by an innocent and injured party, then the elements of fraud are in place. With a finding of fraud, the injured party may seek rescission and/or civil damages. In addition, the state may choose to prosecute the wrongdoer under the penal code. Contract fraud not only sits at the top of the culpability scale but can also be found at the top of the charts on the most popular white-collar criminal list. Its many permutations can be found in virtually all aspects of commercial activity, and as with so many areas of criminal behavior, the consumer pays the ultimate cost of these crimes through passed on costs for insurance, credit, and any number of other services undermined by these sorts of activities.

Another highly sensitive area of mutual assent is found in the law of undue influence. Undue influence involves taking away a person's free will through any manner of physical, emotional, or psychological manipulation. It can happen in any relationship, and where it is alleged, the person claiming to be the victim of undue influence has the burden of proof in showing the alleged duress. One important exception to this general rules involves persons who act in a fiduciary role. Fiduciary is a term derived from the Latin word *fides* meaning faith, honesty, confidence, or trust. A person in a fiduciary role is entrusted with acting for the benefit of another. Most professionals in law, accounting, the healing arts, and business find themselves in fiduciary rules to one degree or another. As for the fiduciary, the burden of proof is now reversed. In dealing with their respective clients, patients, or beneficiaries, a contract is presumed to be under undue influence, and the burden of proof is on the fiduciary to show that the transaction is at arm's length, i.e., fairly arrived at. Most all of the so-called "learned professions" have codes of conduct and canons of ethics designed to address issues of this sort.

Writing Requirement

The origins of the writing requirement for certain contracts are found in two roots: one historical and one practical. The historical root goes back to early English common law as developed under William the Conqueror and his successors. Status in that society was almost entirely measured by how much land one had control over. Being Lord of the Manor meant privilege, power, and rank. Thus contracts involving the transfer of land ownership were of utmost importance because of the bearing they had on social status. Highly ritualized written processes of titled transfers to land evidenced these important contracts. The original title to the land was often traceable to a knight's *fief* or fee for services provided to the sovereign. From this phrase, the highest recognized ownership in land today is still called fee simple absolute.

The second root of the writing requirement is found on a more mundane level, having less to do with knights in shining armor and more with practicality. A writing is considered the best and most neutral evidence of the parties' intent at the time the agreement was entered into. The writing does not lose its memory; it does not take sides. Thus when English lawmakers wrote the Statute of Frauds, they decided the statute would serve them with the best of both worlds--impose a writing requirement on the most important contracts to act as the best evidence in a court of law.

The English version of the Statute of Frauds has been carried over to our legal system virtually intact for over three hundred years with land contracts still being the most important. All U.S. states have adopted their own versions of the statute, and they are virtually uniform in that they require contracts involving interests in land, consideration of marriage, one year plus, third party guarantees, and others to be in writing. The most significant addition to this list came with the adoption of the Uniform Commercial Code. Under the provisions of the UCC, contracts for the sale of goods for more than $500 need to be in writing. Thus, the first question which needs to be answered is: does the statute cover this contract or not?

Once you have decided the contract is covered, what are the effects of having failed to use a writing? Several possibilities may occur at this juncture. The parties may proceed to voluntarily perform the contract. But if one or both decide to assert the statute, its teeth are found in it being used as a defense to enforcement, i.e., if the party against whom contract enforcement is sought has not signed, it may not be enforced against him or her. There are equity-based exceptions to this general rule based on partial performance and promissory estoppel.

Once the contract is finally reduced to writing, the next element of the Statute of Frauds takes hold: the parol evidence rule and the exceptions to it. Think of it the parol evidence rule as being similar to Janus, the figure from Roman mythology, who was guardian of portals and patron of beginnings and endings. He faced both ways at once symbolizing his power to let in or keep out the muses. So too does the parol evidence rule acts as a guardian of the gate of contract. The exclusion face of the rule states that the writing is intended to express the final intent of the parties. All prior or contemporaneous statements must ultimately have been reflected in writing and will remain barred from the interpretation of the instrument. This provision is designed to prevent a fraudulent rewrite of the document after the fact with new evidence.

The converse is found in the exceptions to the parol evidence rule. Long ago, an anonymous legal scholar first said: "The Statute of Frauds should not be used to perpetrate frauds." The exceptions to the parol evidence rule are designed to let in additional information not shown on the original writing in certain limited circumstances. These special circumstances are grounded in public policy and simple practical necessity. Public policy provides an overriding basis in cases involving fraud, misrepresentation, deceit, bad faith, power to avoid based on age or mental capacity, duress, undue influence, and mistake. All these elements are considered in the best interest of public policy and will be allowed into evidence, notwithstanding the statute if the facts warrant it.

The second area of exception to the parol evidence rule involves the explanation of ambiguities. If the contract, as written, contains ambiguous language, parol evidence is allowable to clear the ambiguity as long as it is consistent with the original terms. The nature of the evidence allowable under this rule can range from oral statements made by the parties on up to entire standards of usage and trade used by a particular industry. This exception is particularly important to contracts covered by the UCC. Again the underlying theory is prevention of fraud. Just like Janus, fraud can be prevented by either keeping evidence in or out depending on the individual equities knocking at the contract door.

II. Hypothetical Multi-Issue Essay Question

Refer back to the essay question in Chapter 5. In that question our protagonists were Gabby and Crabby. Gabby was injured in an auto accident involving Mario and the City of Boulder. Remember also what happened to poor Gabby. He died after having been successfully operated on by a very tired Dr. Franklin Pierce. The issues involving possible tort liabilities were reviewed in Chapter 5. We now revisit the problem to review the genuineness of assent issue in the agreement made between Crabby and Dr. Pierce.

Remember the circumstances under which the agreement was entered into. Crabby implored an exhausted Dr. Pierce to perform the operation in spite of his weariness. She promised him $50,000 to perform the operation. Her state of mind at the time of the promise was one of extreme panic and fear for her husband's life. Dr. Pierce professionally completed the operation and seeks the agreed-upon compensation for services rendered. Crabby now claims her assent was not genuine. How do you decide?

III. Outline

Genuineness of Assent
- The requirement that a party's assent to a contract be genuine

Recission
- Recission is an action to undo the contract
- The law permits recission of some contracts made in mistake

Mistakes
- Unilateral mistakes
- Mutual mistakes
  - Mutual mistakes of a past or existing material fact
  - Mutual mistake of value

Misrepresentation
- An assertion that is made that is not in accord with the facts
  - Intentional misrepresentation
  - Innocent misrepresentation

Elements of Fraud
- The wrongdoer made a false representation of material fact
- The wrongdoer intended to deceive the innocent party
- The innocent party justifiably relied on the misrepresentation
- The innocent party was injured

Types of Fraud
- Fraud in the inception (or fraud in the factum)
- Fraud in the inducement
- Fraud by concealment
- Silence as misrepresentation
- Misrepresentation of law
- Innocent misrepresentation

Duress
- Occurs where one party threatens to do a wrongful act unless the other party enters into a contract
- Economic duress

Proving Undue Influence
- A fiduciary or confidential relationship must have existed between the parties
- The dominant party must have unduly used his or her influence to persuade the servient party to enter into a contract

Statute of Frauds
- State statute that requires certain types of contracts to be in writing

Contracts Generally Required to Be in Writing
- Contracts involving interests in land
- Contracts that by their own terms cannot possibly be performed within one year
- Collateral contracts where a person promises to answer for the debt or duty of another
- Promises made in consideration of marriage
- Contracts for the sale of goods for more than $500
- Real estate agents' contracts
- Agents' contracts where the underlying contract must be in writing
- Promises to write a will
- Contracts to pay debts barred by the Statute of Limitations or discharged in bankruptcy
- Contracts to pay compensation for services rendered in negotiating the purchase of a business
- Finder's fee contracts

Contracts Involving Interests in Land
- Real property
  - The land itself as well as buildings, trees, soil, minerals, timber, plants, crops, and other things permanently affixed to the land
- Fixtures
  - Personal property that is permanently affixed to the real property, such as built-in cabinets in a house

Transfer of Other Interests in Real Property
- Mortgages
- Leases
- Life estates
- Easements

Transfer of Other Interests in Real Property
- Mortgages
- Leases
- Life estates
- Easements

One Year Rule
- An executory contract that cannot be performed by its own terms within one year of its formation must be in writing

Equal Dignity Rule
- The equal dignity rule states that agents' contracts to sell property covered by the Statute of Frauds must be in writing to be enforceable

Promissory Estoppel

An equitable doctrine that permits enforcement of oral contracts that should have been in writing. It is applied to avoid injustice.

Transfer of Interests in Real Property Must be in Writing under the Statute of Frauds

This includes:

- Transfers of an ownership interest
- Mortgages
- Leases
- Life estates
- Easements

Sufficiency of the Writing

- Formality
- Required signature
  - Fax can constitute a suitable guarantee
- Integration of several writings
  - Incorporation by reference
- Standards of interpretation

Parol Evidence Rule

Parol evidence is any oral or written words outside the four corners of the written contract

If a written contract is a complete and final statement of the parties' agreement, any prior or contemporaneous oral or written statements that alter, contradict, or are in addition to the terms of the written contract are inadmissible in court regarding a dispute over the contract.

Exceptions to the Parol Evidence Rule

Parol evidence may be admitted in court if it:

- Shows that a contract is void or voidable
- Explains ambiguous language
- Concerns a prior course of dealing or course of performance between the parties or usage of trade
- Fills in the gaps in a contract
- Corrects an obvious clerical or typographical error

IV. Objective Questions

Terms:

1. When only one party is mistaken about some aspect of a contract, this is called ________________ ________________. However, the mistaken party is not usually allowed to rescind the contract.

2. A fact that is important to the subject matter of a contract is known as a ________________ fact.

3. When one party consciously decides to induce another party to rely and act on a misrepresentation, this is called ________________.

4. If a party makes a misrepresentation with knowledge that it is false or without sufficient knowledge of the truth, this is called ________________.

5. When one party threatens to do some wrongful act unless the other party enters a contract, this is called ________________.

6. Land as well as buildings, trees, soil, minerals, timber, and other things permanently affixed to the land are called ________________ property.

7. An interest in land for a person's lifetime that is transferred to another party on that person's death is a ________________ ________________.

8. Where the main purpose of a transaction and an oral collateral contract is to provide pecuniary benefit to the promisor or guarantor, and the collateral contract is treated like an original contract and does not have to be in writing to be enforced, this is called the ________________ ________________ ________________.

9. The rule that requires that all agents' contracts to sell property must be in writing to be enforceable is called ________________ ________________ ________________.

10. When a contract contains some but not all of the terms of the parties' agreement, this is called ________________ ________________.

True/False:

1. ____ A fax has been found to satisfy the signature requirement of the Statute of Frauds.

2. ____ Where a party has been fraudulently induced to enter a contract, the innocent party can either rescind the contract or enforce the contract and sue for damages.

3. ____ To be actionable, a fraudulent misrepresentation must be in spoken or written words.

4. ____ Where there is fraud in the inception, the contract is voidable by the innocent party.

5. ____ Fraud by concealment occurs when one party fails to disclose a fact.

6. ____ Contracts to construct buildings or insure buildings do not have to be in writing under the Statute of Frauds because no ownership interest in the real property is being transferred.

7. ____ Both implied and express easements have to be in writing under the Statute of Frauds.

8. ____ To be an enforceable contract, the entire writing must be in one document.

9. ____ The parties to a written contract may stipulate that their contract is a complete integration and that parol evidence may not be introduced at all.

10. ____ To be legally binding, a written contract does not have to be drafted by a lawyer or formally typed.

Multiple Choice:

1. Ted enticed Fred to enter into a sales contract by intentionally telling Fred certain material facts that Ted knew were untrue. Assume no other relevant facts. On what legal basis can the contract be avoided?
   A. Voidable because of undue influence, which created no real assent.
   B. Voidable because of fraud, where no real assent took place.
   C. Voidable because of duress, which prevented real assent.
   D. Voidable because of mistake, which did not allow real assent to take place.

2. Ed, an auditor for the IRS, became emotionally involved with Rhonda. At the urging of Rhonda, and fearing that she would sever their relationship, Ed reluctantly waived a tax audit settlement that was grossly unfair to the IRS, Ed's employer's best chance for rescinding this outcome is to show that Ed acted under:
   A. A lack of express authority.
   B. Duress.
   C. Undue influence.
   D. There was no consideration.

3. On June 1, Farmer Brown contracted to sell a red barn to Rancher Ron. However, unknown to either party, the barn had burned down on May 20. If Ron sues Brown for breach of contract, Brown's best defense is:
   A. Fraud.
   B. Economic duress.
   C. Mutual mistake of fact.
   D. Misrepresentation.

4. Investor Ivan contracted to buy a large area of land for housing development from Owner Ellen. Before signing the contract, Ivan asked Ellen if she knew of any reason why houses could not be built on the land. Though Ellen knew there was a difficult to discover the soil condition that would make building impossible, she said nothing. Which of the following is correct?
   A. Ellen has committed fraud by concealment.
   B. Ellen has committed an innocent misrepresentation.
   C. Ellen has committed a misrepresentation of law.
   D. Ellen has committed silence as misrepresentation.

5. Assume the facts as in question 4. Which of the following is correct?
   A. Ivan may sue for damages or rescind the contract.
   B. The contract is void.
   C. The contract is enforceable by either Ivan or Ellen.
   D. None of the above.

6. A written agreement was signed by both parties that was intended to be their entire agreement. The parol evidence rule will prevent the admission of evidence that is offered to:
   A. Explain the meaning of an ambiguity in the written contract.
   B. Establish that fraud had been committed in the formation of the contract.
   C. Prove the existence of a contemporaneous oral agreement that modifies the contract.
   D. Prove the existence of a subsequent oral agreement that modifies the contract.

7. Employee Edgar received a verbal offer of employment from Competitive Co. for a year and a half. However, after a few months, Competitive laid off Edgar for economic reasons. Edgar is suing Competitive Co. for lost wages. Which of the following is true?
   A. Edgar will lose because such a contract must be signed by both parties to be enforceable.
   B. Edgar will win because the Statute of Frauds does not cover contracts for employment.
   C. Edgar will win because Edgar's partial performance is a substitute for a written agreement.
   D. Edgar will lose because the contract needed to be in writing.

8. Homeowner Henry contracts with Broker Bob to have Bob sell his house. Which of the following is correct?
   A. The contract must be in writing only if Bob will be employed for more than one year.
   B. Under the equal dignity rule, the contract must be in writing.
   C. Under promissory estoppel, the contract must be in writing.
   D. The contract must be in writing only if Bob's commission will be greater than $500.

9. Supportive Sam phoned Lender Lena and told her that, if she would loan money to Needy Nel, he would pay Lena back if Nel could not. Which of the following is correct?
   A. Sam made a secondary promise.
   B. Sam made a primary promise.
   C. Sam made a joint promise.
   D. None of the above.

10. Under a contract for the sale of land, the Statute of Frauds:
   A. Requires that the entire agreement be in one single document.
   B. Requires the purchase price be fair.
   C. Does not apply if the land's value is less than $500.
   D. Does not require the agreement be signed by all parties.

V. Answers to Objective Questions

1. *Unilateral mistake.* Rescission is allowed only under limited circumstances where the unilateral mistake is tied in with other parties' actual or constructive knowledge of the mistake, where there are clerical or mathematical errors, or where enforcement would lead to an unconscionable result.
2. *Material.* Look at the essential purpose of the contract when trying to identify material facts, i.e., but for this element, would the parties be in this contract at all?
3. *Fraud.* The elements of fraud are intentional false misrepresentation by the wrongdoers coupled with justifiable reliance by the injured innocent party.
4. *Scienter.* This term is derived from the Latin work meaning "to know." In addition to contracts, this term is often used in the definition of numerous crimes.
5. *Duress.* In identifying this element, look for situations where wrongful acts or threats are used to take away another's free will to act or make a contract.
6. *Real property.* This classification is based on traditional case law combined with statutory definitions. These issues will be further detailed in the chapters on real and personal property.
7. *Life estate.* Upon the death of the life tenant, the property passes to the holder of the remainder interest or back to the grantor by reversion. Life estates are a commonly used planning device.
8. *Leading object exception.* This exception to the writing requirement is used in contracts involving promises for the debt of another. It is also known as the "main purpose" rule.
9. *Equal dignity rule.* This is an umbrella-type provision that extends the coverage of the Statute of Frauds requirement to contracts closely related to agreements that are already covered by the statute.

10. *Partial integration.* Parol evidence may be allowed to fill in the gaps if the parties have not fully integrated their agreement in the original document.

True/False:

1. True. An unsigned fax with the company name across the top has been held to constitute an enforceable guarantee.
2. True. The innocent party in such a situation has an option of doing either of these things.
3. False. A misrepresentation may occur by the party's conduct.
4. False. Contracts where there has been fraud in the inception are void rather than just voidable.
5. False. Fraud by concealment occurs when one party takes specific action to conceal a material fact. This goes beyond mere failure to disclose.
6. True. These contracts do not fall under the equal dignity rule requirement, but as a practical matter, they too should be in writing.
7. False. Only express easements must be in writing.
8. False. Several writings may be integrated to form a single written contract.
9. True. Such an agreement is called a merger or integration clause.
10. True. Generally the law only requires that there be a written memorandum containing the essential terms of the parties' agreement. These writings can be, and often are, quite informal.

Multiple Choice:

1. B. Ted made a false misrepresentation of a material fact with an intent to deceive. Fred detrimentally relied on this representation. Therefore, the elements of fraud are in place. A is incorrect because undue influence is the use of some sort of control which makes the decision involuntary. C is incorrect because duress is a threat that is associated with fear. D is incorrect because only mutual mistakes would make this contract voidable, and there was no showing from the facts given that a mutual mistake took place.
2. C. Here, the emotional involvement was so overwhelming that the decision was not voluntary and was the product of undue influence. A is incorrect because Ed probably had express authority to waive the audit. B is incorrect because duress is a threat that induces fear. D is incorrect because there was probably consideration involved in the audit settlement.
3. C. Both parties were unaware of a material fact. A is incorrect because Brown did not fraudulently misrepresent the fact to Ron. B is incorrect because there was no duress.
4. D. By saying nothing when Ivan asked a direct question, Ellen has committed silence as misrepresentation. A is incorrect because fraud by concealment requires specific action to conceal something, not just failure to disclose. B is incorrect because Ellen did not make a statement of fact that she reasonably believed. C is incorrect because the soil condition is not a matter of law.
5. A. Ivan has the option of either suing for damages or rescinding the contract. B is incorrect because Ivan may enforce the contract if he chooses, so it is not void. C is incorrect; Ellen may not enforce the contract.
6. C. If the contract was intended to be their entire agreement, the rule will not allow evidence to show a contemporaneous oral agreement that modifies the contract. A is incorrect because the rule allows evidence to explain the meaning of ambiguous words or phrases. B is incorrect because the rule allows evidence to prove fraud. D is incorrect because subsequent oral agreements may be proven under the rule.
7. D. The employment contract was for more than one year and must therefore be in writing to be enforceable. A is incorrect because the Statute of Frauds does not require that both parties sign the contract. B is incorrect because employment contracts for greater than one year are covered by the

Statute of Frauds. C is incorrect because the doctrine of partial performance is not applicable to employment contracts.

8. B. Agents who sell real property must have a written contract with the other party. A is incorrect because here, the length of Bob's employment is irrelevant. C is incorrect because promissory estoppel does not require writing and is not applicable here. D is incorrect because the amount of the commission is irrelevant when the contract is for such an agent.
9. A. Such a promise to pay the debts of another if that person fails to perform is called a collateral or secondary promise.
10. D. The Statute of Frauds requires contracts for the sale of land to be in writing and to be signed by the party against whom enforcement is sought, not necessarily all parties. A is incorrect because there may be several integrated writings. B is incorrect because the Statute of Frauds does not address the purchase price. C is incorrect because contracts for the sale of land must be in writing regardless of the amount paid.

VI. Answers to the Essay Question

There appears to be no material mistake of fact in this problem on the part of either party. Crabby knew Dr. Pierce was tired and did not really want to operate. He, in turn, did operate in a professional manner in spite of his weariness. The fact that Gabby died anyway was not attributable in any way to how the operation was performed. As was discussed earlier in Chapter 5, the tort issues are to be resolved between Crabby and/or both Mario and the City of Boulder.

There appear to be no indices of misrepresentation, concealment, or fraud in this problem. Both parties entered into the agreement with all cards on the table.

The doctor-patient relationship is clearly classified as a fiduciary relationship. As such Dr. Pierce has the burden of proof to show that any transaction entered into between him and his patient entered into without any undue influence. Because undue influence is presumed, Dr. Pierce must show that the contract was entered with free and genuine consent on the part of Crabby who was acting as Gabby's agent due to Gabby's physical condition at the time. What makes this a difficult problem for Dr. Pierce is that the price offered for the operation by Crabby ($50,000) may be higher than the normal fee charged for a comparable operation. Perhaps Dr. Pierce should have realized that Crabby was in a state of extreme agitation and told her the offer was above and beyond fair market value. It may be argued by Crabby that the excessive fee was a form of economic duress coupled with an abuse of fiduciary position.

Dr. Pierce, in turn, has a legitimate claim to some sort of fair compensation for his services. These services were delivered in a competent manner and it would be inequitable for him not to be paid, notwithstanding the fiduciary relationship. Therefore, Dr. Pierce will likely be paid for his services. The only real issue is whether he will be paid what was originally offered or only the fair market value of comparable services. The probable result lies in the latter based on equity coupled with the concern of courts about contracts that may be overly advantageous to the fiduciary.

Chapter 12
THIRD PARTY RIGHTS AND DISCHARGE

There are only two mistakes one can make along the road to truth: (1) not going all the way, and (2) not starting.

Buddha

I. Overview

Previous chapters on the law of contracts dealt with the establishment of the respective rights and duties of two persons entering into the contracting process. As the old saying goes: "It takes two to tango." Now we are looking at the possibility of a new dance partner cutting in. As at the high school prom, this new participant may not always be a welcome sight! Yet the rules of contract, like dance floor etiquette, require that third parties be allowed to join the dance where proper circumstances are in place. This chapter reviews those situations that allow a new dancer onto the floor and describes what legal tune plays for him.

Third parties can become involved in a contract *ab initio* (from its inception) or after the fact. Contracts with third party involvement from the beginning are generically labeled intended beneficiary contracts. These contracts are broken down into two subcategories: donee and creditor. The donee beneficiary contract is probably the one most students will be familiar with and most likely to be a participant in. Consider what happens when records of newborn infants are filed with local authorities. The insurance industry has long used this source of public information to notify the new parents that the party's over. They must now act like responsible citizens and provide for their young in case any unforeseen calamity befalls them. Thus, insurance is taken out for young Junior as beneficiary of his parent's insurance policy even though Junior has not paid for the policy, does not yet have the personal capacity to enter into contracts, and is really only concerned with staying dry and his next meal. He is now a third party intended (donee) contract beneficiary. And if his rights have vested under the contract, the contract cannot be altered, canceled, or rescinded without his consent. Other examples of third party donee beneficiaries can be found in trusts and contracts to make a will.

The second category of intended third party beneficiaries is found in the law of creditors rights. When there is a preexisting debtor/creditor relationship, that relationship may act as the basis for a second contract where the named creditor beneficiary may be protected as an intended beneficiary. For example, suppose A borrows money from B to buy a car, and the car is used as collateral for the loan. B can now be named as a loss-payee (a person named in the policy to be paid in case of loss) up to the amount owed, even though B has not paid for or been a signatory to the insurance policy between A and his or her insurance company.

The third category, incidental beneficiaries, is not really intended beneficiaries at all but rather wallflowers in the contract dance. The law is very straightforward on this point: if you do not qualify as an intended beneficiary based on donee or creditor grounds, you are classified as an incidental beneficiary and will not have any legally recognized standing to sue for protection under the contract. There is one minor exception to this rule in certain government contracts. Ordinarily, a taxpayer who objects to these contracts is classified as an incidental beneficiary, with no standing to sue, unless he or she can show that they belong to a class for whose primary and immediate benefit the government contract was made. For example, a taxpayer who is charged a special assessment for a sidewalk repair in front of his house may have standing to sue if the sidewalk repair was defective.

The second major category of third party involvement in contracts is the introduction of a third party after the contract has already been formed. This third party joins the dance by way of assignment or novation. In assignment, one of the original parties transfers rights or duties to a new third party participant. Compare this with a novation, derived from the Latin *nova*, meaning new. In a novation, the original contract with A and B is ended when a new contract is entered into between one of the original parties and a new party, C. The most common example is found in the assumption of mortgage obligations. In these forms of after-formation involvement of third parties, the law of assignment is far more important. Public policy in the law of contracts favors the transferability of contract rights and duties. There are, however, situations where this general rule does not apply and consent of the other parties is required for a valid transfer. These include situations involving personal services, creditworthiness, trust, and/or fiduciary relationships. In addition, such transfers may be barred or limited by the agreement itself or by statute. In any event, if there is a material change of risk in the obligations covered by the agreement, limitations are likely to apply vis-à-vis the transferability of rights and duties under the contract. Consider the world of finance, commodities, and the like. The transfer of contract rights from one person or business entity to another facilitates all of these commercially critical practices.

The second objective of this chapter is to examine the basic rules with regard to the discharge of contracts. The rules of performance and breach of contract are rooted in common sense. Most contracts are completed legally when the parties have lived up to their reciprocal obligations under that contract. Conversely, a breach is found when a failure of performance is not somehow excused by law. We are expected to live up to our performance obligations and no more. If those obligations are not met, breach of contract is the result.

The evaluation process of contract performance issues is best broken down into time sequence subparts: precontract, during the contract, and postcontract. In precontract issues, what are the covenants entered into before performance is to be initiated? Were there any conditions that may affect the rights and duties of the parties to contract? Conditions are certain events that have a triggering effect on the obligations of contract. The timing of conditions can be superimposed upon the contract. A precondition or condition precedent calls for the event to take place before the contract goes into effect. For example: "I will buy this car if my mechanic signs off on the engine inspection." A concurrent condition calls for two or more events to coincide in time. Consider an escrow where a third party is used as a holder of property and is instructed to act vis-à-vis that property only upon satisfaction of mutually dependent acts of third parties. This is a common form of property transfer used in the sale of real estate. The escrow holds the deed to the property from the seller until the buyer has delivered the purchase price in a form acceptable to both parties. A condition subsequent is found where performance may be excused by a certain event after the contract was entered into. For example, a parolee is allowed to stay out of prison as long as the conditions of the parole release are met.

There are certain circumstances that will act to excuse nonperformance. These circumstances are also based in common sense. Can it really be reasonable to expect personal service contracts to be enforced after death or disability? Or does it make sense to accept performance after destruction of a unique subject matter of the contract? For example, if you bid millions of dollars for a Van Gogh's "Irises" at an auction, and it was destroyed by lightning, is it reasonable to expect the seller to come up with an exact duplicate? A third form of excuse is found in subsequent illegality. If a contract was legal at the time it was formed, but subsequent events have made its enforcement illegal, courts will no longer enforce its performance covenants based on the new illegality.

In addition to excused nonperformance, there are a number of possible circumstances that may result in a discharge from any further contractual performance. These fall into two main categories: discharge by acts of the parties or by operation of law. Discharge by acts of the parties is voluntary post contract formation events such as mutual recission, reformation, accord and satisfaction, a substituted contract, or novation. In all these scenarios, the parties have, in effect, reentered the bargaining and created a new deal.

In an operation of law discharge, something has happened where the court steps in and declares this contract performance obligation can no longer be enforced. Examples of such legal impediments to enforcement would include the running of a statute of limitation or bankruptcy. In both cases, any further performance under the contract has been legally ended. If, however, the contract duty has not been discharged, excused, or performed, and the absolute duty to perform has been breached, one must examine what remedies are available to the nonbreaching party.

II. Hypothetical Multi-issue Essay Question

Mr. Daro Cepheus was a star in the world of high finance and was in line for a big promotion at Bailout Bank when he found out he was passed over by Ms. Mydras Volans. Her star shone even brighter, and Daro decided not to get mad, but get even. Daro devised an intricate method of selling the same corporate bond all over the world to multiple investors.

He would make multiple sales by having the original certificates held at the bank and send out facsimiles and tell investors the reasons for the delay in sending the originals was due to the postal service. In the meantime, investors from Chad, Mongolia, Japan, France, and Hong Kong, in that order, all sent him $1 million each, for the same bond. Daro is now watching sunsets in Rio while the investor in Hong Kong was the first to smell a rat and contact Bailout Bank. Assuming Daro cannot be found, what are the legal consequences to the remaining parties involved?

III. Outline

Form of Assignment
- Assignor—the obligee who transfers the right
- Assignee—the party to whom the right has been transferred

Rights That Cannot Be Assigned
- Personal service contracts
- Assignment of future rights
- Contracts where assignment would materially alter the risk
- Assignment of legal actions

Rights of Successive Assignees
- American Rule—First assignee in time prevails, regardless of notice
- English Rule—First assignee to give notice prevails
- Possession of Tangible Token Rule—First assignee who receives delivery of the tangible token prevails over subsequent assignees

Delegation of Duties
- Delegation of duties—a transfer of contractual duties by the obligor to another party for performance
- Delegator—the obligor who transferred his or her duty
- Delegatee—the party to whom the duty has been transferred

Duties That Cannot Be Delegated
- Personal service contracts calling for the exercise of personal skills, discretion, or expertise
- Contracts whose performance would materially vary if the obligor's duties were delegated

Third-Party Beneficiaries
- Intended beneficiaries
  - Donee beneficiaries
  - Creditor beneficiaries
- Incidental beneficiaries
  - Have no rights to enforce or to sue under other people's contracts

Covenant
- A covenant is an unconditional promise to perform

Types of Conditions

Discharge of Contracts
- By agreement
  - Mutual rescission
  - Substituted contract
  - Novation
  - Accord and satisfaction
- By excuse
  - Impossibility of performance
  - Commercial impracticability
  - Frustration of purpose
- By operation of law
  - Statutes of limitations
  - Bankruptcy
  - Alteration of the contract

IV. Objective Questions

Terms:

1. The termination of an enforceable contract by full performance of the duties is called __________________.

2. A contract entered into with the intent to confer a benefit or gift on to intended third party is called a _______________.

3. A third party who benefits unintentionally when the contract is performed is called a(n) _______________ beneficiary.

4. The transfer of contractual rights is called an _______________, while the transfer of contractual duties is called a _______________.

5. The party owed a right under the contract is called the _______________, while the party who owes a duty of performance is called the _______________.

6. A clause that prohibits the assignment of rights under the contract is known as a _______________ clause.

7. A clause that permits the assignment of the contract only upon receipt of an obligator's approval called a _______________ clause.

8. A statement that is a constructive notice to all later assignees of the filling assignee's rights in the contract is known as a _______________ statement.

9. An agreement that substitutes a third party for one of the original contracting parties and makes the new party obligated to perform the contract is called a(n) _______________.

10. If the delegatee has not assumed the duties under a contract and is not legally liable to the obligee for nonperformance, this is called _______________ _______________ _______________.

True/False:

1. _____ An intended third party beneficiary can enforce the contract against the party who promised to render performance.

2. _____ Generally, any member of the public can sue as an incidental beneficiary on contracts entered into by the government.

3. _____ If a third party intended beneficiary tries to enforce his rights under a contract, the promisor can assert only personal defenses he has against the third party beneficiary.

4. _____ The rights of an intended third party beneficiary must be vested before the beneficiary can enforce the contract against the promisor.

5. _____ Generally, if a contract is conditional upon meeting a third party's satisfaction, the courts will apply a personal satisfaction test.

6. _____ Legal actions involving both personal rights and contract rights are assignable.

7. _____ An assignor cannot partially assign his contract rights. Either the entire contract is assigned, or none of it is.

8. _____ Any successive assignee who does not obtain the assigned contractual right can sue the assignor for damages.
9. _____ A firm can delegate duties to any of its qualified employees even if the obligee places special trust in the obligor.

10. _____ A temporary impossibility excuses a party from performance if he withdraws during the time the contract performance is suspended.

Multiple Choice:

1. Ted Harley sold his moped factory to Fred Davidson. As part of the contract, Fred assumed the existing mortgage on the property held by Brewer Bank. Regarding the rights and duties of the parties, which of the following is correct?
   A. The promise by Fred need not be in writing to be enforceable by Brewer.
   B. Brewer is a creditor beneficiary of Fred's promise and may recover against him personally in the event of default.
   C. Brewer is not in privity of contract since it was not a party to the assignment.
   D. Ted has no further liability to Brewer and cannot ride off into the sunset worry free.

2. Which of the following is *not* true of donee beneficiary contracts?
   A. The donee beneficiary must be identified under the contract to be conferred with a gift or benefit.
   B. The promisee is the contracting party who directs the benefit.
   C. The promisor agrees to confer performance for the benefit of the donee beneficiary.
   D. The donee beneficiary must have the capacity to enter into contracts.

3. Public policy disfavors the assignment of certain contract rights. Generally, which of the following is freely assignable?
   A. Personal services contracts.
   B. Assignment of nonexistent future rights.
   C. Assignment of existing future contract rights.
   D. Assignment of legal actions.

4. Parent owned an insurance policy on her life, on which she paid all the premiums. Daughter was named the beneficiary. Parent died and the insurance company refused to pay the insurance proceeds to Daughter. An action by Daughter against the insurance company for the insurance proceeds will be:
   A. Successful because Daughter is a third party donee beneficiary.
   B. Successful because Daughter is a proper assignee of Parent's rights under the insurance policy.
   C. Unsuccessful because Daughter was not the owner of the policy.
   D. Unsuccessful because Daughter did not pay any of the premiums.

B. Nagel and Fields entered into a contract in which Nagel was obligated to deliver certain goods to Fields by September 10. On September 3, Nagel told Fields that he had no intention of delivering the goods required. Prior to September 10, Fields may successfully sue Nagel under the doctrine of:
   A. Promissory estoppel.
   B. Accord and satisfaction.
   C. Anticipatory repudiation.
   D. Substantial performance.

6. Andy and Bertha make a contract where Charles is the intended beneficiary. Charles sues Bertha to enforce his rights under the contract. Which of the following is correct?
   A. Bertha can assert any defenses she has against Andy.
   B. Bertha can assert any defenses she has against Charles.
   C. Both of the above.
   D. None of the above.

7. Assume the same facts as is question 6. Which of the following is correct?
   A. Charles' interest in the contract vested when he learned of and consented to the contract.
   B. Charles' interest in the contract vested as soon as Andy and Bertha formed the contract, before he knew of it.
   C. For Charles' interest to vest, Charles must change his position in reliance on the contract.
   D. None of the above.

8. Employee Ellen contracts with Calloway Co. to perform certain services for three years. However, within six months, Ellen sells her house and must move away. Which of the following is correct?
   A. Ellen is excused from performance of the contract because of the objective impossibility of performance.
   B. Ellen is excused from performance of the contract because of frustration of purpose.
   C. Ellen is not excused from the contract and is liable for damages.
   D. None of the above.

9. Developer Dan contracts with Arnie Architect to have Arnie draw up plans for a new golf course clubhouse for $40,000. Which of the following is correct?
   A. This is a personal services contract and is therefore never assignable.
   B. Arnie can assign his right to the $40,000.
   C. Dan can prevent Arnie from assigning Arnie's right to the $40,000 as this is a personal services contract.
   D. Arnie can assign the contract since it is a future right.

10. Ellen and Fred have a contract where Ellen is to bake 100 pies for Fred's restaurant for payment of $500. Ellen bakes and delivers the 100 pies. Ellen then assigns her right to the $500 payment to Grant. Which of the following is correct?
    A. In assigning this right, Ellen can no longer sue Fred for nonperformance.
    B. Ellen and Grant can now both sue for nonperformance.
    C. Fred can sue Ellen for assigning this right.
    D. None of the above.

V. Answers to Objective Questions

Terms:

1. *Discharge.* A discharge canceis any further obligation under a contract.
2. *Donee beneficiary.* Look for the lack of consideration.
3. *Incidental.* Most people who do not qualify as a donee or creditor beneficiary fall into this third category.
4. *Assignment, delegation.* These and the answers to question 5 are dichotomous terms that can be used as a learning tool.
5. *Obligee, obligor.* Proper identification of the terms is essential to the study of third party rights.

6. *Antiassignment*. Courts generally interpret these clauses narrowly.
7. *Approval*. Courts limit the use of such clauses.
8. *Financing*. These are provided for under the UCC.
9. *Novation*. This term is derived from the Latin term meaning "new."
10. *Declaration of duties*. Note the requirement of notice to the obligee.

True/False:

1. True. The intended beneficiary has standing to sue.
2. False. Usually, members of the public cannot sue on contracts entered into by the government since incidental beneficiaries cannot sue as third party beneficiaries.
3. False. The promisor can assert rights he has against the promisee as well as the third party beneficiary.
4. True. Sometimes the donor can change his or her mind vis-à-vis the benefits given under the contract.
5. False. If a contract is conditioned upon meeting a third party's satisfaction, courts will generally apply the objective, reasonable person test.
6. False. Legal actions involving personal rights are not assignable.
7. False. An assignor can partially assign his contract rights to one or more people.
8. True. The assignee steps into the legal shoes of the assignor and is entitled to the transfer of benefits formerly owned by the assignor.
9. True. The rule is subject to the limitations imposed by the rules of fiduciaries and personal service contracts.
10. False. A temporary impossibility excuses a party from performance only during that time. Once this circumstance no longer exists, the party must perform.

Multiple Choice:

1. B. When a purchaser assumes an existing mortgage, the purchaser's promise to pay the mortgage creates a creditor-beneficiary relationship. A is incorrect because a mortgage is within the Statute of Frauds. C is incorrect because an intended beneficiary of a contract may sue to enforce the contract. D is incorrect because Ted is still liable to Brewer after the assignment in which there was no novation.
2. D. Often, donee beneficiaries are too young or incompetent to enter into contracts for themselves. Legal capacity is not one of the requirements of becoming a donee beneficiary.
3. C. Assignment of future rights in existing contracts are generally assignable, subject to any statutory restrictions such as laws relating to garnishment. The others are all generally limited by public policy from assignment.
4. A. Daughter was the intended third party donee beneficiary *ab initio*. There was no assignment here, which makes B incorrect. C and D are incorrect because the donee beneficiary does not have to pay for or own the policy created for her benefit.
5. C. The doctrine of anticipatory repudiation is in effect because Nagel told Fields that Nagel had no intention of delivering the goods prior to the date of performance. A is incorrect because promissory estoppel acts as a substitute for consideration and is not relevant here. B is incorrect because accord and satisfaction is an agreement where a party with an existing duty of performance under a contract promises to do something other than perform the duty originally promised in the contract. D is incorrect because the doctrine of substantial performance would allow for a contract obligation to be discharged even though the performance tendered was not in complete conformity with the terms of the agreement. Here Fields is suing Nagel for breach of contract.

6. C. A is correct because Bertha can raise any defenses she has against the original party to the contract. B is correct because Bertha can raise any defenses she has against the third party beneficiary when he sues to enforce the contract.
7. A. Charles' interest can vest when he learns and consents to the contract, when he changes his position in reliance on the contract, or when he sues on the contract. B is incorrect because at this point Charles knows nothing and cannot have a vested interest. C is incorrect because changing his position in reliance on the contract is only one way for his interest to vest, not the only way.
8. C. Ellen has breached the contract and is liable for damages. A is incorrect because Ellen's inability to complete the contract is subjective, not objective. B is incorrect because the doctrine of frustration of purpose is not applicable here.
9. B. If payment of the $40,000 is all that is left to complete the contract, Arnie is free to assign this to whomever he chooses. A is incorrect because personal services contracts can be assigned if payment is all that is required or both parties agree. C is incorrect because Dan cannot control whom Arnie assigns payment to. D is incorrect because this is not a future right.
10. A. Once Ellen assigns this right, Grant stands in her shoes, and she gives up the right to sue Fred. B is incorrect because Ellen cannot sue Fred. C is incorrect because Fred cannot sue Ellen for assigning the right to payment.

VI. Answers to Essay Question

This contract did not identify any intended third party beneficiaries at the time of its creation. It did, however, anticipate that third parties would eventually become involved. The market for corporate bond obligations is such that it can be readily expected that these notes will be transferred a number of times before they have fully matured. Thus, third party involvement will come after the original bond is sold to the first buyer by way of assignment.

This sort of contract is freely assignable as a financial obligation, and any attempts to limit its transferability would be narrowly construed by the courts. The technical procedures for a required writing were complied with, but there was obvious fraud because of multiple sales of the same asset by Mr. Cepheus.

Because we have a multiple assignment here, a number of options are available to the court depending on which rule is being used in its jurisdiction. Under the American Rule, the investor from Chad would be first in line for relief. Under the British rule, the investor from Hong Kong would prevail because of his notice being given first to Bailout Bank. In addition, he can argue that the investor from Chad failed to but should have demanded physical delivery of the tangle token, the bond itself. In all likelihood, the investor from Hong Kong would therefore win. The other investors may want to consider going after Bailout Bank under an agency theory of liability for the fraud committed by its former employee.

The basic performance obligation was to sell only what was owned. Here, there was no act of a party or operation of law excuse for the actions taken by Mr. Cepheus.

Chapter 13
REMEDIES FOR BREACH OF TRADITIONAL AND ON-LINE CONTRACTS

The common law is nothing else but statutes worn out by time.
Sir John Eardley Wilmot

I. Overview

We are expected to live up to our performance obligations. If these obligations are not met or somehow excused, a breach of contract is the result in the eyes of the law. There are a number of possible circumstances that may result in a discharge from any further contractual performance. These fall into two main categories: discharge by acts of the parties or by operation of law. Discharge by acts of the parties are voluntary post contract formation events such as mutual recission, reformation, accord and satisfaction, a substituted contract, or novation. In all these scenarios, the parties have, in effect, reentered the bargaining and created a new deal.

In addition, the contract obligation may have been met and breached with less than full performance. In substantial performance, 100 percent of the performance was not provided. If the breach was not material or intentional, the nonbreaching party may sue for damages but not recission. For example, if a $100,000 house has the wrong doorknob on it, and replacement would cost $100, the original contract can still be enforced less the price of the correct doorknob. Compare this with a house with no doors. The failure to provide doors may be construed as only partial performance of a house-building contract because of lack of security. The nonbreaching party may then sue for recission or recover damages.

The first objective of this chapter is to introduce students to remedies in contract after a breach has taken place. An overview of the remedies available to the nonbreaching party in contract can be compared to the various directions posted on a map when your car has broken down. Because your trip has been brought to an abrupt halt, you must decide which options make the most sense to you. Do you try to go back where you started? Can the car be fixed where you are, or do you need a tow to your original destination? By analogy, contracts can also be brought back (recission), fixed (damages or reformation), or taken forward to their original destination (specific performance). The majority of remedies provided by the courts for contract breach fall into the category of a repair by way of monetary damages. In this repair process, there is an attempt to provide the innocent party with a financial replacement for the benefits that he or she had under the original contract.

Think of monetary damages as an ascending staircase, starting at the bottom with token or nominal damages and going all the way to punitive damages. Nominal damages are applicable where there is little real economic consequence arising from the breach. These damages are sometimes called a Pyrrhic victory, named after the victory of Pyrrhus, King of Epirus, over the Romans at Asculum in 279 B.C. The victory was so costly that the war was ultimately lost. In a nominal damage award, the winner wins in principle but not in significant economic terms.

The next step is found in compensatory or actual damages. These damages seek to restore the benefit of the bargain by providing a monetary substitute for what was lost due to the breach. For example, if Mario had a deal with Zoomo Motors for a new car priced at $20,000 and had to buy a comparable model from Zamay Motors for $30,000 due to Zoomo's breach, the actual measure of compensatory damages would be $10,000. If Mario had a resale contract for the car to A.J. for $40,000 and Zoomo knew of this second contract, Zoomo may also be liable for an additional $10,000 in consequential damages to Mario. In addition, had Zoomo's breach been related to a bad faith tort within the contract setting, the court might grant punitive damages. Punitive damages are a form of court imposed civil punishment. Normally, punitive damages are not granted for a contract breach alone, but if the bad faith involved shocks the conscience of the court, tort damages may be applicable. One other damage measure possibility lies in the area of liquidated or agreed upon damages set by the parties. A court will examine these damages to make sure that they are not a disguised penalty.

The other two paths available to the nonbreaching party are going backwards or going forward with the contract. In going backwards, a court is asked to return the parties to their precontract position by way of recission or restitution, i.e., undo the contract. A common example of recission is found in return of deposit clauses in purchase contracts or in consumer protection statutes which provide for a three day cooling off period after the contract is signed. These statutes generally allow for a unilateral right of recission within the three-day period.

The last category of remedies involves going forward with the original terms of the agreement. These remedies are classified as equitable remedies because the breach cannot be adequately compensated by normal economic damage measures. The underlying theory is that the contract must somehow be enforced as a matter of equity and fair play rather than substituted by money alone. Equitable remedies are traceable to the English legal history of the Court of Chancery and are found in modern day versions of specific performance, quasi-contract, and injunctive relief.

The second objective of this chapter is to look at those situations where a civil wrong, defined as a tort, is somehow associated with the contract setting in question. Ordinarily, recovery for breach of contract alone is limited to damages. However, with the addition of a contract-related tort to the equation, courts may also add tort damages. The major areas of concern in the contract setting are:

1. Intentional interference with contract relations.
2. Breach of implied covenants of good faith and dealing.

If the elements of these torts are found, courts may impose punitive damages far in excess of actual contract damages. Be sure to examine the *Pennzoil v. Texaco* case for a graphic illustration of this principle.

II. Hypothetical Multi-issue Essay Question

Ms. Dee Minimis entered into a contract with Mr. Lex Loci to sell her roller skate instruction business, Derbytown Rollerball, Inc., on January 1, 2001 for $10,000. Roller-skating was on the wane, and Dee wanted to get out while the getting was good. Under the terms of the contract, Dee agreed not to compete with Lex for a reasonable period of one year within a five-mile radius of Derbytown's business location. In addition, the contract had a liquidated damages clause of $5000 in case Ms. Minimis breached the noncompetition clause. Dee took the proceeds of the sale and lost it all on the Derbytown lottery. She also realized that she made a big mistake in selling the business because of the advent of the in-line skate. She decided to establish Derbytown Superskate next door to her former place of business. She is not worried about the liquidated damages clause because of her insolvency. Lex Loci wants to sue Ms. Dee Minimis for breach of performance and asks the court for all possible remedies provided by law. What result?

III. Outline

Types of Performance and Their Legal Consequences

Anticipatory Breach
- Also called anticipatory repudiation
- A breach that occurs when one contracting party informs the other that he or she will not perform his or her contractual duties when due
- Nonbreaching party's obligations under the contract are discharged immediately
- Nonbreaching party has the right to sue

Types of Monetary Damages
- Compensatory—Compensates a nonbreaching party for the loss of the bargain, placing him or her in the same position as if the contract had been fully performed
- Consequential—Compensates a nonbreaching party for foreseeable special damages
- Liquidated—Agreement by the parties in advance that set the amounts of damages recoverable in case of breach
- Nominal—Damages awarded against the breaching party even though the nonbreaching party has suffered no actual damages because of the breach

Mitigation of Damages
- A nonbreaching party is under a legal duty to avoid or reduce damages caused by a breach of contract

Rescission and Restitution
- Rescission
  - An act to undo a contract
  - Available if there has been a material breach of contract, fraud, duress, undue influence, or mistake
- Restitution
  - Returning of goods or property received from the other party in order to rescind a contract
  - If the actual goods or property is not available, a case equivalent must be made

Equitable Remedies
- Specific performance
  - Court orders the breaching party to perform the acts promised in the contract; subject matter of the contract must be unique
- Reformation
  - Court rewrites a contract to express the parties' true intentions; usually used to correct clerical errors
- Quasi contract
  - Permits the recovery of damages for breach of an implied-in-law contract where no actual contract exists between the parties
- Injunction
  - Court order that prohibits a party from doing a certain act

Torts Associated with Contracts
- Intentional interference with contractual relations
- Breach of the implied covenant of good faith and fair dealing

Punitive Damages
- Damages awarded
  - To punish the defendant

To deter the defendant from similar conduct in the future
To set an example for others

IV. Objective Questions

Terms:

1. Damages based on contract-related torts are generally stated as _________________ damages.

2. This covenant is implied in certain types of contracts when determining the objective of the contract. It is called the covenant of _______________ _______________ _______________ _______________ _______________.

3. When a party to a contract renders performance exactly as required by the contract, thereby discharging the contract, it is called _______________ _______________.

4. When one contracting party informs the other that he will not perform his contractual duties when due, it is known as _______________ _______________.

5. The nonbreaching party may recover _______________ _______________ compensation for injuries in the form of money.

6. When a remedy is based on the concept of fairness and includes specific performance, reformation, quasi-contract, and injunction, it is referred to as an _______________ remedy.

7. This remedy compensates the nonbreaching party for foreseeable damages. The nonbreaching party must know or have reason to know that the breach will cause special damages. This remedy is called _______________ _______________.

8. This remedy provides that the parties may agree in advance. However, the actual damages must be difficult to determine, and the amount must be a reasonable estimate of the harm that would result from a breach. It is called _______________ _______________.

9. The avoidance or reduction of damages caused by a breach of contract is called _______________.

10. The process of undoing a contract is called _______________.

True/False:

1. _____ Punitive damages are generally rewardable for breach of contract.

2. _____ Concurrent conditions arise when there is both a condition precedent and a condition subsequent.

3. _____ An Internet contract cannot be breached.

4. _____ An accord is a completed contract to settle a contract dispute which discharges the original contract.

5. ____ The nonbreaching party may only recover monetary damages from the breaching party if the breach was material.

6. ____ Generally, only equitable remedies are available for breach of contract.

7. ____ To be liable for special damages, the breaching party must know or have reason to know that the breach will cause special damages to the other party.

8. ____ In certain circumstances, a liquidated damages clause may be considered a penalty and therefore not be enforceable.

9. ____ The nonbreaching party is under no obligation to avoid damages after a contract breach.

10. ____ If a party breaches the implied covenant of good faith and fair dealing, the nonbreaching party may be able to recover punitive, as well as other, damages.

Multiple Choice:

1. Which of the following is *not* an element of the tort of intentional interference with contract relations?
   A. A valid enforceable contract between contracting parties.
   B. Money damages must also be awarded.
   C. Third-party knowledge of the contract.
   D. Third-party inducement to breach the contract.

2. Which of the following is *not* classified as an equitable remedy?
   A. Mitigation of damages.
   B. Injunction.
   C. Specific performance.
   D. Reformation.

3. Feuding Fran and Fighting Fred have a dispute over a contract. Eventually, Fran and Fred agree that Fran will pay a reduced amount for Fred's performance of reduced duties. Which of the following is correct?
   A. Fran and Fred have resolved the dispute by reaching accord and satisfaction.
   B. Fran and Fred have resolved the dispute by reaching only an accord.
   C. Fran and Fred have not resolved the dispute, and either party can rescind the contract.
   D. Fran and Fred have not resolved the dispute, and both parties must fully perform the contract.

4. Competent Construction Co. (CCC) contracts to build a house for Yuppie Muppie. Yuppie requests that a specific type of tile be installed in the floor of the entry way. However, CCC mistakenly uses the wrong tile. Which of the following is correct?
   A. CCC has committed a minor breach of the contract.
   B. Yuppie can deduct the cost of installing the right tile from the contract price.
   C. Yuppie can sue CCC to recover the cost to install the right tile if Yuppie has already paid CCC.
   D. All of the above.

5. Builder Boris contracts to construct a large office building for Dr. Dan at a contract price of $1 million. The contract provides that all trim and doors will be made of top quality oak. However, Boris cannot tell the difference and mistakenly uses a lower quality oak. Which of the following is correct?
   A. Dan can do nothing since this is only a minor breach.
   B. Dan can get monetary damages even though this is a minor breach.
   C. Dan can get an equitable remedy even though this is a minor breach.
   D. None of the above.

6. Assume the same facts in question 5 except that before Boris begins any work, Dan rescinds the contract. Also, assume Boris's cost to complete the contract would have been $750,000. Which of the following is correct?
   A. Boris can recover nothing since there has been no performance, and therefore, Boris has lost nothing.
   B. Boris can recover $250,000.
   C. Boris can recover $1 million.
   D. Boris can only recover what he has already spent on materials.

7. Restaurateur Ralph contracts to have Custom Construction build a restaurant. The contract states that the restaurant will be completed by June 1 and has a liquidated damages clause in case the deadline is not met. However, Custom Construction falls further behind schedule, and the restaurant is not completed until September 10. Which of the following is correct?
   A. The liquidated damages clause will be enforced regardless of the amount, since actual damages would be extremely difficult to determine.
   B. The liquidated damages clause will not be enforced if actual damages are determined later.
   C. The liquidated damages clause will not be enforced because courts almost never enforce such clauses since they are considered penalties.
   D. None of the above.

8. Max contracts with Ned to sell 300 kites to Ned's Kite Shop for $1,000. At the last minute, Max calls Ned and rescinds the contract. Luckily, after much searching, Ned is able to find Orville who also sells kites. Orville agrees to sell 300 kites for $1000. If Ned sues Max, which of the following is correct?
   A. Ned will be awarded consequential damages.
   B. Ned will be awarded compensatory damages.
   C. Ned will be awarded liquidated damages.
   D. Ned will be awarded nominal damages.

9. George's Garage contracts to buy $5,000 worth of car parts from Susan Supplier. However, Carl's Car Shop hears of the contract and calls Susan. Carl offers Susan $6,000 if she will breach the contract with George. Which of the following is correct?
   A. Carl is liable for intentional interference with contractual relations.
   B. Carl is liable for a breach of the covenant of good faith and fair dealing.
   C. Carl is not liable for any torts, but is liable for contract damages.
   D. None of the above.

10. Specific performance is used as a remedy when:
    A. The nonbreaching party cannot specifically request it.
    B. The breaching party requests it in lieu of money damages.
    C. Money damages will not suffice.
    D. None of the above.

V. Answers to Objective Questions:

Terms:

1. *Punitive.* These damages are a form of civil punishment and are used in tort law to deter the defendant from engaging in similar conduct in the future.
2. *Good faith and fair dealing.* Breach of this implied covenant results in a contract-related tort.
3. *Strict performance (or complete performance).* This performance is what is called for under the contract.
4. *Anticipatory breach.* This type of material breach triggers rights in the other party to stop his own performance and seek remedies. It is designed to mitigate harm to the nonbreaching party.
5. *Monetary damages.* This is the most common of the remedies provided in contract cases. They are designed to give the nonbreaching party the economic benefit of the bargain.
6. *Equitable.* Remedies in equity are historically traceable to the English Court of Chancery and were originally based on appeals to the King's conscience.
7. *Consequential damages.* To be liable, the breaching party must know or have reason to know that the breach would cause these damages to the other party.
8. *Liquidated damages.* Courts want to be sure that these damages are fairly arrived at and do not act as a penalty against the breaching party.
9. *Mitigation.* This doctrine applies to all measures of damages.
10. *Recission.* This remedy tries to put the parties back into the position they were in before the contracting process began.

True/False:

1. False. In the absence of a contract-related tort, such as intentional interference with contract relations, punitive damages are not generally awarded for contract breach.
2. False. Concurrent conditions arise when parties to a contract must render performance simultaneously.
3. False. A contract is a contract, even if it is over the Internet.
4. False. An accord is only an agreement of how to resolve a contract dispute. It is not a completed contract.
5. False. The nonbreaching party may recover damages even if the breach is only minor.
6. False. Generally, only monetary damages are available for breach of contract. Equitable remedies are only used if monetary damages are insufficient.
7. True. The key element is foreseeability.
8. True. If the actual damages are determinable or the liquidated damages clause provides for a large recovery, the clause may not be enforced.
9. False. The nonbreaching party is under an obligation to avoid or reduce any further damages after the breach. This is called mitigation.
10. True. Such a breach is also a tort. Therefore, tort damages, such as punitive damages, may be awarded.

Multiple Choice:

1. B. Money damages are not a required element of this contract-related tort. A, C, and D are required elements.
2. A. Mitigation is used in damage awards, and requires the nonbreaching party to take reasonable steps to reduce damages. B, C, and D are all equitable remedies.

3. B. Where there is only an agreement, but not completion of the agreement, this is accord, but not satisfaction. A is incorrect for the reason above. C is incorrect because this is a resolution and neither party can rescind the contract without being held liable. D is incorrect because if an alternative agreement has been reached, the parties do not have to fully perform the original contract.
4. D. A is correct because though this is a breach, it is minor. B and C are correct because these are the remedies that Yuppie has.
5. B. Monetary damages are available even if the breach is minor. A is incorrect since Dan does have some recourse. C is incorrect because an equitable remedy is probably not available since monetary damages would suffice.
6. B. Boris can recover the amount of profit he would have had on the contract. A is incorrect because Boris has lost his expected profit and can recover this amount. C is incorrect because Boris cannot recover the whole contract price, but only his expected profit. D is incorrect because Boris can still recover his expected profit even if he has spent money on materials.
7. A. In such a situation, damages are difficult to calculate. If the amount of liquidated damages is reasonable, the clause will be enforced. Be is incorrect because if the liquidated damages clause is enforceable, actual damages will not be awarded. C is incorrect because if certain requirements are met, such clauses will be enforced.
8. D. Although Max breached this contract, Ned was able to completely mitigate damages and cannot recover any significant amount of money. Therefore, Ned did not suffer any monetary damage. However, if Ned wants to sue Max on principle, he will be awarded nominal damages.
9. A. Since Carl knew of this contract and intentionally induced Susan to breach it, he is liable for intentional interference with contractual relations. B is incorrect because since Carl was not a party to the contract, he did not have a duty of good faith and fair dealing. C is incorrect because since Carl was not a party to the contract, he could not breach it.
10. C. This is the only element of specific performance that applies. It is sought where money will not suffice.

VI. Answers to Essay Question

Ms. Dee Minimis had a number of performance obligations under the terms of her contract with Mr. Lex Loci. The most obvious, of course, was to pay the agreed upon price for the business. The most important performance obligation covenant was the condition subsequent involving her duty not to compete for a period of one year within a five mile radius of Derbytown Rollerball.

Ms. Dee Minimis does not appear to qualify for any of the excuses from her performance obligations, such as accord and satisfaction or novation, based either on the actions of the parties or on legal excuses such as a statute of limitation.

Ms. Dee Minimis had a definite duty to honor the noncompetition covenant in her contract with Mr. Lex Loci. The clause appears to be a reasonable provision connected with the sale of a business and does not create an undue limitation on competition. In addition, this obligation is not excused or discharged by reason of her teaching in-line skating, rather than roller-skating, instruction. The nature of both enterprises is essentially the same.

There is a specific provision in this contract for liquidated damages. Normally, where such a provision provides a true alternative performance and is not a penalty, courts will honor the provision. Here the liquidated damages clause appears fair on its face and was openly bargained for by both parties. The problem is that even if Ms. Minimis were willing to pay the $5,000, she is not in a financial position to do so because of her lottery habit. Her insolvency is a product of voluntary acts.

Because the liquidated damages clause is rendered ineffective due to Ms. Minimis' insolvency, the imposition of any money damages would not provide Mr. Loci with any effective law-based remedy in this case. Equitable remedies must therefore be examined.

Because personal services are involved in this contract, affirmative enforcement cannot be granted. This bar to enforcement of personal services contracts is rooted in the Thirteenth Amendment to the U.S. Constitution that prohibits forced or coerced labor. Thus, Ms. Minimis cannot be forced to teach in-line skating. She can, however, be subjected to a negative injunction from providing those services in violation of the noncompetition clause. Because the money damages in this case, liquidated or otherwise, are illusory, the injunctive remedy can be granted as a remedy for breach of the noncompetition clause.

Ms. Dee Minimis may have engaged in both intentional interference with the contract between Mr. Lex Loci and his new clients as well as in bad faith and dealing with Mr. Loci at the outset. If the elements of these torts are found, the court may impose punitive damages.

Chapter 14
IINTELLECTUAL PROPERTY AND INTERNET LAW

There is no surer way to misread any document than to read it literally.
Learned Hand

I. Overview

One of the great debates currently being waged in the halls of the U.S. Congress and state legislatures is to what extent a government entity should act as a partner in the promotion of new business opportunities such as e-commerce. As a matter of public policy and related tax policy, there is no question that government sanctioned jump-starts to business can bring long-term benefits to the larger community by way of jobs and additional tax revenues. This chapter focuses on several areas of well-accepted legal partnerships, i.e., areas of law with long established rules which provide innovative, creative, and new methods of special business legal protections in the name of the larger public good. These protected areas revolve around the law of patents, copyrights, and trademarks. They are designed not only to reward original effort but also to foster and encourage continued new contributions to the common good by creating an economically protected relationship between the writer, inventor, and others, and the government.

The areas of patent, copyright, and trademark law are all individually complex and call for use of highly specialized law practitioners. They have a common denominator in that each of these areas provides a legally protected mechanism for rewarding the creator of work in the economic marketplace. This reward system revolves around two basic principles:

1. Exclusive use of the economic benefits that result from the protected activity.
2. Legal recourse against those who somehow infringe upon this exclusive economic protection.

Where proper statutory procedures are used, the benefits can be great. For example, once a patent is granted, the inventor can reap the exclusive economic benefits for fourteen or seventeen years depending on the type of patent issued. The protection provided under copyright laws is even longer. Since 1976, U.S. copyright laws have extended protections for authors and other creative persons for their lives plus fifty years. The protection can be even longer for copyrights registered by business entities. In the area of trademarks and related marks, the continued use of the registered mark, coupled with renewals can extend the protections of law indefinitely. The most important caveat in trademark law is: "Use it or lose it." Failure to maintain legal protection of trademarks, service marks, or related marks can lead to a loss of protection for them. If the mark is allowed to become a common term for a product, it may be deemed generic and may no longer be entitled to protections that come with exclusive use of the term. One example of such a generic term is found in the word aspirin. Because the original marketers of the product did not protect the name vigorously enough, the name aspirin has fallen into the generic pool of language used to describe products with those chemical properties found in that product, i.e., aspirin.

The best way for a student to approach these areas of the law is to recognize that each of these legally recognized areas of protection carries with it a benefit/burden dichotomy—the benefits of the protective statute will pass to those who know how to use the statute. Proper use of the statute calls for all the individual elements required under the statute to be complied with. Once that burden has been met, the benefits will follow. Students should, therefore, familiarize themselves with respective steps required to obtain patent, copyright, or trademark exclusive use protection as described in the chapter. The multi-issue essay question will illustrate the process and how it is used to protect intellectual property rights.

II. Hypothetical Multi-issue Essay Question

Ted Huie, Fred Dewie, and Ned Louie are all recent graduates of the prestigious law school, I OWE U, located in scenic Backwater, Massazona. Having all recently passed the tough Massazona State Bar Examination, they decided to forego the riches of Wall Street and start their own law practice in Backyard, which is just down the road from Backwater. Their chosen area of specialization in law practice is patents, copyrights, and trademarks. They want to use their proper surnames in advertising their new law firm but are concerned that the use of the law firm name Huie, Dewie, and Louie may raise a few concerns at Walt Disney Studios, Inc. Walt Disney Studios has long held the copyright to the cartoon duck characters named Huey, Dewey, and Louie, and our newly-admitted law practitioners don't want to start their law careers by running "afowl" of the law! What advice would you give them on this issue?

III. Outline

Common Types of Unfair Competition
- Palming off
- Disparagement
- Misappropriation of trade secrets
- Misappropriation of the right to publicity

What Is Intellectual Property?
- Intellectual property is objects such as inventions, writings, trademarks, etc., which are often a business's most valuable asset

Types of Intellectual Property
- Patents
- Copyrights
- Trademarks and other marks
  - Service marks
  - Certification marks
  - Collective marks

Patents
- To be patented, the invention must be:
  - Novel
  - Useful
  - Nonobvious

What Can Be Patented?
- Machines
- Processes
- Compositions of matter
- Improvements to existing machines, processes, or compositions of matter
- Designs for an article of manufacturer
- Asexually reproduced plants
- Living material invented by man

Public Use Doctrine
- A patent may not be granted if the invention was used by the public for more than one year prior to the filing of the patent application

Patent Infringement
- Unauthorized use of another's patent
- A patent holder may recover damages and other remedies against a patent infringer

Copyright
- Only tangible writings, i.e., writings that can be physically seen, are subject to copyright registration and protection
- Examples include:
  - Books
  - Periodicals
  - Newspapers
  - Lectures
  - Sermons
  - Addresses
  - Musical compositions
  - Plays
  - Motion pictures
  - Radio and television productions
  - Maps
  - Works of art, including paintings, drawings, sculpture, jewelry, glassware, tapestry, and lithographs
  - Architectural drawings
  - Photographs

Copyright Infringement and Fair Use
- Copyright infringement occurs when a party copies a substantial part of the plaintiff's copyrighted work without permission
- The Fair Use Doctrine permits certain limited use of a copyright by someone other than the copyright holder without permission of the copyright holder

Trademark
- Trademark law is intended to:
  - Protect the owner's investment and goodwill in a mark
  - Prevent consumers from being confused as to the origin of goods and services
- Trademarks can be registered for 10 years and can be renewed for an unlimited number of 10 year periods

Marks That Can Be Trademarked
Trademark
A distinctive mark, symbol, name, work, motto or device that identifies the goods of a particular business, e.g., Xerox
Service mark
Mark used to distinguish the services of the holder from those of its competitors, e.g., United Airlines
Certification mark
Mark that is used to certify that goods and services are of a certain quality, e.g., wines from Napa Valley
Collective mark
Mark used by cooperatives, associations, and fraternal organizations, e.g. Boy Scouts of America

Length of Terms

IV. Objective Questions

Terms:

1. A type of unfair competition that occurs when a small company tries to pass off one of its products as those of a larger rival is _______________ _______________ .

2. A product formula, pattern, design, compilation of data, customer list, or other business secret is a _______________ _______________.

3. False statements about a competitor's products in comparative advertising are called _______________ _______________.

4. The tort of _______________ is an attempt by another person to appropriate a living person's name or identity for commercial purposes.

5 Objects such as inventions, writings, trademarks, etc. that are often a business's most valuable asset include _______________ property.

6. A patent may not be granted if the public used the invention for more than one year prior to the filing of the patent application. This is known as the _______________ _______________ doctrine.

7. Writings that are capable of visual perception are _______________ _______________.

8. A requirement for a mark to be under federal protection when an ordinary term has become a brand name is a _______________ _______________.

9. A distinctive mark, symbol, name, word, motto, or device, which identifies the goods of a particular business and distinguishes the business and products of the holder from those of its competitors are all ways of identifying a _______________.

10. State laws that allow persons and companies to register trademarks and service marks are _______________ statutes.

True/False:

1. _____ If the owner of a trade secret does not take all reasonable precautions to protect a trade secret, the secret will no longer be protected under unfair competition laws.

2. _____ A disparaging statement is any statement, true or false, which is made by one party about the products, services, property, or reputation of another business.

3. _____ When a plaintiff proves disparagement, he can recover both actual and general damages.

4. _____ Suits brought under the Federal Patent Statute of 1952, Copyright Revision Act of 1976, and the Lanham Trademark Act are properly brought in federal court, not state court.

5. _____ Design patents are valid for fourteen years and can be renewed at the end of that time.

6. _____ Corporations and businesses are given copyright protection for either 75 or 100 years while individuals are protected for the life of the author plus fifty years.

7. _____ Cyber business plans are patentable.

8. _____ The registration of a trademark is given nationwide effect as notice that the mark is the registrant's property.

9. _____ If the registrant of a trademark does not use the registered trademark symbol, the registrant will lose all protection and have no remedy against infringements.

10. _____ Only the party to whom a trademark was originally registered can seek cancellation of the trademark.

Multiple Choice:

1. All of the following are necessary to prove the tort of palming off except:
   A. The defendant used the plaintiff's logo, symbol, mark, etc.
   B. There is a likelihood of confusion as to the source of the product.
   C. At least one consumer purchased the product under the mistaken belief that the product was that of the well-established company.
   D. All of the above are necessary to prove the tort of palming off.

2. If a defendant is liable for the tort of palming off, the plaintiff can:
   A. Recover the profits made by the defendant from use of the trade secret.
   B. Recover damages suffered.
   C. Either A or B, but not both.
   D. Both A and B.

3. Individual Ivan, an amateur scientist, discovers the trademarked trade secret of TechniCorp. Which of the following is correct?
   A. Ivan is liable for misappropriation of a trade secret regardless of how he discovered it.
   B. Ivan is liable for misappropriation of a trade secret only if he used unlawful means to obtain it.
   C. Ivan is not liable for misappropriation of a trade secret because only corporate competitors are liable for this tort.
   D. None of the above.

4. Entrepreneur Ellen develops a line of hair products and markets the products as the "Michael Jackson Line." Assume Michael Jackson did not authorize Ellen to develop and market this line, but that she did so on her own. Which of the following is correct?
   A. Ellen has committed the tort of appropriation.
   B. Ellen has committed the tort of palming off.
   C. Ellen has committed the tort of disparagement.
   D. None of the above.

5. Inventor Elizabeth invented a new process for making sunglasses that are 100 percent scratch resistant. Inventor Elizabeth patented the process under the Federal Patent and Trademark Office in Washington, D.C. Which of the following is correct?
   A. The design is protected under both federal and state law.
   B. The patent is good for seventeen years.
   C. The patent can be renewed at the end of the patent period.
   D. All of the above are true.

6. Author Allen and Author Ann worked as coauthors on a book. Both agreed that the book is a masterpiece and should be copyrighted. Which of the following is correct?
   A. Allen and Ann must agree to copyright the book under either one of their names, but not both.
   B. Allen and Ann can each copyright their own work.
   C. Allen and Ann are protected under copyright law for 75 years.
   D. Allen and Ann cannot copyright the book.

7. Assume now that Allen and Ann are journalists. They work together for several weeks writing news briefs and stories on an upcoming presidential election. Which of the following is correct?
   A. Allen and Ann must agree to copyright the news under either one of their names, but not both.
   B. Allen and Ann can each copyright their own work.
   C. Allen and Ann are protected under copyright law for 75 years.
   D. Allen and Ann cannot copyright the news.

8. Sneaky Sam tries to copy the copyrighted work of Famous Fran. Which of the following is correct?
   A. Sam is not liable unless he copies the entire document.
   B. Sam is not liable unless the portion he copies is word for word.
   C. Sam is liable even if he copies only part of the document and it is not word for word.
   D. Sam is liable even if he only copies a brief quotation for comment in a book review.

9. To register a mark, which of the following must be met?
   A. The mark has actually been used in commerce.
   B. The applicant verifies an intention to use the mark in commerce and actually does so within six months of registering.
   C. Either A or B.
   D. None of the above.

10. Trader Tom is involved in the commerce of his state, Nelow. However, Tom is never involved in commerce outside Nelow or with other states. Which of the following is correct?
    A. Tom can only register his mark under state law.
    B. Tom can register his mark under either state or federal law.
    C. Tom can only register his mark under federal law as there are no state trademark laws.
    D. None of the above.

V. Answers to Objective Questions

Terms:

1. *Palming off.* These torts are particularly aimed at luxury goods manufacturers.
2. *Trade secret.* Determine the efforts taken by the company to keep it secret.
3. *Product disparagement.* Comparative advertisements are OK if they are truthful.
4. *Appropriation.* We are all entitled to reap the benefits of our personal identity.
5. *Intellectual.* Patents, copyrights, etc., qualify as intellectual property.
6. *Public use.* The rule seeks to impose a duty on the inventor to act in a timely manner.
7. *Tangible writings.* As hi-tech evolves, expect this definition to be expanded.
8. *Secondary meaning.* This meaning often has tremendous value; consider the value of terms like Coke and Pepsi.
9. *Trademark.* In many ways, these have become a new international language of business.
10. *Antidilution.* The burden is placed on the person seeking the benefit of the law's protection.

True/False:

1. True. Remember the "Use or lose" maxim.
2. False. A disparaging statement is a false, not a true, statement. True statements, although they may be damaging, are not disparaging statements.
3. False. When a plaintiff proves disparagement, he or she can recover only actual damages.
4. True. These are federally protected intellectual property enactments.
5. False. Although design patents are valid for fourteen years, they cannot be renewed.
6. True. See the Copyright Revision Act of 1976.
7. True. Taking the lead from the State Street Case, many people have filed for and received patents for business and financial models that are used over the Internet.
8. True. The protection is provided under federal statutes.
9. False. Use of the symbol is not mandatory. However, certain remedies may be lost if the symbol is not used.
10. False. A third party is free to seek cancellation of a previously registered trademark.

Multiple Choice:

1. C. Only A and B must be shown to prove the tort of palming off. Actual consumer confusion does not have to be shown.
2. D. The plaintiff can recover both profits and damages suffered and is not limited to choosing either one or the other.
3. B. To be liable, Ivan must have obtained the trade secret through some unlawful means. A is incorrect because if Ivan used lawful means, he is not liable. C is incorrect because individuals as well as corporations can be liable for trademark infringement.
4. A. Unauthorized commercial use of another person's name constitutes the tort of appropriation. B is incorrect because palming off involves one merchant trying to sell his goods or services as those of a well-known competitor. Here, there was exploitation of a famous person's name. C is incorrect because disparagement involves making false statements about a competitors product or services.
5. B. The process patent is protected for seventeen years. A is incorrect because the design is protected under only federal, not state, law. C is incorrect because the patent cannot be renewed at all.
6. B. Allen and Ann can each copyright his or her own work since they are the original authors of the work. A is incorrect because each can copyright his or her own work. C is incorrect because Allen and Ann are protected for their lives plus fifty years under copyright law. D is incorrect because such original works can be copyrighted.
7. D. News cannot be copyrighted.
8. C. If Sam copies a substantial and material part of Fran's work, he is liable for copyright infringement even if it is not word for word. A is incorrect because Sam need only copy a substantial and material part of the work to be liable. B is incorrect because the work need not be copied word for word for Sam to be liable. D is incorrect because using a brief quotation for comment in a book review falls under the fair use doctrine and is permissible.
9. C. Either A or B can be shown to register a mark.
10. A. Since Tom's business is purely intrastate, he can only register his mark under state law. B is incorrect because federal law requires that the commerce be interstate to be federally protected. C is incorrect because Tom must register under state law, not federal law.

VI. Answers to Essay Questions

The use of a company name may be subject to scrutiny under any number of protected areas of the law including patents, copyrights, and/or trademarks or related marks. The most likely areas of concern are copyright and trademarks. There is already a legally recognized copyright protection accorded to Walt Disney Studios, Inc. for the creative cartoon characters known as Huey, Dewey, and Louie. They have long been very popular with children and adults alike and have been vigorously protected by the owners of the copyright. In addition, trademark law may come into play where the owner of a cartoon character has marketed cartoon-related products using the likeness of the three ducks in question.

Is a surname copyrightable or subject to trademark protection? Normally a surname, standing alone, with no creative work attached thereto or without any secondary distinctive meaning, may not be protected. Here both elements—creative work and secondary distinctive meaning—can be found in the names associated with the cartoon characters. Thus Walt Disney Studios, Inc. would have proper grounds for seeking protection for the continued exclusive use of these names as used for the cartoon characters in the public media and in trade.

Does it accord protection in all situations or can legitimate distinctions be made where there is no dilution of the rights and protections provided for under the statute? Consider *Mead Data Central, Inc. v. Toyota Motor Sales, U.S.A., Inc.*, 875 F.2d 1026 (2nd. Cir. 1989). In addition, is it fair use to list one's own surname as a business entity that coincidentally sounds and looks the same as another person's protected intellectual property? The need for specialized professional counsel is readily apparent.

Chapter 15
ELECTRONIC COMMERCE AND INFORMATION TECHNOLOGY LICENSING

Internet is so big, so powerful and pointless that for some people it is a complete substitute for life.
Andrew Brown

I. Overview

The key features of this chapter seek to introduce the student to the basic history of e-commerce and the Internet. The first key objective of this chapter is to describe the most current efforts to tailor the laws of contract, intellectual property, and free speech protections to the realities of cyberspace. The Internet was established in 1969 as a method by which electronic communications were to be used for military and strategic national defense purposes. This basic structure of multiple computers being able to "talk" to each other ended up having many uses beyond its original military origins. Starting in the early 1980's, under the auspices of the National Science Foundation, the Internet was established to facilitate high-speed communications among research and academic centers on a global scale. The efficiencies realized by those centers were not lost on business, and by the late 1990's, the Internet ushered in the Information Age as we know it today.

Among many early legal issues that arose from the widespread use of the Internet were the constitutional interpretations of free speech. An early landmark case on this issue is found in *Reno v. American Civil Liberties Union* (1997) where the U.S. Supreme Court struck down two early attempts to limit or prohibit certain types of speech transmitted over the Internet. The court held that communications transmitted over the Internet were entitled to the same protections as ordinary commercial speech so long as it was not classified as "invasive" in the same sense as radio or television. In addition, the court was concerned about the lack of precision in the statute that sought to limit "indecent" and "patently offensive" communications on the Internet. This decision clearly sent an early signal to users of the Internet that it is a medium of communication in which basic free speech rights will be protected.

With the advent of e-commerce over the Internet, the second key objective of this chapter illustrates how policy makers at all levels of government and industry are seeking to evolve the law to foster and protect commerce while not intruding on basic privacy rights. The Federal Electronic Communications Privacy Act is one such example. That statute makes it a crime to engage in unauthorized interceptions of electronic communications. Other examples of policy-making laws include the intellectual property world of domain names and the Federal Anticybersquatting Consumer Protection Act, which seeks to protect commerce from wrongful or bad faith intrusions of company or individuals' names. Tradenames have enormous value. Such mechanisms such as the Uniform Dispute Resolution Policy established by the Internet Corporation for Assigned Names and Numbers are designed to expedite disputes over domain names. In addition, a number of states are proposing the Uniform Electronic Transactions Act and the Computer Information Transactions Act as models to be used to set up structures for licensing agreements for Internet-based intellectual property and Internet commercial transactions.

What is most interesting about these acts is the fact that they are very heavily patterned after long-existing provisions of the Uniform Commercial Code. Since the early 1950's the U.C.C. has provided the basic model for conducting commerce in this country. It has been adopted, at least in part, by every state in the union. It has been kept up to date by a group known as the National Conference of commissioners on Uniform State Laws. This group is made up of academics, jurists, law and business practitioners and economists. They formulate model laws that are, in turn, proposed to the individual states. Each state may adopt its own particular version of the law, but in the vast majority of cases, the uniform laws are enacted intact. The goal of this process is to create a uniform set of interconnected and common legislation becomes part of well-functioning and predictable state laws on commerce. It is too early to tell how well these new uniform laws on e-commerce will be received by the respective states. It does seem certain, however, that such laws will be quite necessary to foster and protect the ever-growing scales of commerce now being conducted over the Internet.

Some of the key issues covered in these proposed laws include the electronic formation of contracts, warranty protections provided in electronic contracts, licensing rights and duties, and remedies for breach of electronic commerce contracts. All in all, it is a work in progress that makes it this the most exciting time ever to be a student of business law.

II. Hypothetical Multi-Issue Essay Question

Everybody's on Line (Everybody's) provides free e-mail on the World Wide Web. It generates income from pop up ads, paid for by web site advertisers. Currently, there are over five million contracted subscribers to Everybody's service. One of the services provided by Everybody's allows its subscribers to send messages to any other e-mail address in the world.

Spam is defined as unsolicited bulk e-mail. Everybody's is concerned that its subscribers do not use Everbody's facilities and services to spam other e-mail users. Under its "Terms of Service" subscription contract, Everybody's specifically states that subscribers agree not to send spam. Failure to abide by this contract will result in termination of the contract and free e-mail services by Everybody's.

Joe Blow is a subscriber to Everybody's services. He was notified of the "Terms of Service" by way of "shrink-wrap" conditions on the face of his contract and by way of his "point-and-click" consent to those terms every time he logged on to Everybody's web site. Joe likes to think of himself as a free thinker and decided that the whole world should get to see the "real Joe." Joe sent countless nude photos of himself to millions of viewers on web. These photos had no redeeming artistic value and would be classified as pornography. He used Everybody's e-mail return address.

Everybody's is quite upset and comes to you to review its options. What constitutional, intellectual property, and contract issues should be taken into account? What remedies does Everybody's have?

III. Outline

E-Commerce and the Internet
- E-commerce
  - The sale of goods and services by computer over the Internet
- Internet
  - A collection of millions of computers that provide a network of electronic connections between computers

Electronic Mail
- Electronic written communication between individuals and businesses using computers connected to the Internet

World Wide Web
- An electronic connection of millions of computers that support a standard set of rules for the exchange of information

Domain Name
- A unique name that identifies an individual's or company's web site
- Domain names can be registered

Uniform Computer Information Transactions Act
- A model act issued by the National Conference of Commissioners on Uniform State Laws
- Establishes a uniform and comprehensive set of rules that regulate the creation, performance, and enforcement of computer information transactions

Licenses
- Licenses are contracts that transfer limited rights in intellectual property and informational rights
- Licensor
  - The owner of intellectual property or informational rights who transfers rights in the property or information to the licensee
- Licensee
  - The party who is given limited rights in or access to intellectual property or informational rights owned by the licensor

How a Licensing Arrangement Works

Formation of a Contract
- A contract can be formed in nay manner that shows agreement, including an offer and acceptance, or conduct of both parties, or operation of electronic agents
- A contract is formed when an electronic acceptance is received
- Keep in mind that all the normal rules of contracts apply to e-contracts

Warranties
- Certain warranties may attach to a license of informational rights either because the licensor has made express warranties or the law has imposed one or more implied warranties
  - Types of warranties
    - Warranties of noninterference and noninfringement
    - Express warranty
    - Implied warranty of merchantability of the computer program
    - Implied warranty of informational content

Implied warranty of fitness for a particular purpose

IV. Objective Questions

Terms:

1. Model acts designed to provide uniform and comprehensive rules for contracts involving computer information transactions and software and information licenses have been developed by the _______________ _______________ of _______________ on _______________ _______________ _______________.

2. Title V of the 1996 Telecommunications Act of 1996 sought to prohibit certain offensive or indecent transmissions over the Internet. This legislation is generally known as _______________ _______________ Act of 1996.

3. A number of state courts have held that the writing requirements of the _______________ of _______________ are satisfied by e-mail electronic signatures.

A. Web sites are identified by a unique Internet _______________ _______________ _______________.

5. In November 1999, the U.S. Congress passed the _______________ _______________ _______________ Act to guard against improper registrations of domain names on the Internet.

A. A computer information transaction is an agreement to create, transfer, or license computer information or informational rights under Section 102(a) (11) of the _______________ _______________ _______________ _______________ Act.

7. The National Conference of Commissioners on Uniform State Laws has proposed an act to create uniform laws for electronic records and signatures. That proposed act is called _______________ _______________ _______________ Act.

8. Under the U.C.I.T.A, a licensor of informational rights warrants that no third person holds any claim or interest in the information or claim of infringement of misappropriation. These warranties are known as the warranties of _______________ and _______________.

9. If a party repudiates the contract before performance is due, under the U.C.I.T.A., this is known as _______________ _______________.

A. Under the U.C.I.T.A., when a licensor breaches a contract, the licensee can engage in a commercially reasonable substitute transaction. This remedy is called _______________.

True/False:

1. _____ Because communications made over the Internet are not classified as speech, they are not protected by the First Amendment to the U. S. Constitution.

2. _____ The Internet is an international network of interconnected television networks.

3. _____ The Internet was first begun in 1969 by the Department of Defense.

4. _____ In the 1997 case of *Reno v. American Civil Liberties Union,* the Supreme Court upheld the Uniform Electronic Transaction act.

5. _____ Electronic signatures have been held to satisfy the Statute of Fraud's writing requirements in a number of states.

6. _____ The Electronic Communications Privacy Act makes it a crime to intercept electronic communications except in cases of law enforcement or permission by the party providing the electronic communication service.

7. _____ E-mail contracts are enforceable.

8. _____ The Internet Corporation for Assigned Names and Numbers is a private court that can adjudicate disputes over domain names with jury trials.

9. _____ Under the Uniform Computer Information Transactions Act, title to a copy of software or electronic information is determined by license.

10. _____ An implied warranty of merchantability for a computer program warrants that the program is fit for the ordinary purposes for which the program is used.

Multiple Choice:

1. In the case of *Reno v. American Civil Liberties Union*, the US Supreme Court stated that the Internet is not as "invasive" as:
   A The U.S. Army.
   B. The Russian Army.
   C. Radio and television.
   D. Rap music.

2. Under the Electronic Communications Privacy Act, certain stored electronic communication may be accessed without violating the law if it is used for:
   A. White House files.
   A. Law enforcement.
   B. E-mail sent from private homes.
   C. University records.

3. China has agreed to open up its Internet markets with all of the following restrictions except:
   A. Foreign companies cannot own more then 50 percent of any Internet business in China.
   B. Foreign companies must obtain a special license from the Chinese government for such investments.
   C. China will regulate the content of Internet Web sites, including advertising placed on the Web.
   A. Foreign companies can own more than 50 percent of any Internet business in China.

4. Under a contract with the U.S. government, which of the following organizations is responsible for conducting arbitration proceedings for challenging cybersquatting?
   A. The Federal Fair Trade Commission.
   B. The U.S. Justice Department.
   C. The Internet Corporation for Assigned names and Numbers.
   D. The Internet Antisquatting Commission.

5. An agreement to transfer limited rights to parties for specified purposes and limited duration as defined in Sec. 102(a)(40) of the U.C.I.T.A. is called a:
   C. Copyright.
   D. Patent.
   E. Trademark.
   F. License.

6. ABC Corporation wants to enter into a contract with XYZ Corporation to allow XYZ to use its propriety informational rights. Under the terms of the U.C.I T.A., an authentication is required if the dollar amount of the contract is:
   A. $500 or more.
   B. $1000 or more.
   G. $5000 or more.
   H. $10,000 or more.

7. Joe Dooks delivers a copy to his licensee, May Bee. Under the terms of the contract, this copy was delivered by FedeEx and was, in fact, left at May Bee's home address. May Bee never opened the package and tossed it out thinking it was "junk mail." Under the U.C.I.T.A., the risk of loss of the copy will fall on:
   A. FedEx.
   A. Joe Dooks.
   B. May Bee.
   C. U.S. Postal Service.

8. Under Section 217 of the U.C.I.T.A., consumers are *not* bound by unilateral electronic errors if the consumer does all except:
   A. Upon learning of the error promptly notifies the other party of the error and does not use or receive any benefit from the information, or make the information or benefit available to a third party.
   A. Delivers all copies of the information to the third party or destroys all copies of the information pursuant to reasonable instructions from the other party.
   C. Pays all shipping, reshipping, and processing costs of the other party [UCITA § 217].
   D. Keeps the copies because they are defective and have no value.

9. Implied warranties are imposed automatically by the U.C.I.T.A. unless they are expressly disclaimed by the licensor under Section 406(c). The implied warranties can include all <u>except</u>:
   A. Merchantability of the computer program.
   B. Informational content.
   C. Express statement of quality.
   D. Fitness for a particular purpose.

10. A licensee can get a specific performance remedy under Section 811 of the U.C.I.T.A. for all except the following:
    A. The license of a digital music video program.
    B. The agreement to a Rock Star to sing in the video.
    C. The right to re-sell the video.
    D. The right to make copies of the video.

V. Answers to Objective Questions

Terms:

1. *National Conference of Commissioners on Uniform State Laws.* This group is made up of jurists, academics and practioners who advise the states on the enactment of uniform laws.
2. *Communications Decency Act of 1996.* Certain provisions of this act were declared to be in violation of the 1st Amendment to the U.S. Constitution by the U.S. Supreme Court in the case of *Reno v. American Civil Liberties Union.*
3. *Statute of frauds.* This is a common law-based writing requirement that has been carried over into certain contracts conducted over e-mail.
4. *Domain name.* This is based on the name registered for each site by filing the appropriate form with the domain name registration service and paying the appropriate fee.
5. *Anticybersquatting Consumer Act.* The Anticybersquatting Consumer Act is designed to protect individuals and businesses from bad faith infringements on trade names and/or famous individual names.
6. *Uniform Computer Information Transactions Act.* The National Conference of Commissioners on Uniform State Laws is currently proposing this act to states as of 1999.
7. *Uniform Electronic Transactions Act.* Its goal is to place electronic records and signatures on the same level as paper contracts and written signatures.
8. *Noninterference and noninfringement.* These warranties are patterned after the implied warranties already found in the U.C.C. with regard to the sales of goods.
9. *Anticipatory repudiation.* This remedy is patterned after similar remedies allowed for in the U.C.C. The non-breaching party may suspend his or her own performance, wait for a commercially reasonable time and then seek any remedy for breach of contract.
10. *Cover.* Again, this remedy is patterned after the buyers right to cover after a breach by the seller, in a U.C.C. contract.

True/False:

1. False. Communications made over the medium of the Internet are classified as speech and thus subject to the protections offered by the 1st amendment to the U.S. Constitution.
2. False. The Internet is an instructional network of interconnected computers.
3. True. The Department of Defense designed the Internet in 1969 to create electronic communications for military and national defense purposes.
4. False. The Supreme Court declared certain portions of the Communications Decency Act of 1996 to be in violation of the 1st amendment to the U.S. Constitution.
5. True. While not yet uniformly decided upon, a number of state courts have upheld electronic signatures as sufficient to satisfy the writing requirements of the Statute of Frauds.
6. True. The Electronic Communications Privacy Act generally makes interceptions of electronic communications illegal except with consent of the owner or for law enforcement purposes.
7. True. E-mail contracts meet the writing requirements for enforceable contracts.
8. False. The Internet Corporation for assigned Names and Numbers is a corporation under contract

with the U.S. government and is responsible for regulating the issuance of domain names in the Internet. It is not a court.
9. True. Under section 5O2(a) of the Uniform Computer Information Transactions Act, the license agreement determines when, where, and how title to software is controlled.
10. True. Section 403 of the Uniform Computer Information Act generally follows the warranty precedents on merchantability as first used in the Uniform Commercial Code.

Multiple Choice:

1. C. Radio and T.V. The court held that communications over the Internet do not include an individual's home or appear in one's computer screen unbidden.
2. B. Law enforcement. Government and law enforcement entities that are investigating suspected illegal activity may access stored electronic communications provided they are complying with all required procedural safeguards such as warrants.
3. D. Foreign companies may not own more than 50 percent of any Internet business in China.
4. C. The Internet Corporation for Assigned Names and Numbers is responsible for regulating the issuance of domain names on the Internet.
5. D. License. This is a contract that authorizes limited access to and use of informational rights.
6. C. Section 201(a) of the U.C.I.T.A. requires authentication if the contract fee is $5,000 or more
7. C. May Bee is liable for the risk of loss even though she did not yet inspect the package. She had that right under Section 608 of the U.C.I.T.A. but failed to exercise it. When she assumed control of the copy with receipt of the same, the risk of loss passed to her under Section 607 of the U.C.I.T.A.
8. D. A, B, and C are all required under Section 217. The consumer may not unilaterally just keep the copies even if they are defective.
9. C. Express warranties are not implied and if made, they may not be disclaimed under Section 406(a). Implied warranties can be disclaimed under section 406(c).
10. B. Specific performance cannot be ordered for a personal service contract. This is based on the principle that an individual cannot be forced to perform a contract against his or her will. They cannot be prevented from performing for others in breach of a contract by way of injunctive relief.

VI. Answer to Essay Question

This case raises numerous possible legal issues and is based, in part, on a famous 1998 case entitled *Hotmail Corp. v. Von$ Money Pie, Inc.* In that case, the defendants were found to be in violation of their subscription contract terms with Hotmail Corp. Specifically, they had engaged in the transmission of pornography spam. Defendants abused Hotmail's "Terms of Service." The U.S. Supreme Court has long held that pornography is not protected speech, and even if it were held to be Joe Blow's copyrighted property, it could not be used in this manner.

Because of the harm done to the plaintiff's reputation, goodwill, and commercial computer facilities by Joe's practices, the court will most likely follow the precedent set in the Hotmail case and issue an injunction against Joe Blow. His actions are in breach of his contract with Everybody's and his speech will not be protected in this case.

Chapter 16
FORMATION OF SALES, LEASE, AND INTERNET CONTRACTS

Any intelligent fool can make things bigger, more complex, and more violent. It takes a touch of genius-and a lot of courage-to move in the opposite direction.
E.F. Schumacher

I. Overview

The residents of the Hawaiian Islands have witnessed one of nature's most interesting phenomenons: a total solar eclipse. As obscured as the sun's light may have been, there still shone around the shadow of the moon a ring of light called a *penumbra*. When you begin to examine the Uniform Commercial Code (UCC) as a specialized body of contract law, it appears to have eclipsed the common law of contracts, although the light of the common law still shines on. The UCC is intended to cover a number of areas of contract formerly resolved by common law, but the basic elements for both are the same. The four elements of contract are still in place just as the light continues to surround the moon's shadow. What is vastly different is the implementation of how those elements are arrived at in light of commercial realities.

Early on, the Law Merchant of England set up special rules for commercial contracts with the realization that the law should be written to foster and encourage commerce rather than encumber it. The early *faire* courts were established by and for merchants. They were designed to have law reflect the needs of commerce. Some of the principles that evolved included:

1. Holding merchants to a higher standard established by their peers.
2. Providing for a uniformity of interpretation for commercial contracts.
3. Providing for a uniform flexibility in the formation, modification, and termination of commercial contracts so as to allow for the realities of the marketplace.
4. Retaining the common law essential ingredients of equity, fair play, good faith, and conscionability in commercial dealings.

The UCC is both the literal and spiritual descendent of the Law Merchant. Its predecessor, the 1906 Uniform Sales Act, was ultimately adopted by thirty-seven states. The UCC eclipsed it in 1952. One of the key authors of the UCC was Professor Karl Llewellyn whose hallmark was legal realism. The law should reflect the realities of commerce that are already working rather than impose unnecessary obstacles or impediments to business. The UCC has been adopted, at least in part, in all fifty states. It continues to be one of the single most important legislative enactments in American legal history. It is updated and revised in order to keep up with the changing realities of the marketplace.

II. Hypothetical Multi-Issue Essay Question

Benny, a bakery owner, entered into an oral contract with Alfa, a "natural" farmer, to buy five tons of whole grain unprocessed wheat germ to be used in Benny's famous "Wowee" brand muffins. The contract price was $17,500 per ton and the wheat germ was to be shipped on Lionel Lines, Inc. The contract was for one year.

After two tons had already been shipped, Alfa informed Benny that he might not be able to deliver more because he fears the threat of the greenhouse effect on his crops. This notice took place on June 3, 2001, when the market price for comparable wheat germ was $19,000 per ton.

Benny went to Alfa's competitor, Romeo, on December 1, 1990 and purchased the remaining three tons of wheat germ for $20,500 per ton. Benny now sues Alfa for the difference between the contract amount and the price he paid Romeo for the wheat germ. What result?

III. Outline

Overview of the Uniform Commercial Code

Article 1	General provisions
Article 2	Sales
Article 2A	Leases
Article 3	Commercial paper
Rev. Article 3	Negotiable instruments
Article 4	Bank deposits and collections
Article 4A	Wire transfers
Article 5	Letters of credit
Article 6	Bulk transfers
Article 7	Documents of title
Article 8	Investment securities
Article 9	Secured transactions

Article 2
- Sale
  - Passing of title from a seller to a buyer for a price
- Goods
  - Tangible things that are movable at the time of their identification in the contract
- Merchant
  - A person who deals in the goods of the kind involved in the transaction, or
  - A person who by his or her occupation holds him or herself out as having knowledge or skill particular to the goods involved in the transaction

Article 2A (Leases)
- Lease
  - A transfer of the right to the possession and use of the named goods for a set term in return for certain consideration
- Lessor
  - The person who transfers the right of possession and use of goods under the lease
- Lessee
  - The person who acquired the right to possession and use of goods under the lease

Terms of an Offer
- Open terms
  - Open price term
  - Open payment term
  - Open delivery term
  - Open time term
  - Open assortment term

Acceptance
- Any reasonable manner or method of communication is considered acceptance under the UCC

Consideration
- Under the UCC, modifying a sales or lease contract needs no consideration to be binding
- If the parties agree that any modification of a sales or lease contract must be in writing, then oral modification is not enforceable

IV. Objective Questions

Terms:

1. The doctrine that states that courts have the power to determine whether an entire contract or a contract clause is unfair or inequitable is called the _______________ doctrine.

2. Tangible things that are moveable at the time of their identification in the contract are called _______________.

3. A person who deals in goods of the kind involved in the transaction, or by his occupation holds himself out as having knowledge or skill peculiar to the goods involved in the transaction, is called a _______________ under the UCC.

4. The rule that says if a merchant offers to buy or sell goods and gives a written and signed assurance on a separate form that the offer will be held open, he cannot revoke the offer for the time stated or if not time is stated, for a reasonable time is called the _______________ _______________ rule.

5. The rule that says a yet to be determined term can be read into a contract is called the _______________ _______________ rule.

6. Goods that by nature or usage of trade are equivalent to any like unit goods of the same grade and quality are called _______________.

7. A person who in good faith and without knowledge that the sale violates the ownership or security interests of a third party buys the goods in the ordinary course of business from a person in the business of selling goods of that kind is called a _______________ _______________ _______________ _______________ _______________.

8. A sale that involves the provision of both a service and a good in the same transaction is called a _______________ _______________.

9. A shipment that is offered to the buyer as a replacement for their original shipment, when the original shipment cannot be filled is known as an _______________ _______________.

10. The rule that says an open term can be "read into" a UCC contract is generally known as the ___________________ ____________________.

True/False:

1. ____ Disparity of bargaining power alone may make a contract unconscionable.

2. ____ Article 2 applies to all mixed sales that include a good and a service.

3. ____ If a confirmation contains additional or different terms than those in the offer, it extinguishes the offer and acts as a counteroffer under the UCC.

4. ____ Under the UCC, as in common law, an agreement modifying a sales contract must be supported by consideration to be binding.

5. ____ Under the UCC, output contracts and requirements contracts cannot have an open quantity term since the quantity of goods that the parties intend to buy or sell cannot be implied.

6. ____ Nonmerchants are held to a subjective standard of honesty in contracting while merchants are held to an objective standard of fair dealing in the trade.

7. ____ Growing crops, timber, and other things are considered "goods" under the UCC, Article 2 even if they are not severed from the realty.

8. ____ Rules established for sales contracts by Article 2 of the UCC take precedent over the common law of contracts.

9. ____ If a seller ships nonconforming goods in response to a buyer's offer, this constitutes both an acceptance of the offer and a breach of the sales contract.

10. ____ Under the UCC, as in common law, an agreement modifying a sales contract must be supported by consideration to be binding.

Multiple Choice:

1. Mega Widgets Supply offered to sell to Racing Widgets Corporation 20 units of widget subassemblies at $1,000 per unit under specified delivery terms. Racing Widgets accepted the offer as follows: "We accept your offer for 20 units at $1,000 per unit per inspection by USAC." Which of the following is correct?
   A. A contract was formed on Mega Widgets' terms.
   B. Racing Widgets' reply constitutes a conditional acceptance, not a counteroffer.
   C. Racing Widgets' reply constitutes a counteroffer, and no contract was formed.
   D. A contract was formed on Racing Widgets' terms.

2. With regard to a contract by the UCC Sales Article, which one of the following statements is correct?
   A. Merchants and nonmerchants are treated alike.
   B. The contract may involve the sale of any type of personal property.
   C. The obligations of the parties to the contract must be performed in good faith.
   D. The contract must involve the sale of goods for a price of more than $500.

3. Office Wholesalers writes a letter offering to sell 20 desks to Fantastic Furniture for $300 each. Fantastic Furniture writes back, "We accept your offer for 20 desks, but we would really like 30 desks. Please ship 10 more desks at the same price." Which of the following is correct?
   A. Fantastic Furniture made a counteroffer, so no contract has been formed.
   B. Fantastic Furniture's acceptance was effective and established new terms for the contract.
   C. This contract is void because it does not meet the requirements of the Statute of Frauds.
   D. Fantastic Furniture accepted Office Wholesalers' offer, and there is a contract for at least 20 desks.

4. Fine Jewelry orders 30 pairs of earrings from Artist Anita. Anita accepts and agrees to supply the earrings. However, no mention is made of price. Which of the following is correct?
   A. Fine Jewelry must pay Anita a reasonable (probably market) price.
   B. The contract fails for indefiniteness since the parties never agreed.
   C. Either party can void the contract before performance without being liable for damages.
   D. None of the above.

5. Merchant Marty and Nonmerchant Nancy negotiate a contract for the sale of goods that is covered by Article 2 of the UCC. Which of the following is correct?
   A. Merchant Marty and Nonmerchant Nancy are both held to the same standard of good faith since the contract is governed by the UCC.
   B. Merchant Marty and Nonmerchant Nancy are both held to an objective standard since a standard applies if one party is a merchant.
   C. Merchant Marty is held to an objective standard of good faith while Nonmerchant Nancy is held to a subjective standard of good faith.
   D. None of the above.

6. Article 2 of the UCC governs only sales for goods, not real estate. Which of the following is *not* covered by Article 2?
   A. Minerals, oil, or gas that are removed by either the buyer of the seller.
   B. Growing crops or timber that are removed by either the buyer or the seller.
   C. Structures or materials that are removed by the seller.
   D. All of the above are covered by Article 2.

7. Under the UCC Sales Article, a firm offer is created only if the:
   A. Offeree is a merchant.
   B. Offeree gives some form of consideration.
   C. Offer states the time period during which it will remain open.
   D. Offer is made by a merchant in a signed writing.

8. Merchant Mike and Merchant Mary negotiate a sales contract by exchanging preprinted forms. If the acceptance contains additional terms, those terms will not become part of the sales contract if:
   A. The offer expressly limits the acceptance to the terms of the offer.
   B. The additional terms materially alter the terms of the original contract.
   C. Offeror notifies the offeree that he objects to the additional terms within a reasonable time.
   D. None of the above will become part of the sales contract.

9. Which of the following is *not* an exception to the writing requirements of the UCC Statute of Frauds?
   A. Contracts for the sale of goods over $500.
   B. Contracts for specially manufactured goods.
   C. Admissions in pleadings or in court.
   D. Part payment and acceptance.

10. Which of the following states has excluded Article 2 of the UCC in its legislative enactments:
   A. Colorado.
   B. New Mexico.
   C. Louisiana.
   D. New York.

V. Answers to Objective Questions

Terms:

1. *Unconscionability.* Section 2-302 of the UCC allows courts to refuse enforcement of a sales contract if it finds either all or part of the contract to be unconscionable.
2. *Goods.* Because of the broad application of this definition, the UCC specifically seeks to avoid listing all the possible items that may be classified as goods. Rather, it seeks to specifically exclude certain items such as money, stocks, bonds, and patents.
3. *Merchant.* Even though this provision has been broadly interpreted so as to include all sorts of part-time or hobby sellers as merchants, courts have disagreed about whether farmers are merchants within the UCC definition.
4. *Firm offer.* The net effect of this rule is to create a common law option contract result without having to provide the consideration required by the common law. The maximum period allowed under this rule is three months.
5. *Open terms.* The UCC seeks to recognize commercial realities by allowing the parties to leave major terms of the contract open at the time of its formation. Under the common law rule of indefiniteness, many of these open terms would not be allowed.
6. *Fungible.* This term is derived from the old English term "fundas," meaning common land. In the UCC, it is not land, but rather goods, that are identical with others and can be replaced by similar weight, measure, and number.
7. *Good faith purchaser for value.* A good faith purchaser for value is a person to whom good title can be transferred from a person with a voidable title. The real owner cannot reclaim goods from a good faith purchaser for value.
8. *Mixed sale.* Look at the main objective of the transaction to see if it is covered by the UCC or the common law of contracts.
9. *Accommodation shipment.* The accommodation is a counteroffer from the seller to the buyer. The buyer is free to either accept or reject the counteroffer.
10. *Gap-filling rule.* The UCC is much more tolerant of open terms than the common law. These provisions reflect the "legal realism" of the UCC in recognizing the realities of the marketplace.

True/False:

1. False. Gross disparity in bargaining power may make a contract unconscionable. However, some disparity alone is not usually enough to declare a contract unconscionable.
2. False. Mixed sales may or may not be governed by Article 2. This determination is made on a case-by-case basis.

3. False. Except under specific circumstances, confirmations with additional or different terms do not extinguish the offer. The UCC is more liberal in this respect than the common law.
4. False. New consideration is not required to support a contract modification under the UCC.
5. False. Under the UCC, output contracts and requirements contracts are the only types of contracts where the quantity term can be left open.
6. True. These standards are set out Section 2-103(1)(b) of the UCC. This standard is designed to reflect the so-called "higher" standards on merchants who are expected to know the rules of the business they are in.
7. True. This rule is set out in Section 2-107(2) of the UCC. These items are goods if they can be severed from the realty without causing material harm to it.
8. True. This reflects the "eclipse" factor discussed in the overview.
9. True. See Section 2-206(1)(b) of the UCC.
10. False. Consideration is not required to support a modification under the UCC. See Section 2-209(1).

Multiple Choice:

1. D. A definite expression of acceptance sent within a reasonable time operates as an acceptance even though it states additional or different terms. Between merchants, such terms become part of the contract unless the offer expressly limits acceptance to its terms, they materially alter it, or notification of objection is given within a reasonable time. A is incorrect because the offer did not limit acceptance to its terms, and Mega Widgets made no objection to the additional terms. B is incorrect because the reply with additional terms was not conditioned upon acceptance of those terms. C is incorrect because the reply was not a counteroffer. It was a definite expression of acceptance that formed a contract.
2. C. Both parties, whether merchants or nonmerchants must perform the contract in good faith. A is incorrect because merchants and nonmerchants are treated differently in different circumstances. B is incorrect because certain types of personal property are not covered by the UCC. D is incorrect because the contract may be for any price if the other requirements are met.
3. D. Fantastic Furniture's acceptance of the offer for 20 desks was definite and unconditional. Fantastic Furniture did, however, request an additional 10 desks. Whether the extra 10 desks will be sold is a separate issue. A is incorrect because the acceptance for 20 desks was definite and unconditional. B is incorrect because the 10 extra desks is not a new term of the original contract. The 20 desks will be sold. However, the extra 10 desks are separate. C is incorrect because the contract is written in the form of letters exchanged between the parties.
4. A. A contract with an open price term is enforceable under the UCC. When no price is specified, the buyer pays a reasonable price. B and C are incorrect because such a contract is valid and enforceable.
5. C. Since Merchant Marty is a merchant, he is held to a higher, objective standard of good faith. Nonmerchant Nancy is not a merchant and is therefore held to a lower, subjective standard of good faith. A is incorrect because merchants and nonmerchants are held to different standards under the UCC. B is incorrect for the same reason.
6. A. Minerals, oil, or gas that is removed by the buyer is not considered a good. However, if such resources are first removed from the land by the seller, they are considered goods. B and C are incorrect because both are considered goods under the UCC.
7. D. Both these elements need to be in place in order to constitute a firm offer, which makes A incorrect. B is incorrect because no consideration is required and C is incorrect because if no time period is stated, the UCC allows for a statutory limit of three months for the offer to be kept open.
8. D. A, B, and C are all excluded from becoming additional terms if they are part of the acceptance.
9. A. B, C, and D are all specifically provided for in UCC as exceptions to the writing requirements of the UCC State of Frauds.

10. C. Because of the French-based civil law tradition historically used in Louisiana, only parts of the UCC have been adopted by that state. Article 2, 6, and 9 have been excluded from its adoption of the UCC.

VI. Answer to Essay Question

The main purpose of this transaction was the sale of goods. A sale consists of passing title from seller, Alfa, to buyer, Benny for a price, $17,500 per ton of wheat germ. Alfa may have provided some services by way of harvesting the grain, but those services were only incidental to the main purpose of the contract. Growing crops are classified as goods under the UCC regardless of whether the buyer or the seller severs the crop from the realty. Both the subject matter and purpose of the contract bring it under the purview of the UCC.

Benny would definitely be classified as a merchant in that a baker is expected to have knowledge and skills related to ingredients used in making bread products. The courts are divided as to whether farmers are merchants within the UCC definition. Your assumption on the merchant status of Alfa should be made clear at this point because it may impact the remainder of your answer. For purposes of resolving the problem, assume here that Alfa is a merchant based on his knowledge of grains.

This contract does not raise any issues of capacity, legality, or consideration. The real problem lies in the formation of mutual assent and failure to put it into writing. The UCC is much more open-ended than the common law in its requirements for formation of contract. A contract for sale of goods may be made in any manner sufficient to show intent to be bound, including conduct. Here there was an oral offer to buy and an acceptance by shipment of conforming goods.

The contract was for more than $500 and should have been in writing. There are three exceptions to the writing requirements of the UCC Statute of Frauds:

1. Specially manufactured goods.
2. Admissions in pleading in courts.
3. Party payment or acceptance.

Here we have partial acceptance based on the first two tons shipped. There was no refusal to accept those goods, so the contract is enforceable to that extent. The remainder of the contract may not be enforceable by Benny because he failed to get a writing. Had there been a written confirmation sent by Alfa, the writing requirement would have been met if no objection was made to the terms within 10 days after the date of receipt. Had the writing requirement been met this way, Alfa would be in breach of his duty to deliver the remaining three tons.

Because there was no writing, the best method available for interpretation is course of performance since the contract was entered into. If it is decided that the writing requirement was met by way of a written confirmation, that writing would act as the express terms of the contract.

Alfa seeks a modification mid-term in the contract. It appears that an unsubstantiated fear of the greenhouse effect would not be a commercially reasonable basis upon which to seek excuse from continued performance. Benny, in turn, may have waited too long to buy the goods elsewhere. The market price at the time of the alleged breach was $19,000 and his actual damages may be limited to the difference between his contract price of $17,500 and $19,000 for each of the remaining three tons. This assumes that Alfa was bound by the contract. If the Statute of Frauds defense applied, Alfa would be discharged from any further obligations. The most likely result here, however, will be that the Statute of Fraud defense will not prevail due to the part payment or part performance exception to that rule. As such, Alfa will most likely be bound by the contract and have to pay for his alleged breach.

Chapter 17
PERFORMANCE OF SALES, LEASE, AND INTERNET CONTRACTS

When everything has to be right, something isn't.
Stanislaw J. Lee

I. Overview

Under the common law of contracts, ownership title has always played a critical role. It has been the linchpin that holds together the benefit/burden dynamics of rights and duties towards property interest. The UCC has greatly modified these rights and duties in contracts covered by it. By placing less emphasis on title alone, it has allowed controversies involving title, risk of loss, and insurable interest to be more independently resolved from each other. The underlying purpose behind allowing this sort of "separate path" resolution of each of these issues is found in the gaps created by the common law emphasis on title alone. For example, under the common law, title was generally required to obtain an insurable interest on property. Assume A was selling to B. Only A could insure the property until such time as B took title. But what about B's non-title economic interests in A's property? If B already had a contract of resale for the property, should not the economic benefits anticipated from that resale contract also be insurable? The UCC recognized that in order to protect and foster commerce, the traditional rules based on title were simple inadequate to protect the realistic needs of commerce.

Passage of title under the UCC begins with the process of identification of the goods. Once the goods are identified as being part of the proper subject matter of a UCC contract, determine what the parties have expressly stated vis-à-vis that issue. The UCC is similar to many other areas of the law in that it tries to balance freedom of contract with a need for clearly ascertainable ground rules. Title can pass from seller to buyer in any legal manner expressly agreed upon by the parties. However, if the parties are silent or ambiguous on this issue, the UCC will impose its own title-transfer rules upon the parties. What makes these rules particularly problematic is the possible intervention of third party rights. In its efforts to protect reliance on the enforceability of commercial contracts, the UCC has given innocent third parties much more protection than was provided under the common law of contracts. That protection has often been provided at the expense of prior title-holders. These modified rules will be outlines in the key question checklist.

The common law risk of loss rules have also been substantially amended by the UCC. Under the traditional rules of common law, the risk of loss stayed with the title. Here too, the UCC allows the parties to address the issues explicitly, and if they fail to do so, the statute will impose its own set of riskshifting rules. Once the goods are identified, the UCC will allocate the risk of loss based on the overall nature of the contract. Note that this identification of contract, be it conditional, shipment, delivery, or consignment, is what determines the risk of loss, not who has title. Students must first, therefore, be able to identify the essence of the contract in order to be able to ascertain which rights and duties the UCC attributes to the respective parties arising from that designation. As in the case of the UCC title rules, the possible infusion of third party rights always looms overhead when risk of loss is attributed.

Compare all three issues: title, risk of loss, and insurability to the lines indicated on a medical monitoring device. Each line is independently giving the reader critical information yet they are components of a larger issue--the overall health of the patient. In UCC contract resolution, each of these issues is independently answered as part of the process of resolving the overall rights and duties under the contract. They are all separate issues but yet are mutually dependent as part of one legal being.

II. Hypothetical Multi-Issue Essay Question

Lancelot Loefric and his wife Mercia were shopping around for a new luxury car. They test drove a number of machines and were concerned with finding the right combination of style, luxury, snob appeal, and safety. Price was no object because Lancelot made millions selling scripts to TV soaps. He was constantly thinking of brilliant new ways to revive dead characters through the use of dream sequences and shower scenes.

Last week Lancelot and Mercia signed a sales contract with Hot Air Autos, Inc., to buy, on approval, a new model Zoroastermobile number 1234567. It is equipped with all the latest features including airbags for all occupants of the car. Under the terms of the sale, Lancelot and Mercia could return the car anytime up to a week after the contract was signed. One day before the week expired, Lancelot hit a pothole and six airbags blew up at once inside the car. He lost control, and the car was wrecked. Lancelot tows the wrecked car back to Hot Air, and they refuse to take it back. What results?

III. Outline

Passage of Title
- A title is legal, tangible evidence of ownership of goods

Shipment and Destination Contracts
- Shipment contract
  - A contract that requires the seller to ship the goods to the buyer via a common carrier
- Destination contract
  - A contract that requires the seller to deliver the goods either to the buyer's place of business or to another destination specified in the sales contract

Document of Title
- An actual piece of paper, such as a warehouse receipt or bill of lading, that is required in some transaction of pick up and delivery

Carrier Cases—Movement of Goods
- Shipment contract
  - The buyer bears the risk of loss during transportation
- Destination contract
  - The seller bears the risk of loss during transportation

Noncarrier Cases—No Movement of Goods
- Merchant-seller
  - Does not pass risk of loss to the buyer until the goods are received
- Nonmerchant-seller
  - Passes the risk of loss to the buyer upon "tender of delivery" of the goods

Conditional Sales
- Sale on approval
  - A type of sale in which there is no actual sale unless and until the buyer accepts the goods
- Sale or return
  - A contract that says the seller delivers goods to a buyer with the understanding that the buyer may return them if they are not used or resold within a stated or reasonable time
- Consignment
  - An arrangement where a seller (the consignor) delivers goods to a buyer (the consignee) for sale

Risk of Loss—Breach of Sales Contract
- Seller in breach
  - If he or she tenders or delivers nonconforming goods to the buyer
- Buyer in breach
  - If he or she refuses to take delivery of conforming goods
  - If he or she repudiates the contract
  - If he or she otherwise breaches the contract

Passage of Title in Sales by Nonowner Third Parties and Sales of Goods Subject to Security Agreements

IV. Objective Questions

Terms:

1. The principle of distinguishing the goods named in the contract from the seller's other goods is called ________________ ________________ ________________.

2. When a merchant allows a customer to take goods for a specified time in order to see if the goods fit their needs, this is called a ________________ ________________ ________________.

3. A contract that requires the seller to ship the goods to the buyer via a common carrier is a ________________ ________________.

4. The right to be able to reclaim goods sold with an imperfect title from the purchasers is called ________________ ________________, while a ________________ ________________ is a title that a purchaser has if the goods were obtained by fraud or some other misrepresentation.

5. A person who in good faith and without knowledge that the sale violates the ownership or security interests of a third party buys the goods in the ordinary course of business from a person in the business of selling goods of that kind is a ________________ ________________ ________________ ________________ ________________ ________________ ________________.

6. A delivery term that requires the seller to arrange to ship the goods and put the goods in the carrier's possession is ________________ ________________ ________________ ________________ ________________ ________________. The seller bears the expense and risk of loss until this is done.

7. A delivery term that requires the seller to deliver and tender the goods alongside the named vessel or on the dock designated and provided by the buyer is ________________ ________________ ________________ ________________ ________________. The seller bears the expense and risk of loss until this is done.

8. A sales contract that requires the seller to deliver conforming goods to a specific destination is a ________________ contract.

9. A delivery term that requires the seller to bear the expense and risk of loss until the goods are unloaded from the ship at its port of destination is called ________________.

10. A contract that requires the seller to bear the expense and risk of loss of the goods during transportation is a ________________ ________________, ________________ ________________ contract. The seller is under no duty to deliver replacement goods to the buyer since there is no contractual stipulation that the goods will arrive at the appointed destination.

True/False:

1. ____ The seller retains title and risk of loss of both existing and future goods until she identifies them to a sales contract.

2. ____ Under a destination contract, title passes to the buyer as soon as the goods are identified in the contract.

3. ____ If a buyer rejects or refuses to receive goods, or justifiably revokes acceptance of the goods, title to the goods revests in the seller.

4. ____ If a person with a voidable title to goods transfers title to a good faith purchaser for value, the real owner cannot reclaim goods from such a purchaser.

5. ____ In entrusting cases, the buyer of the goods bears the risk of loss even though he is an innocent party. The purpose of this principle is to promote confidence in commercial transactions.

6. ____ Under shipment contracts, the risk of loss passes to the buyer when the seller delivers the goods to the carrier. Therefore, the buyer bears the risk of loss during transportation.

7. ____ In a sale on approval, the sale is effective as soon as the buyer takes possession of the goods.

8. ____ Goods sold pursuant to a sale or return contract are subject to the claims of the seller's creditors while the buyer possesses them since the title to the goods still belongs to the seller.

9. ____ The buyer may obtain an insurable interest in goods before the title or risk of loss passes to him.

10. ____ The buyer and seller may never have an insurable interest in the same goods at the same time.

Multiple Choice:

1. Which of the following is *not* classified by the UCC as a conditional sale:
   A. Sale on approval.
   B. Sale with open terms.
   C. Sale or return.
   D. Sale on consignment.

2. Which of the following is false?
   A. "F.O.B." means the seller is responsible for delivering the goods to the carrier. The seller bears the risk of loss to this point.
   B. "F.A.S." means the seller must deliver the goods alongside a vessel or dock. In addition, the seller bears the risk of loss until the buyer has possession.
   C. "C.I.F." means the seller is responsible for certain costs, including cost of insurance and freight. The seller bears the risk of loss until the goods are on the carrier.
   D. "No arrival, no sale" means the seller bears the risk of loss during transportation.

3. In July, Santa's Station agreed to buy 5,000 bushels of pears from Holiday Pear Growers. The pears were to be delivered weekly beginning in late November. Which of the following is correct?
   A. The pears are future goods when the contract is made, but the contract is still enforceable.
   B. The pears are future goods when the contract is made, and therefore the contract is unenforceable.
   C. The pears are not identified until Holiday Pear Growers ships them.
   D. None of the above.

4. Assume the same facts as in question 3. Also, assume Holiday Pear Growers and Santa's Station agree this will be a destination contract. Which of the following is correct?
   A. Holiday Pear Growers must deliver the pears to Santa's Station or some other designated place, and Holiday Pear Growers bears the risk of loss in transportation.
   B. Holiday Pear Growers must ship the pears via a common carrier, and Santa's Station bears the risk of loss in transportation.
   C. Holiday Pear Growers must deliver the pears to Santa's Station or some other designated place and Santa's Station bears the risk of loss in transportation.
   D. Holiday Pear Growers bears the risk of loss until Santa's Station takes possession of the pears.

5. Lucrative Larry contracts to buy a new Porsche at Pam's Dealership. Larry agrees to pay $70,000 for the new car. Pam's dealership deals extensively and exclusively in Porsches. However, Pam bought the car from another dealer, Paul, before selling it to Larry. In addition, Pam wrote a check for the car that later bounced. Which of the following is correct?
   A. Paul can reclaim the car from Larry.
   B. Larry was a good faith purchaser for value and can keep the car.
   C. Paul entrusted the car to Pam so he bears the risk of loss.
   D. Pam possessed good title at the time she sold the car.

6. Assume the same facts as in question 5, except instead of $70,000, Larry paid $700 for the new car.
   A. Paul can reclaim the car from Larry.
   B. Larry was a good faith purchaser for value and can keep the car.
   C. Paul entrusted the car to Pam so he bears the risk of loss.
   D. Pam possessed good title at the time she sold the car.

7. On May 1, Home Designs contracts to buy fifty sofas from Furniture Manufacturer for $200 per sofa for a total contract price of $10,000. Home Designs and Furniture Manufacturer agree that Home Designs will bear the risk of loss beginning on May 1 even though the sofas are still in Furniture Manufacturer's warehouse. Which of the following is correct?
   A. The risk of loss agreement is unenforceable since Furniture Manufacturer must bear the risk while the goods are in its possession.
   B. The risk of loss agreement is unenforceable since who bears the loss is governed by the UCC, not the parties.
   C. The risk of loss agreement is enforceable since the parties agreed to it and the goods exist and have been identified to the contract.
   D. None of the above.

8. Assume the same facts as in question 7 except there is no risk of loss agreement. In addition, assume Furniture Manufacturers breaches the contract by tendering nonconforming goods and Home Designs revokes acceptance when the goods are received. Before the sofas are returned, a fire destroys them. Which of the following is correct?
   A. Furniture Manufacturers is liable for damage beyond the amount that Home Design's insurance will cover.
   B. Furniture Manufacturers is liable for the total contract price of $10,000.
   C. Home Designs must bear the entire loss of $10,000.
   D. Home Designs must bear the loss from damage beyond the amount which Furniture Manufacturers' insurance will cover.

9. Vacuum Valerie Co. allows Bachelor Bob to take home and use a SuperDuper Vacuum for two weeks to see if he likes the vacuum without buying it. If, at the end of two weeks, Bob wants to buy the vacuum, Valerie will sell it to him for "a great price." Which of the following is correct?
   A. This is a sale on consignment.
   B. This is a sale or return.
   C. This is a sale on approval.
   D. None of the above.

10. Assume the same facts as in question 9. Also, assume at the end of the first week, Bob's house burns down, through no fault of his own, destroying the vacuum. Which of the following is correct?
    A. Bob bears the risk of loss and is liable to Vacuum Valerie Co. for the cost of the vacuum.
    B. Vacuum Valerie Co. bears the risk of loss and cannot recover from Bob.
    C. Vacuum Valerie Co. and Bob must split the cost of the vacuum.
    D. None of the above.

V. Answers to Objective Questions

Terms:

1. *Identification of goods*. This process is used for a number of reasons throughout the UCC, including allocating title, risk and insurable interest.

2. *Sale on approval.* This is one of several types of conditional sale transactions recognized by the UCC.
3. *Shipment contract.* See Section 2-401(2)(a) of the UCC. Under this section the seller is required to make proper shipping arrangements and deliver the goods into the carrier's hands.
4. *Void title, voidable title.* These doctrines are designed to protect the reliance factor of good faith purchasers in the stream of commerce.
5. *Buyer in the ordinary course of business.* See Section 1-201(a) of the UCC. If a person qualifies under this section, he takes the goods free of any third-party security interests in the goods.
6. *Free on board point of shipment.* See Section 2-319(1)(a) of the UCC.
7. *Free alongside point of shipment.* See Section 2-319(2)(a) of the UCC.
8. *Destination.* Such contracts require the seller to bear the risk of loss of the goods during their transportation.
9. *Ex-ship.* See Sections 2-322(1) and (2)(b). This term requires the seller to bear the expense and risk of loss until the goods are unloaded from the ship at its point of destination.
10. *No-arrival, no-sale.* See Section 2-324(a) and (b) of the UCC.

True/False:

1. True. Identification is the first step in the allocation of risk of loss process.
2. False. Under a destination contract, title passes to the buyer when the seller tenders delivery of the goods to the specified location.
3. True. See Section 2-401(4) of the UCC.
4. True. This rule is designed to assist the innocent third party.
5. False. In entrusting cases, the entrusting party bears the risk of loss even though both she and the buyer are both innocent parties.
6. True. Generally, see Section 2-319 for this set of rules in shipment contracts.
7. False. In a sale on approval, the sale is effective when the buyer accepts the goods. Acceptance of the goods requires acts beyond merely taking possession.
8. False. Goods sold pursuant to a sale or return contract are only subject to the claims of the buyer's creditors while the buyer possesses them. In addition, while the buyer has possession of the goods, he also has title to the goods.
9. True. This is one of the changes the UCC made from the common law of contracts that relied on title.
10. False. A buyer and seller may have an insurable interest in the same goods at the same time.

Multiple Choice:

1. B. Open terms are allowed under the UCC in certain contracts and are not classified as conditional sales. A, C, and D are all classified as conditional sales under the UCC.
2. B. While the seller must deliver the goods alongside a vessel or dock, the seller only bears the risk of loss until the goods are delivered, not until the buyer takes possession. A is correct. C is correct. D is correct.
3. A. Since the pears are not grown yet, they are future goods. However, the pears are identified in the contract, and the contract is enforceable. B is incorrect because the contract is enforceable. C is incorrect because the pears are identified in the contract since the pear trees are growing crops.
4. A. A destination contract requires the seller to deliver the goods to the buyer's place of business or some other designated place. B is incorrect because the seller is not required to ship the goods via common carrier under a destination contract. C is incorrect because the buyer does not bear the risk of loss while the seller is delivering the goods. D is incorrect because the seller only bears the risk of loss until the goods are delivered to the designated place.

5. B. Since Larry acted honestly in this contract and paid value of $70,000 for the car, he is a good faith purchaser for value. Under such circumstances, Paul cannot reclaim the car. A is incorrect for the reason discussed above. C is incorrect because Paul did not entrust the car to Pam, but thought he was selling it to her. D is incorrect because Pam had a voidable title at the time she sold the car.
6. A. Here, Larry did not pay "value" or sufficient consideration for the car and he is not a good faith purchaser for value. Therefore, Paul can reclaim the car. B is incorrect for the reason discussed above. C is incorrect for the reason discussed in question 5. D is incorrect for the reason discussed in question 5.
7. C. The parties may agree about who will bear the risk of loss if the goods under the contract are damaged as long as the goods exist and are identified and the obligation of good faith is met. A and B are incorrect for the reason discussed above.
8. A. B is incorrect because the breaching seller is only liable for the amount the buyer's insurance does not cover. C is incorrect because the buyer can recover the amount not covered by insurance from the breaching seller. D is incorrect for the reason discussed in answer A.
9. C. A is incorrect because a sale on consignment is usually between two merchants for retail sale of the seller's goods. B is incorrect because a sale or return is usually between two merchants for retail sale of the seller's goods. However, the goods may be returned to the seller if the buyer does not resell them.
10. B. Under a sale on approval, the seller bears the risk of loss until the buyer accepts the goods. Here, Bob had not yet accepted the vacuum but was still using it on a trial basis. A and C are incorrect for the reason discussed above.

VI. Answers to Essay Question

The main purpose of this transaction was the sale of goods. Hot Air Autos may have provided some services by way of maintenance and repair of the auto for purposes of preparation for sale or after the fact by way of warranty. Those services, however, would be considered only incidental to the main purpose of the contract. Automobiles are classified as goods under the UCC. Thus both the subject matter and the main purpose of the contract fall under the purview of the UCC.

Hot Air Autos is definitely classified as a merchant and is expected to have the knowledge and skills associated with the business of preparing autos for retail sales. Lancelot and Mercia, in turn, would be classified as consumers. It does not appear from the facts presented that they had any intent to engage in resale of the auto. Their main purpose was to purchase the car for personal use.

This contract does not raise any issues or capacity, legality, or consideration. The real issue revolves around the formation of mutual assent and the interpretation of the terms of the contract. Under the UCC, a contract for the sale of goods may be made in any manner sufficient to show intent to be bound, including conduct. The four elements of contract appear to be in place here.

Where there are express written terms, those will be examined first as a source of interpretation. If there were no express written terms, a court could move to course of performance, course of dealing, and usage of trade in the particular industry involved.

The goods involved in this contract are covered by the UCC. An automobile is a personal good, and it is standard practice to identify autos in sales contracts by the manufacturer's serial number affixed to the body of the car. The serial number was used in this contract.

The parties here have explicitly agreed to certain conditions as part of their contract. Those express conditions are tied to a sale on approval to be finalized within one week after the contract was signed. The buyers had the right to return the automobile at any time prior to the expiration of that one-week period.

There is no third party involvement in this case except for the possibility of an insurance company paying one of the parties for their loss and then stepping into their legal shoes by subrogation. There was no evidence of any third party involvement as a good faith purchaser for value or buyer in the ordinary course of business. Had Lancelot and Mercia sold the car to their neighbor Figaro three days after the contract, then Figaro may have taken better title to the car than Lancelot and Mercia had at the time.

Because the parties did not explicitly agree on who assumed the risk of loss in this contract, the UCC general rules with regard to conditional sales must be examined. Under the UCC, the risk of loss remains with the seller on approval contracts until the buyer accepts. There is an exception to this rule if harm to the goods was caused by the buyer's negligence. Here it appears that Lancelot's loss of control was due to a defect in the product, not his own negligence. Thus the loss falls on the seller.

Both parties had an insurable interest in the car based on their respective economic contract interests as recognized by the UCC regardless of title.

Chapter 18
PRODUCT LIABILITY AND WARRANTIES

Security is mostly a superstition. It does not exist in nature ... Life is either a daring adventure or nothing.

Helen Keller

I. Overview

Warranties

Buyer beware! Throughout the history of common law, the law's expectations of buyer protection started and ended with the buyers themselves. A person entering the marketplace was expected to personally know the seller and to resolve any differences directly with him or her. The role of government and its courts as a source of buyer protection was expected to be minimal. That worked well enough in the pre-industrial age populated by small communities with direct dealings between the buyer and seller. What are the realistic chances today of an aggrieved buyer working out a problem directly with a scion of Mr. Ford or Mr. Honda? The postindustrial global marketplace has become too complex to expect a buyer to resolve these issues directly with the seller. The law has stepped in with numerous remedial measures to help alleviate the shortcomings created by *caveat emptor*.

The UCC rules of warranty are one such statutory set of buyer protections. Remember that there are five common denominators applicable to all warranties. First, determine whether any sort of warranty may be in existence either by acts of the parties or by imposition of law. Examine the scope and nature of the promises or assurances made under the alleged warranty. Second, decide if there has been any nonconformance with the terms of the warranty in the time frame covered by the transaction. This time frame may be limited to just the initial sale, or the period of performance may have been extended by way of post-sale promises. Third, if there has been a breach of warranty by reason of nonconformity to the promises made, has this breach caused an injury of any sort? Fourth, ascertain the measure of the alleged injury. Finally, are there any defenses, contractual or otherwise, applicable in this case? Remember, warranties represent only one path to rectification of harm. Other avenues may include tort law including the various permutations of product liability reviewed in the next chapter, consumer protection statutes, and equitable remedies.

Product Liability

Traditional negligence doctrines have failed to adequately protect consumers' hurt by defective products. Courts led the way to newer avenues of recourse by way of the doctrine of strict liability. A sampling of arguments from the two sides of the public policy debate surrounding this doctrine is set out below. The actual mechanics of the doctrine are found in the Restatement (Second) of Torts at Section 402A.

On the side of product users, the following points are worth considering:

1. Traditional tort law doctrines based on fault and defenses related thereto have not always adequately served the injured person. The evolution of strict liability doctrines is a logical consequence of having this deficiency in the law.

2. Contract law, both common and UCC, has also failed to provide adequate assurance to the victims of product harm. Consumers traditionally have had less real bargaining power in contracting when it comes to attaching responsibility for harm created by a product.

3. Various legislative enactments at both local and national levels designed to protect consumers tend to be reactive rather than proactive. Bans on products are enacted only after so many injuries have occurred that the product's continued existence in the marketplace can no longer be tolerated. For example, consider how long it took to get three-wheel all-terrain vehicles and lawn darts off the market. Or do semiautomatic "street sweepers" really serve any legitimate societal purpose?

4. As a practical matter, manufacturers of products should provide protections against defects in products. Compared to the consumer, manufacturers have the resources to research, develop, and test against harm. Can you as a buyer of an automobile really test the airbag before you buy the car?

5. The sanctions imposed by law for defective products should act as a deterrent to further introduction of faulty products into the marketplace. If the sellers of goods know this, they will try harder to make products safer in the first place.

6. Manufacturers and other members of the chain of distribution have traditionally had the deep pockets to fight off the little guy by using so-called hard ball tactics in fighting all claims regardless of the equities of the individual claims.

7. There is an extreme inconsistency among the states regarding the rules of evidence and the like that make, or otherwise legitimate, a crap shoot when going after nationally based business organizations who can seek out the more business friendly forum.

As persuasive as some of these arguments may be, the other side of the coin has its own convincing points. Some key arguments for the seller against the current products liability laws are as follows:

1. The cost of the present system has simply run amok and is a model of inefficiency. For every dollar that is spent on paying for the cost of the harm done by defective products, nearly fifty percent is spent on the transfer cost without reaching the victim.

2. Technology never has been and cannot be expected to be one hundred percent precise. The potential harm created by products is dependent on the state of the art at the time, and to require more is to impose 20/20 retroactive hindsight.

3. The present day procedural rules have allowed a deep pocket mentality to set in. Rules like joint and several liability of cotortfeasors allow an entity with only a small percentage of responsibility to be liable out of proportion to that level of responsibility based only on its financial resources.

4. Many socially beneficial products are kept out of the U.S. marketplace because of fears raised by our product liability system in the eyes of potential importers. Conversely, the enhanced costs of U.S. products based on built-in liability insurance costs makes U.S. products less competitive overseas. This diminished participation in the worldwide marketplace hurts all of us. The rules of product liability in the U.S. compared to those of Japan illustrate this point.

5. There is a prevailing lottery mentality as a result of large damage awards from product liability cases. The harm done may be minimal, but the pain and suffering losses coupled with potential punitive damages have inspired too many consumers (and their attorneys) to go on a treasure hunt in the courts.

6. Some products, by definition, are involved in high risk and must be assured of some legal protections in order to be financially viable in the marketplace. Consider, for example, such hi-tech medicinal devices such as heart pacemakers.

One unfortunate aspect of having all these protections has been the creation of a false comfort. That mind-set says: "If there is harm, the blame falls on everyone but me!" The law is there to be used by the innocent, good faith victim of harm. If, however, a person harmed suffers from legal myopia, a tendency to see only what he wants to see, he should not be surprised if the law does not protect him in all events. Just as it was wrong to expect the buyer to find his own way in *caveat emptor*, it is also wrong to expect the UCC, the government, and its courts to cover buyers in all instances. Buyer beware may be outdated in the legal sense, but it makes just as much sense as ever in reality that all product liability claims can somehow be expected to be a sure bet. Yet when you consider real cases like the one used in the hypothetical multi-issue essay question, you can see why the law must act to restrain reckless and callous design, manufacture, and sale of truly harmful products.

II. Hypothetical Multi-Issue Essay Question:

Clint Wayne and his fiancée, Molly Oakley, of Boulder, were as infatuated with their guns as they were with each other. They collected all sorts of new and antique weaponry and were NRA certified expert marksmen. One of Clint's prized possessions is his 1953 Big Bang Long Barrel (a.k.a. the tumescent tool that won the West!). This model was a deliberate copy of the 1873 Colt Peacemaker. The Long Barrel was manufactured by the Big Bang Company from 1953 to 1973, when the Saturday Night Bobbitt replaced it. Current gross sales are better than ever at over $90 million per year with a net profit of $22 million. The owner of the Big Bang Company is Mr. I.M.A. Big-Bang IV.

One of the reasons for the replacement of the original Long Barrel model in 1973 was a pesky little problem of the gun going off when it was accidentally bumped. When the gun was fully loaded and uncocked, the hammer rested against the end of the firing pin, which in turn was in contact with the bullet. Thus, even a slight contact with the hammer could set off a chain reaction that fired the gun. To avoid this possibility, the Big Bang Company instruction manual specifically advised owners to keep the chamber under the hammer empty by loading only five bullets. The company has refused, however, to formally recall the gun and could not be forced to do so under current gun control laws.

On a fateful day last month, Clint was picking up Molly to go to the shooting range. He had his Big Bang Long Barrel on the front seat of his pickup truck and pushed it aside to clear the seat for Molly. The gun fell to the floor, discharged, and killed Molly instantly.

Assume your court has jurisdiction over all the parties and issues presented in this controversy. What are consequences of this accident to Clint, Molly's estate, the Big Bang Company, and Mr. I.M.A. Big-Bang IV.

III. Outline

Warranties of Title and No Infringements
- Good title
- No security interests
- No infringements
- No interference

Warranty of Quality
- Seller's or lessor's assurance to buyer or lessee that the goods meet certain standards of quality
- Warranties may be expressed or implied

Express Warranty
- Affirmation that the goods meet certain standards of quality, description, performance, or condition

Implied Warranty of Merchantability
- Goods must be fit for the ordinary purposes for which they are used
- Goods must be adequately contained, packaged, and labeled
- Goods must be of an even kind, quality, and quantity within each unit
- Goods must conform to any promise or affirmation of fact made on the container or label
- Quality of the goods must pass without objection in the trade
- Fungible goods must meet a fair average or middle range of quality

Implied Warranty of Fitness for Human Consumption
- A warranty that applies to food or drink consumed on or off the premises of restaurants, grocery stores, fast-food outlets, and vending machines

Implied Warranty of Fitness for a Particular Purpose
- Implies that the goods are fit for the purpose for which the buyer or lessee acquired the goods if:
  - The seller or lessor has reason to know the particular purpose for which the goods will be used
  - The seller or lessor makes a statement that the goods will serve that purpose
  - The buyer or lessee relies on the statement and buys or leases the goods

Implied Warranty Arising from a Course of Dealing or Usage of Trade
- Implies that goods comply with:
  - Course of dealing (prior conduct between the parties) or
  - Usage of trade (customs of the industry or market)

Warranty Disclaimers
- Statements that negate express or implied warranties
- Written disclaimers must be conspicuously displayed to be valid
- Unconscionable disclaimers will not be enforced by a court

Other Warranty Issues
- Overlapping and inconsistent warranties
- Third-party beneficiaries of warranties
- Damages recoverable for breach of warranty
- Statute of limitations

Products Liability
The liability of manufacturers, sellers, and others for the injuries caused by defective products

Tort Liability Based on Fault
Negligence
A tort related to defective products where the defendant has breached a duty of due care and caused harm to the plaintiff
Misrepresentation
When a seller or lessor fraudulently misrepresents the quality of a product and a buyer is injured thereby

Strict Liability
Strict liability is imposed irrespective of fault
All parties in the chain of distribution of a defective product are strictly liable for injuries caused by that product
Privity of contract between plaintiff and defendant is not required
The doctrine applies even if the injured party had no contractual relations with the defendant

Defects in Manufacture
Manufacturer fails to properly assemble a product
Manufacturer fails to properly test a product
Manufacturer fails to adequately check the quality of a product

Defects in Design
Defects that occur when a product is improperly designed

Crashworthiness Doctrine
Automobile manufacturers are under a duty to design automobiles so they take into account the possibility of harm from a person's body striking something inside the automobile in the case of a car accident

Defect in Packaging
Occurs when a product has been placed in packaging that is insufficiently tamperproof

Failure to Warn
A defect that occurs when a manufacturer does not place a warning on the packaging of products that could cause injury if the danger is unknown

Other Product Defects
Failure to provide adequate instructions
Inadequate testing of products
Inadequate selection of component parts or materials
Improper certification of the safety of a product

Defenses to Product Liability
- Supervening or intervening event
- Generally known dangers
- Government contractor defense
- Assumption of risk
- Misuse of the product
- Statutes of limitation and statutes of repose
- Contributory negligence
- Comparative negligence

IV. Objective Questions:

Terms:

1. A statute that limits the sellers' liability to a certain number of years from the date when the products was first sold is a statute of _______________.

2. A defect that occurs when a manufacturer does not place a warning on the packaging of products that could cause injury if unknown is called a _______________ _______________ _______________ defect.

3. The defense to a product liability action in which the defendant must prove that (1) the plaintiff knew and appreciated the risk, (2) the plaintiff voluntarily assumed the risk, and (3) the plaintiff's undertaking of the risk was unreasonable is known as the _______________ _______________ _______________ defense.

4. The liability defense that says the plaintiff abused the product that in ordinary amounts would not be dangerous is called _______________ _______________.

5. A legal theory that says all manufacturers of fungible products who have been sued by a plaintiff using those products must be able to prove their innocence or be assessed damages based on the number of units produced at the time that the plaintiff purchased the product is called _______________ _______________ _______________.

6. Under a warranty against _______________, the lessor warrants that no person holds a claim or interest in the goods that arose from an act or omission of the lessor that will interfere with the lessee's enjoyment of his leasehold interest.

7. Under the theory of _______________ _______________ _______________ _______________, buyers and lessees can recover for breach of an express warranty if the warranty was a contributing factor that induced the buyer to purchase the product or the lessee to lease the product.

8. A test to determine merchantability based on what the average consumer can expect *not* to find in food products is the _______________ _______________ test.

9. A warranty that governs the actions of a seller or lessor with whom the customer has never dealt is a(n) _______________ _______________ _______________ _______________ _______________ _______________ _______________.

10. A warranty that governs the actions of a seller or lessor based on previous dealings with the same seller or lessor is a(n) _______________ _______________ _______________ _______________ _______________ _______________ _______________ _______________.

True/False:

1. ____ Compliance with federal or state safety standards for product design automatically absolves a defendant from product liability.

2. ____ In the case of a supervening event, only the party who caused the defendant harm and all sellers after him are liable, but prior sellers are absolved from strict liability.

3. ____ Manufacturers are liable for damages caused by both foreseeable and unforeseeable misuse of products.

4. ____ Generally, to receive punitive damages, the plaintiff must prove the defendant intentionally injured him or acted with reckless disregard.

5. ____ Strict liability, like negligence, requires the injured person to prove the defendant breached a duty of care.

6. ____ A buyer who furnishes specifications for goods to a seller can hold the seller liable if a third party claim arises out of compliance with the specifications. This is because the seller gives a warranty of no infringement.

7. ____ Express warranties may be written, oral, or inferred from the seller's conduct.

8. ____ Manufacturers are not liable for express warranties made by wholesalers and retailers unless the manufacturer authorizes or ratifies the warranty.

9. ____ In effect, to impose liability for a breach of the implied warranty of merchantability, a plaintiff must show that the product may injure an appreciable number of people.

10. ____ In establishing a breach of the implied warranty of fitness for a particular purpose, the seller or lessor must have actual knowledge of the buyer's or lessee's particular purpose and rely on this knowledge.

Multiple Choice:

1. Pharmaceutical Company of America (PCA) produces and packages aspirin for sale to stores including Smith's Corner Store. Patsy Prepared buys a bottle of PCA aspirin at Smith's. Two weeks later Patsy's friend, Sarah Sickly takes two aspirin for a headache while she is at Patsy's house. Within minutes, Sarah is deathly ill and is rushed to the hospital. Sarah suffered extensive injuries from a toxic substance in the aspirin. Which of the following is correct?
   A. Sarah cannot sue PCA or Smith's because she did not buy the aspirin and is therefore not in privity with PCA.
   B. Sarah cannot sue PCA or Smith's because they are protected by the defense of unavoidably dangerous products.
   C. Sarah cannot sue PCA or Smith's because she assumed the risk of taking the aspirin.
   D. Sarah can sue PCA or Smith's even though she did not buy the aspirin and is not in privity.

2. Neighbor Nel sells her lawn mower to Homebody Homer because she is moving away. Homer is injured by a defective part in the lawn mower. Which of the following is correct?
   A. Nel is strictly liable for the defect in the mower since she was in the chain of distribution.
   B. Nel is strictly liable for failing to provide adequate instructions.
   C. Nel is strictly liable because she is not a merchant and this was a casual sale.
   D. Nel is not strictly liable because only manufacturers can be held liable.

3. Pedestrian Pat is injured while standing on the corner of an intersection when Biker Becky's new motorcycle speeds out of control because of a defective part. Which of the following is correct?
   A. In most jurisdictions, Pat can sue the motorcycle manufacturer even though he was only a bystander.
   B. In most jurisdictions, Pat cannot sue the motorcycle manufacturer since he was only a bystander.
   C. Biker Becky is strictly liable for the defect.
   D. None of the above.

4. Wilfred takes a medication for his heart condition. Seven different pharmaceutical companies produce the medication. Unfortunately, Wilfred experiences seriously damaging side effects about which he was not warned by any of the pharmaceutical companies. Wilfred wants to sue the medication manufacturer. However, the medication is fungible, and Wilfred cannot identify the pharmaceutical company responsible for the drug he took. Assume Wilfred is in a state that allows suits based on market share liability. Which of the following is correct?
   A. Each of the seven defendant pharmaceutical companies must prove they did not make the medication Wilfred took.
   B. Wilfred must sue each company based on the number of units of the medication they produce.
   C. Wilfred cannot sue unless he identifies the pharmaceutical company most likely to have made the medication he took.
   D. None of the above.

5. Which of the following is correct? To recover for strict liability, the injured party must show:
   A. That the product was defective.
   B. That the product was the cause of the injury.
   C. Who caused the product to become defective?
   D. All of the above must be shown to recover for strict liability.

6. Merchant Melinda, a merchant who regularly sells stereo equipment to the public, sells a Sonic stereo to Customer Cathy. Melinda states that this stereo will serve Cathy's needs because it is "guaranteed to last for ten years." However, Sonic only guarantees the stereo will last for five years. Which of the following is correct?
   A. If the stereo breaks down in three years, both the warranties of merchantability and fitness have been breached.
   B. If the stereo breaks down in six years, the warranty of fitness has been breached, but the warranty of merchantability has not.
   C. Both of the above.
   D. None of the above. The warranties of merchantability and fitness are only breached if the stereo breaks down before the ten years has expired.

7. Which of the following is not required to prove a breach of the implied warranty of merchantability?
   A. The product was not merchantable at the time of the sale.
   B. The product proximately caused the plaintiff's injury.
   C. The seller-merchant gave an express warranty of merchantability at the time of sale.
   D. All of the above are required to prove a breach.

8. Restaurateur Rudolf owns a large chain of restaurants. For ten years, Rudolf has purchased new restaurant dishwashers from Dealer Dan. Although it is not customary in his business, Dan has always assembled the internal parts of the dishwasher before delivering it to Rudolf. However, on his last sale to Rudolf, Dan did not assemble the internal parts and when Rudolf turned on the dishwasher, extensive damage was done. Which of the following is correct?
   A. Dan has breached the implied warranty arising from usage of trade.
   B. Dan has breached the implied warranty arising from a course of dealing.
   C. Both of the above.
   D. None of the above. Dan was not required to assemble the internal parts and therefore cannot be held liable for his failure to do so.

9. Which of the following is not necessary to show an implied warranty of fitness for a particular purpose?
   A. The seller was a merchant.
   1. The seller has reason to know the particular purpose for which the buyer is purchasing the goods.
   C. The seller states that the goods will serve the buyer's purpose.
   D. The buyer relies on the seller's skill and judgment and purchases the goods.

10. Merchant Mary sells a used car to Customer Carl. The contract states that Carl purchases the car "as is." Which of the following is correct?
   A. Such a disclaimer is unconscionable and will not be enforced.
   B. Such a disclaimer is only effective if Carl has examined the car.
   C. Such a disclaimer is an expression that makes clear that there are no implied warranties.
   D. Such a disclaimer is only effective if this is the usual course of dealing in the trade.

V. Answers to Objective Questions

Terms:

1. *Repose*. These statutes remain controversial because the harm done by certain products may be a long time in coming as illustrated by some cancer-causing toxins found in the food chain.

2. *Failure to warn.* Because certain products are inherently dangerous, manufacturers and sellers of such products have a duty to warn users about those dangers. Failure to warn about these inherent dangers can support a product liability action.
3. *Assumption of risk.* We all assume numerous risks in everyday activities, such as driving, that cannot always be avoided. This defense works best where the seller or lessor can show that the plaintiff's undertaking of the risk was unreasonable under all the circumstances involved in the particular case.
4. *Abnormal misuse.* Some commentators have criticized this defense for being too narrow. Note that it only protects the seller or lessor against abnormal misuse. One must guard against normal misuse. Critics would argue that any misuse is improper and should be allowed as a defense.
5. *Market share liability.* This theory is also sometimes called enterprise liability or horizontal liability. Because of the social engineering aspects of this theory, it remains controversial in public policy debates on products liability.
6. *Interference.* See Section 2A-211(1). This provision is similar to the warranty of no infringements and is implied by the UCC.
7. *Basis of the bargain.* Look for a fact pattern wherein the express warranty was a contributing factor (not necessary the only factor) in inducing the sale. These factors can include statements made before, during and/or even after the bargain was entered into (in cases modifying an existing contract).
8. *Foreign substance.* This test has been adopted by a few states and makes it easier to adjudicate warranty claims involving food products. Some consumer groups are opposed to this test because they argue that it gives too much protection to the seller in cases where the substance is not foreign yet still defective.
9. *Implied warranty arising from usage of trade.* This warranty is looks at what is "par for the course." It is designed to assure that persons entering into a trade or business are held to the prevailing standards imposed by that trade or business.
10. *Implied warranty arising from a course of dealing.* This warranty looks at the established practice between the parties as a base-line expectation level of performance, i.e., the course of dealing may have established a precedent in setting a standard for later dealings between the same parties.

True/False:

1. False. While noncompliance with federal or state safety standards for product design is *prima facie* evidence of a defect in design, compliance alone does not necessarily absolve a defendant from product liability.
2. True. The supervening event breaks the chain of liability tracing.
3. False. Manufacturers are liable for damages caused by foreseeable misuse of a product, but not unforeseeable misuse.
4. True. Note these damages are commonly used in product liability cases in order to send a signal to the manufacturer.
5. False. Under strict liability, the injured person does not have to prove the defendant breached the duty of care.
6. False. If a buyer furnishes specifications for certain goods, the seller cannot be held liable if compliance results in a third party claim.
7. True. Sections 2-312(1) and 2A-210(1) are not limited just to written affirmations.
8. True. This concept comes from the law of agency.
9. True. This rule is intended to protect the seller from being liable for a user's unique idiosyncrasies where the harm was caused solely by that idiosyncrasy.
10. True. These are all elements of this warranty as set out in Section 2-315 and 2A-213 of the UCC.

Multiple Choice:

1. D. Sarah does not have to be in privity with the manufacturer or retailer of the aspirin since strict liability is a tort doctrine. Under strict liability, sellers are liable to the ultimate user or consumer. A is incorrect because Sarah does not have to be in privity. B is incorrect because aspirin is not an unavoidably dangerous product. C is incorrect because taking aspirin is not generally considered an unreasonable risk.
2. C. Nonmerchants who are involved in casual sales cannot be held strictly liable. A is incorrect because this was a casual sale, and Nel is not a merchant regularly involved in the sale of lawn mowers. B is incorrect because this was a casual sale, and she was not responsible for giving instructions on the use of the lawn mower. D is incorrect because wholesalers and retailers in the line of distribution can also be held strictly liable.
3. A. Most jurisdictions provide protection for people who are injured even if they are only bystanders. B is incorrect for the reason stated above. C is incorrect because Becky is a consumer and not strictly liable for a defect which is the fault of the manufacturer.
4. A. In a suit based on market share liability, the plaintiff sues all manufacturers and they, in turn, must prove they did not produce the product that caused injury. B is incorrect because Wilfred must sue all seven manufacturers. Whether and how much each manufacturer is liable is dependent on whether they can prove they did not produce the medication. C is incorrect because under market share liability, Wilfred need not first identify the pharmaceutical company most likely to have produced the medication he took.
5. C. Under strict liability, the injured party does not have to prove who caused the product to become defective. However, that party must show that the product was defective and that the product was the cause of injury. A is required. B is required. D is incorrect because the injured party does not have to show who caused the defect.
6. B. It is unrealistic to expect cars not to crash, and when they do, it is scientifically provable that secondary crashes will take place inside the car. Thus, the crashworthiness doctrine forces the manufacturers to take that reality into account when they are designing a car.
7. C. The manufacturer and all other merchants in the chain of distribution can be held liable for defective packaging. A is incorrect because SSS, as a retailer, is also strictly liable for defective packaging. B is incorrect because there is no evidence of fraud here.
8. D. A, B, and C are all elements of the traditional doctrine of assumption of risk. In practice, the courts narrowly apply the defense.
9. D. Although this drug is unsafe in certain ways, it apparently serves a valuable social function. In addition, DMI adequately warned all of the side effects Paul experienced. Therefore, under the principle of unavoidably dangerous products, DMI is not liable. B is incorrect because DMI did warn of all the side effects. C is incorrect because such a drug would not be classified as a generally known danger, but an avoidably dangerous product.
10. B. Trimming hedges is an obvious misuse of the lawnmower. In addition, this is arguably an unforeseeable misuse. Therefore, HHI would not be liable. A is incorrect because there was no supervening event which caused a defect in the lawn mower. C is incorrect because lawnmowers are a frequently used product and not considered unavoidably dangerous. D is incorrect because under the circumstances, misuse of the product is a more applicable defense.

VI. Answers to Essay Question:

This case involves the sale of a product that is classified as a good under the UCC. Some services may be incidentally involved by way of repairs and the like, but the dominant purpose of the transaction is the sale of a good. The user of the product here was the buyer-owner, Mr. Clint Wayne, and the injured party was a user-bystander, Ms. Molly Oakley.

Of the possible theories listed under the traditional law of tort, abnormally dangerous activities is most likely to be raised and will most likely fail. As dangerous as guns are, their use is far from unusual in our society. There are over 200 million weapons in the U.S., most of them legally owned. The harm created by these weapons is well documented on the news every night. Thus the harm done by them is a "normal" element of every day life in the U.S., like it or not.

There is no question that a theory of strict liability can be applied based on defects in design, manufacture, instruction, inadequate selection of component parts, and more under Section 402A of the Restatement 2d of Torts. It can be reasonably assumed that any weapon should have built into it reasonable safety mechanisms to guard against unintended misfires from a mere incidental contact with the gun. Guns can and do bump against other objects or fall and should not go off like a loaded hand grenade when they do. To design a gun with a hammer resting against the firing pin creates a definite danger to both the user and the bystanders. To assume that the user will keep the gun only partially loaded is myopic at best and callous to the reality of how guns are actually used. Giving warnings or instructions to only partially load is akin to telling drivers not to go over an unreasonably slow speed limit because the car my not be able to stop.

The defendants may argue that the gun falling off the seat was an intervening event and that there is an assumption of risk in using a generally known dangerous product like a gun. In addition, our current gun laws exempt this product from the Consumer Product Safety Commission's power to recall. These defenses should not be allowed to bar recovery. Other gun manufacturers made comparable products with adequate safety mechanisms at the same time as this one, so the state of the art defense should not be allowed. As for risk assumption, the risks associated here are with gun safety, not hand grenade-like attributes of a gun. There was no apparent willful misuse of the product by Mr. Wayne. In related cases, over 40 people have been killed by this product including baggage handlers who were handling bags containing these weapons. As for the statute of limitations, guns have an unusually long life span of use because they do not wear out. Witness the large trade in antique and collectable firearms in this country.

There has been some recent activity on the part of Congress to stem some of the harm done by all these weapons. Consider the Brady Bill and the ban on certain assault weapons. But as a practical matter, annual gun deaths of all sorts will soon outnumber the death toll of auto-related deaths in this country and for those people (as well as Ms. Oakley); it is all too little too late. Because this product is not fungible but unique to the Big Bang Company, only those persons and entities in the chain of distribution should be held liable. As for damages, actual and punitive damages could be allowed.

Chapter 19
CREATION AND TRANSFER OF NEGOTIABLE INSTRUMENTS

To an imagination of any scope, the most far-reaching form of power is not money, it is the command of ideals.
Oliver Wendell Holmes

I. Overview

The law and language of commercial paper contains some of the most befuddling terminology within a profession already beset with too much prattle. Many students find it to be the nadir of their legal studies, and generally place it on a par with a root canal or some sort of highly invasive medical procedure. A colleague once described the practice of law as an attempt by the attorney to lead his or her client through a minefield without either of them being blown up in the process. The law of negotiable instruments is one such minefield. It is tricky, dangerous, and full of unpleasant surprises. Yet the importance of negotiable instruments as a tool of commerce cannot be denied, and the minefield must be crossed.

Because this area of the law is so complex, the study of commercial paper is best begun with an introduction to key terms, the players who use those terms, and the mechanics of the negotiation process. Remember, that a special set of formalized contracts are being used as a substitute for money, extension of credit, and record keeping. Like so many areas of specialized human endeavor, this techno-babble may sound strange and unreasonably complex at first blush. But as you develop an appreciation for the actual scope of commercial transactions carried on with negotiable instruments, it is hoped that you will see that there is a rhyme and reason to it.

One of the main objectives of this chapter centers on how to identify the key elements of a negotiable instrument. There are six components or elements in all UCC negotiable instruments. The compliance formalities are strict. If these elements are found to be in place, a substitute medium for money has been created. This substitute is not just a matter of convenience, but rather a virtual necessity which has evolved as the next stage past money. Currency was first devised as method of providing an alternative method of exchange to barter. Even though barter is still used by many in the underground economy, the vast majority of commercial dealings provide for sale of goods or services in exchange for legal tender or a money substitute.

Out of this process, one can readily see that use of cash, as a form of payment is not always practical or wise. An entire financial services industry, the traveler's check, is premised on the risks inherent in the use of large amounts of cash for travel. Those same concerns apply to the commercial marketplace.

Money simply does not work as well as negotiable instruments for purposes of convenience, safety, record keeping, and extension of credit. Negotiable instruments are important money substitutes, and as such, it is critical that the elements of negotiability be identified and in place before these documents are passed to third parties.

Another problematic misconception held by first-time students of negotiable instruments is the difference between negotiability and legality. Many students proceed to duly memorize the elements of negotiability and learn how to identify them in an instrument. Having done that, they proceed to declare an instrument that lacks a missing critical element of negotiability as legally null and void. Nonnegotiability does not *ipso facto* preclude legality. Even though an instrument may fail to meet the requirements proscribed by the UCC for negotiability, that fact does not render it illegal. Nor is it necessarily barred from being enforced as a contract.

The UCC's technical requirements for the creation of a negotiable instrument are quite detailed. Some cynics have argued that these technicalities were written in order to assure continued employment for lawyers and bankers. Their real underlying motivation is to prepare the document for transfer to, and acceptance by, third parties as a substitute for money. The document is literally put in circulation, and that path of circulation may sometimes circumvent the globe before it comes home to its place of origin. Throughout this entire journey it will be passed on from one party to another under the same basic ground rules--the UCC rules of transfer and negotiability. The key objectives of this chapter center around outlining those ground rules as set out in Article 3 of the UCC.

The two basic methods of third party acquisition of instruments are by assignment and negotiation. The common law of contracts generally covers assignment issues. A negotiable instrument can be assigned as well as negotiated but not vice-versa. An instrument that does not meet the formalities set out in the prior chapter cannot be negotiated. Thus you must first examine the document itself for the elements of a negotiable instrument. If the instrument fails on that score, look to the law of assignment to resolve the issue. If the elements of negotiability are in place, then look at the last element listed: "payable to order or bearer." These are the two methods of negotiable transfer allowed by Article 3. These words provide your first road signs to which direction this instrument may take on its travels through the hands of third parties. Remember it is only a clue to the first step. Subsequent steps in the negotiation process may convert the instrument from one method of transfer to the other. Think of them as the two rails of a railroad track. They are connected and run parallel to each other yet never really meet. The instrument must ride on one or the other, but not both at the same time, in order to stay on its path of negotiation.

The first path of transfer is by physical possession of the instrument if the document is designated as bearer paper. Bearer paper is negotiated by delivery of the document. Think of it as analogous to money. You do not have to sign it, indorse it, or in any way change the "paper" itself to change ownership. It only takes a legal physical transfer. Because this method is so open-ended, substantial risks inure with its use. Witness the use of travelers' checks as testimony for the dangers of using bearer-types of exchange. The same dangers apply to bearer negotiable instruments.

The second method allowed by Article 3 involves order paper. This document has been made payable to a specific payee or indorsed to a specific indorsee. As such, further negotiable transfer to a third party calls for not only delivery as required for bearer paper, but also indorsements. All indorsements have three sub issues and should be examined in light of these three subcomponents: name, qualification, and restriction.

II. Hypothetical Multi-issue Essay Question

Farmer Jake Gothic is fed up. After severe drops in farm commodity prices combined with bad weather, he was almost broke. His only consolation was that he qualified for the simplified income tax form, the 1040SI. The form has only two lines: (1) How much did you make? (2) Send it in!

Unfortunately Jake's problems were not over. After he claimed a medical deduction for his hot tub and a home office use for his privy, the IRS audited him. His revised tax bill was $6,000, due and payable by high noon on April 16th. Failure to do so would have him suffer the same fate that befell Al the Taxdodger Capone.

Jake is so angry at the results of the audit that he decided to pay his tax bill in a unique way. He shaved the side of his cow, Bessie, and wrote the following on her in permanent ink:

	April 16, 2000
To:	Bailout Bank Farmland, U.S.A. Pay to the order of: The IRS $6,000
Signed:	Angry Farmer Jake Gothic
For:	To the Infernal Revenue Sewer for taxation without representation!

Jake walked Bessie into the IRS office where the cow proceeded to leave a natural deposit. What results?

III. Outline

What is a Negotiable Instrument?
- A negotiable instrument is a special form of contract that satisfies the requirements established by Article 3 of the UCC
- Also called commercial paper

Negotiable Instruments
- Functions
  - Substitute for money
  - Credit device
  - Record keeping device
- Types
  - Drafts
  - Checks
  - Promissory notes
  - Certificates of deposit

Creating a Negotiable Instrument
- A negotiable instrument must:
  - Be in writing
  - Be signed by the maker or drawer
  - Be an unconditional promise or order to pay
  - State a fixed amount of money
  - Not require any undertaking in addition to the payment of money
  - Be payable on demand or at a definite time
  - Be payable to order or to bearer

Nonnegotiable Contracts
- Nonnegotiable contracts fail to meet the requirements of a negotiable instrument, and therefore, are not subject to the provisions of UCC Article 3

Transfer of Negotiable Instruments
- Negotiable instruments can be transferred by:
  - Assignment
  - Negotiation
    - Negotiating order paper
    - Negotiating bearer paper
    - Converting order and bearer paper

Assignment
- Assignment
  - The transfer of rights under a contract
- Assignor
  - The transferor in an assignment situation
- Assignee
  - The transferee in an assignment situation

Negotiation of Order Paper

Negotiation of Bearer Paper

Types of Indorsements
- Blank—Does not specify a particular indorsee; creates bearer paper
- Special—Specifies the person to whom the indorser intends the instrument to be payable; creates order paper
- Unqualified—An indorsement that does not disclaim or limit liability
- Qualified—An indorsement that disclaims or limits the liability of the indorsee
  - Special qualified indorsement
  - Blank qualified indorsement
- Nonrestrictive—No instructions or conditions attached to payment of funds
- Restrictive—Conditions or instructions restrict the indorsees rights
  - Conditional inducement
  - Inducement prohibiting further inducement
  - Inducement for deposit or collection
  - Inducement in trust

Other Issues Concerning Indorsements
- Misspelled or wrong name
  - The payee or indorsee can indorse the instrument in the misspelled name, the correct name, or both
- Forged indorsements
  - With few exceptions, unauthorized indorsements are wholly inoperative
  - Imposter rule
- If an imposter forges the indorsement of the named payee, the drawer or maker is liable on the instrument and bears the loss
  - Fictitious payee rule
- The drawer of maker is liable on a forged or unauthorized indorsement of a fictitious payee

IV. Objective Questions

Terms:

1. It is used as a substitute for cash, a credit device, and a record-keeping device used to conduct various business and personal functions. It is called a _______________ _______________.

2. Drafts and checks are considered _______________ _______________ _______________ because the drawer orders the drawee to pay the payee.

3. Commercial paper payable on demand, on sight, on presentation, or at no specifically stated time is called a _______________ instrument.

4. The requirement of negotiable instruments that says they must be able to be easily transported between areas is called the _______________ _______________.

5. A clause in a negotiable instrument that gives the maker the right to pay the amount due prior to the maturity date of the note is called a _______________ _______________.

6. An indorsement that disclaims or limits liability on the instrument is an _______________ _______________.

7. An indorsement that contains some sort of instruction from the indorser is a _______________ _______________.

8. An instrument is _______________ _______________ when it is payable to two or more people with "and" between their names and requires both of their indorsements to negotiate the instrument.

9. An instrument is _______________ _______________ _______________ _______________ when it is payable to two or more people with "or" between their names and requires only one of the named persons to indorse the instrument.

10. A rule that says if an impostor forges the indorsement of the named payee, the drawee or maker is liable on the instrument and bears the loss is called the _______________ _______________.

True/False:

1. ____ A holder in due course takes an instrument subject to any defenses that could be asserted against the payee.

2. ____ Once a drawee accepts a draft and the draft is returned to the drawer, the drawer can no longer transfer the draft as a negotiable instrument.

3. ____ A promissory note is an order to pay and can be freely transferred.

4. ____ To satisfy the writing requirement of negotiability, the writing must be on a preprinted form. Typewritten or handwritten writings are not acceptable.

5. ____ To satisfy the permanency requirement, a writing must be on paper. No other objects are acceptable.

6. ____ A restrictive indorsement, though it purports to prohibit further negotiation, is ineffective to destroy the negotiability of the instrument.

7. ____ If there is a proper indorsement in trust, the indorser is personally liable to the clients or heirs of the fiduciary to whom it was indorsed.

8. ____ If an indorsee does not comply with the instructions of a restrictive indorsement, he is liable to the indorser for losses that result from such noncompliance.

9. ____ If the name of the payee is misspelled on an instrument, the payee cannot indorse the instrument, and it is therefore nonnegotiable.

10. ____ When an instrument contains additional language, words of negotiation take precedence over words of assignment.

Multiple Choice:

1. Generally speaking, the laws of commercial paper and negotiable instruments in effect in the U.S. today are found in:
   A. The Uniform Negotiable Instruments Law.
   B. Bills of Exchange and the Law Merchant.
   C. Article 3 of the Uniform Commercial Code.
   D. Article 10 of the Uniform Commercial Code.

2. On July 1, 2001, Hector agrees to sell Rhonda a cast iron replica of a carnival riding horse for $1000. Hector delivers the horse. Rhonda gives Hector $500 in cash and a written instrument, signed by Rhonda, in which she promises to pay the balance by the end of the Month. The instrument property identifies the riding horse. Under the UCC, Rhonda has given Hector a:
   A. Promissory note.
   B. Negotiable draft.
   C. Trade acceptance.
   D. Negotiable time draft.

3. Shady Sharon received a check from Trusting Tim for goods Tim purchased. Sharon knew the goods were defective and unusable. However, by the time Tim discovered this fact, Sharon had negotiated Tim's check to Builder Bob for a new patio Bob was building for Sharon. Which of the following is correct?
   A. Bob is not a holder in due course because Tim has a claim against Sharon.
   B. Bob is not a holder in due course because he is not the payee.
   C. Bob is a holder in due course.
   D. Sharon is a holder in due course.

4. Which of the following is incorrect?
   A. A negotiable instrument may contain an acceleration clause.
   B. A negotiable instrument may allow the maker to satisfy a note by performance of services or payment of the money.
   C. A negotiable instrument may have an extension clause.
   D. A negotiable may be lacking a specified due date.

5. Budgeting Bill writes a check on June 1 to Craftsman Cameron for goods Bill bought from Cameron. However, Bill postdates the check for June 10. Which of the following is correct?
   A. The check is payable on June 1.
   B. The check is payable on or after June 10.
   C. The check is payable as soon as Cameron has possession of it.
   D. The check is nonnegotiable since it is postdated.

6. Amy Allen writes a check to Cathy Carter. Ben indorses the check as follows:
   Pay Dan Davis
   (signed) Ben Baker
   Ben then delivers the check to Dan. Which of the following is correct?
   A. When Ben signed the check, it was converted to bearer paper.
   B. Ben's indorsement of the check was a special indorsement.
   C. Amy does not have to pay Dan since she only negotiated the check to Ben.
   D. All of the above are correct.

7. Assume the same facts as in question 6. Which of the following is correct?
   A. Ben is liable to Dan if Amy defaults on payment of the check.
   B. Ben's indorsement was an unqualified indorsement.
   C. Amy is liable for nonpayment even after it is negotiated from Ben to Dan.
   D. All of the above are correct.

8. Image Insurance Co. makes a check payable to Attorney Ann. The check is for a claim settlement with Ann's client, Client Courtney. Ann indorses the check as follows:
   (signed) Attorney Ann
   without recourse
   Ann then delivers the check to Courtney. Which of the following is correct?
   A. This indorsement guarantees payment of the check.
   B. This is an unqualified indorsement.
   C. This is a blank qualified indorsement.
   D. This is a special qualified indorsement.

9. Assume the same facts as in question 8. Also assume that after Courtney receives the check, she indorses it as follows:
   Pay to Furniture Fred
   (signed) Client Courtney.
   Courtney then delivers the check to Fred in exchange for new furniture. Which of the following is correct?
   A. Courtney is not liable if the insurance company does not pay the check.
   B. Fred has no recourse if the insurance company does not pay the check.
   C. Ann is liable if the insurance company does not pay the check.
   D. Courtney is liable if the insurance company does not pay the check.

10. Which of the following is not a type of restrictive indorsement?
    A. A qualified indorsement.
    B. A conditional indorsement.
    C. An indorsement in trust.
    D. An indorsement for deposit.

V. Answers to Objective Questions

Terms:

1. *Negotiable instrument.* This term was adopted in the revision of Article 3 of the UCC. The term formerly used as a generic description of negotiable instruments was commercial paper.
2. *Orders to pay.* These are distinguished from promissory notes and certificates of deposit that are classified under the UCC as promises to pay. Remember, because the bank is holding your money in your checking account, you have the power to decide how that money is to be disbursed.
3. *Demand.* These instruments are payable at any time after they are issued. This term can be used as part of a draft, check, or promissory note. It is not usually found on a certificate of deposit because of an agreed upon period of time which must expire before payment is made with interest.
4. *Portability requirement.* This rule is designed to ensure free transfer of the instrument. Even though unusual methods may be used as illustrated by the essay question, as a practical matter the best practice is to use traditional paper so that the instrument will be readily accepted.
5. *Prepayment clause.* The use of prepayment, acceleration, or extension clauses in a negotiable instrument does not affect its negotiability. Such clauses are commonly found in promissory notes.
6. *Qualified indorsement.* Look for the term "without recourse" to identify this kind of indorsement. It does not stop further negotiation.
7. *Restrictive indorsement.* Four types of restrictive indorsements are allowed under Sec. 3-205 of the UCC.
8. *Payable jointly.* This is allowed under Sec. 3-116 of the UCC.
9. *Payable in the alternative.* As compared to "payable jointly," under this sort of instrument, either party can indorse and negotiate the instrument.
10. *Impostor rule.* The reason behind this rule is to make sure that the drawer and maker know their indorsee when he or she further negotiates the instrument.

True/False:

1. False. A holder in due course who takes an instrument has many defenses that could be asserted against the payee.
2. False. The drawer can freely transfer the draft as a negotiable instrument.
3. False. A promissory note, while it can be freely transferred, is a promise to pay, not an order to pay.
4. False. A writing may be in preprinted, typewritten, or handwritten form and will thus satisfy the writing requirement for negotiability.
5. False. A writing must be on some type of permanent object. However, it does not have to be on paper, and several other objects have been found to satisfy the requirement.
6. True. Under Sec 3-206(1), it is treated as a special indorsement.
7. False. A proper indorsement in trust releases the indorser from personal liability to clients or heirs of the fiduciary.
8. True. Restrictive indorsements are effective against all holders except intermediary banks in the collection process.
9. False. If the name of the payee is misspelled on an instrument, the payee can indorse the instrument with the misspelled name, her correct name, or both. The negotiability of the instrument is not destroyed.
10. True. See Sec. 3-204(4) of the UCC. Extra words with an indorsement do not affect the indorsement.

Multiple Choice:

1. C. Article 3 of the UCC provides the main body of law used in the U.S. today with regard to commercial paper and negotiable instruments. A and B were precursors to the UCC, and D is incorrect because the UCC has only 9 main articles
2. A. A promissory note is a two party instrument in which the maker promises to pay the payee a specified sum of money. Rhonda is the maker, and Hector is the payee. A promissory note may make reference to the underlying transaction, such as the sale of the horse, without destroying negotiability. Answers B and D are incorrect because a draft, whether negotiable or not, is a three party instrument. A draft is an instrument in which one party orders a second party to pay a third party a sum of money. C is incorrect because a trade acceptance is a type of draft that contains an order by a seller of goods directing the buyer to pay the seller of the goods.
3. C. Bob is a holder in due course since he negotiated the instrument without notice that Sharon had fraudulently induced Tim to pay for the goods. A is incorrect because Bob is a holder in due course even though Tim has a claim against Sharon. B is incorrect because Bob can be a holder in due course without being a payee. D is incorrect because Sharon no longer holds the instrument.
4. B. An instrument that may be satisfied by the performance of services does not meet the sum certain requirement and is therefore nonnegotiable. A, C, and D are all permissible clauses in negotiable instruments.
5. B. A postdated check is payable on or after the date on the check. A and C are incorrect because the check is not payable before June 10. D is incorrect because a postdate does not render a check nonnegotiable.
6. B. Since Ben indorsed the check payable to Dan Davis, the check had a special indorsement, not a blank indorsement. A is incorrect because a special indorsement makes the check order paper, not bearer paper. C is incorrect because Amy must pay the holder of the check. Ben was free to negotiate the paper to Dan or anyone else he chose.
7. D. A is correct because indorsers are liable to indorsees. The fact that Amy wrote the check originally does not release Ben from liability. In fact, Ben's indorsement of the check is a guarantee of payment. B is correct because Ben's indorsement was unqualified since he did not sign it "without recourse." C is correct because Amy is ultimately liable for payment on the check. Even if Amy defaults on payment of the check and Dan receives payment from Ben, Ben can then recover from Amy.
8. C. This indorsement is blank because Ann did not make the check payable to a specific indorsee. A is incorrect because Ann indorsed the check "without recourse." Therefore, Ann does not guarantee payment of the check. B is incorrect because an indorsement without recourse is a qualified indorsement. D is incorrect because a special indorsement names a specific indorsee.
9. D. Courtney is liable if the insurance company does not pay the check since her indorsement was unqualified. The fact that Ann's indorsement to Courtney was qualified does not protect Courtney from liability on the check. To be protected from liability, Courtney would have had to make her own qualified indorsement on the check. A is incorrect because Courtney is liable for the reasons discussed above. B is incorrect because Fred has recourse against both the insurance company and Courtney. C is incorrect because Ann's qualified indorsement protects her from liability if the insurance company does not pay the check.
10. A. A qualified indorsement protects the indorser from liability but is not a type of restrictive indorsement. B, C and D are all types of restrictive indorsements.

VI. Answers to Essay Question:

Farmer Jake is involved with processing of a payment for an amount owed to the IRS arising out of the audit. In spite of the entire body of federal law generated by the Internal Revenue Code and its various regulations, the main focus of this question will center on the UCC rules as they apply to this instrument used by Farmer Jake to pay his tax deficiency.

This instrument is purported to be a check as defined by the UCC. The drawer is Farmer Jake. The drawee is Bailout Bank of Farmland, U.S.A., and the payee is the IRS.

In order to qualify as a negotiable instrument under the UCC, the document must incorporate the six essential ingredients. These six elements are a writing, signature, unconditional promise or order to pay in a certain sum of money on demand or at a definite time. The document must incorporate the magic words to order or bearer as required by the UCC.

First, the writing used here is handwriting on the side of the cow. Jake used permanent ink and the instrument is portable. The IRS can walk Bessie down the street to the Federal Reserve Bank or to Jake's bank to process the draft. Second, Jake may use any form of informal name or nickname he chooses as long as his identity is clearly identified. Third, he is also allowed to add informational items on the check if he chooses. Usually these negotiations are found in the lower left-hand corner and have no effect on negotiability. The order to pay written on the side of Bessie is not conditioned in any way. Fourth, the amount stated is a sum certain, six thousand U.S. dollars, and is not commingled with any other sort of tender of payment. Fifth, because there is not language written on Bessie that attempts to set any time frame for payment, the instrument can be considered to be a demand instrument. Finally, this instrument has the one of the required forms of magic words: "Pay to the Order of." Just because Jake has no love lost for the IRS and adds colorful descriptors, those items will not render the instrument nonnegotiable. In any event, the IRS is not coy about how it is paid as long as it is paid. It regularly cashes shirts written in blood and the like. Jake should just make sure he gets a better tax adviser next year so he won't have to give the IRS his bull, too.

Chapter 20
HOLDER IN DUE COURSE AND LIABILITY

The ways by which you may get money almost without exception lead downward.
Henry David Thoreau

I. Overview

This chapter has two main objectives: first, to introduce the student to the unique UCC concept of holder in due course (HDC) and second, to illustrate the basic liabilities (and defenses thereto) that arise out of the use of negotiable instruments. The focus of these two objectives, therefore, now turns to the third party who becomes involved with a negotiable instrument. How, when, and where will the UCC protect that third party and when will it not? Because any discussion of negotiable instruments is interwoven with their role as a substitute for money, a comparison with money may be useful. If someone hands you a dollar that is clearly a photocopy of currency, should you accept it in lieu of currency? Or should you be put on notice that there is a problem and act accordingly? The doctrine of holder in due course is designed to protect good faith third party takers of money substitutes. It is designed to create justifiable reliance on instruments where the elements of HDC are in place. It is not designed to protect those who think play money is real and do not bother to learn the rules. Like so many protections afforded by the law, it is really only a matter of degree, not absolute immunity from one's own ignorance.

The degrees of protection afforded to third party takers of instruments begin with a comparison between the common law of contracts and the UCC. Under the common law, an assignee steps into the shoes of the assignor. Generally, he or she undertakes the assignor's obligations and benefits from his rights. The problem with this "what you see is what you get" philosophy is that it may not provide enough protection for third parties who may not be aware of the history behind the document. As a buyer of commercial paper, you want to buy only what you see and not get ambushed by potential lawsuits lurking in the bushes. The law of assignment simply does not cut off enough history and often allows third party takers of documents to be stuck with legal entanglements that were not of their making.

The HDC concept allows for a fundamental shift in the law in that, where its elements are in place, the taker will be free of all claims and most of the defenses which could otherwise be asserted by other parties having contact with the instrument. These elements are set out in the key question checklist. Remember, as radical as this shift is from the common law of contract assignment, the metamorphosis is not complete. The HDC still takes the instrument subject to a number of defenses called real defenses. In addition, past abuses of HDC principles have led to some statutory overlay on the part of Congress through its delegation of regulatory authority to the Federal Trade Commission (FTC). In the area of consumer credit transactions, the FTC has stepped in by preempting the HDC rule. The law of contract under this rule subjects the HDC of consumer credit paper to all the defenses and claims of the consumer. Various states have also promulgated similar rules as outlined in the Uniform Consumer Credit Code or by case law.

The second objective of this chapter is to review the liability, defense, and discharge aspects of negotiable instruments. No contract is forever. It may last a long time, but it will end. Negotiable instruments are a highly formalized and technically complex set of contracts used primarily as a money substitute. Paper money eventually wears out and needs to be taken out of circulation. So too must a negotiable instrument ultimately run its course. That course may be just a matter of days in the case of a check, or many years as seen in long term notes.

There are two main theories for attaching liability in negotiable instruments. The first is found in the rights and duties arising out of the signatures of parties to the instrument. These rights and duties are essentially derived from contract law as modified by the UCC. The second source of liability is found in a set of warranties imposed by the UCC. Warranties are promises imposed by the statute upon those who want to be players in this game. These promises are not always actually expressed by the party against whom the warranties are imposed, but the responsibility to live up to them is still there. Think of them as duties that arise out of using the road of negotiation: if you go down this road, you must follow the rules of the road.

In addition, remember that by their very nature, negotiable instruments are handled by many people. The adjudication of the rights and duties of numerous takers of these instruments vis-à-vis each other is of critical importance through the entire spectrum of UCC studies. How those respective rights and duties finally fall into place at the point of discharge of a negotiable instrument is the second major objective of this chapter. In the end, the law must provide some form of resolution. A judge cannot sit on the bench and just mutter: "Comme ci, comme sa?" The UCC rules on discharge help make the judge's job a little easier.

II. Hypothetical Multi-Issue Essay Question

Tippee Tonic was a store manager for Acme Pig-Pens, the world's largest supplier of sanitary hog housing. Their bilevel Calistoga model, equipped with a state of the art mud bath, is a runaway best seller. As the manager of Acme's anchor store, Tippee was authorized by his employer to indorse and deposit all checks payable to Acme into the corporation's accounts at Hogsbreath Bank.

Tippee went hog crazy over Blossom Foxfire, who never seemed to have enough money. Blossom wanted to be swept away on a romantic interlude and said she didn't care if they went to "Myorca or Youroca as long as we get out of Hogsbreath." Tippee couldn't resist the offer. He indorsed several checks made out to Acme's order and deposited them in his own account at Hogsbreath Bank. The checks totaled $20,000, and now Tippee and Blossom are nowhere to be found.

Acme demands that Hogsbreath Bank credit its account and the bank defends by stating that Tippee had the power to indorse. What result?

III. Outline

- Holder in Due Course
  - Holder in due course
    - Takes a negotiable instrument
    - For value
    - In good faith
    - Without notice of defect
    - The instrument bears no apparent evidence of forgery, alterations, or irregularity

- Holder in Due Course
  - Signature Liability of Parties
    - A person cannot be held contractually liable on a negotiable instrument unless his or her signature appears on the instrument
  - This is also called contract liability
  - A signature is any name, word, or mark used in lieu of a written signature

- Primary v. Secondary Liability
  - Primary liability
    - Absolute liability to pay a negotiable instrument subject to certain real defenses
  - Secondary liability
    - Liability on a negotiable instrument that is imposed on a party only when the party primarily liable on the instrument defaults and fails to pay the instrument when due

- Liability of Accommodation Makers and Accommodation Indorsers Compared

- Warranty Liability of Parties
  - The law implies certain warranties on transferors of a negotiable instrument
    - Transfer warranties
    - Presentment warranties

- Real and Personal Defenses
  - Real defenses can be raised against a holder in due course
    - Minority
    - Extreme duress
    - Mental incapacity
    - Illegality
    - Discharge in bankruptcy
    - Fraud in the inception
    - Forgery
    - Material alteration
    - Notice of discharge
  - Personal defenses cannot be raised against a holder in due course
    - Breach of contract
    - Fraud in the inducement
    - Ordinary duress

- Discharge
  - Actions or events that relieve certain parties from liability on negotiable instruments
    - Payment of the instrument
    - Cancellation

Impairment of the right of recourse

IV. Objective Questions

Terms:

1. When this occurs in two situations: (1) value is given if the holder pays money for a negotiable instrument; and (2) value is given if the holder performs an agreed upon act in return for a negotiable instrument, it is called ________ ________ ________ ________ ________.

2. A person who has purchased a limited interest in a negotiable instrument who is a holder in due course only to the extent of the interest purchased is called a ________ ________ ________ ________ ________.

3. The doctrine that says a holder cannot qualify as an HDC if she has notice that there is any adverse claim against or defense to its payment is called the ________ ________ doctrine.

4. The rule that says a holder who does not qualify as a holder in due course in his own right becomes a holder in due course if he acquires the instrument through a holder in due course is called the ________ ________.

5. A defense that can be raised against both holders and holders in due course is called a ________ ________.

6. A defense that can be raised against enforcement of a negotiable instrument by an ordinary holder is called a ________ ________.

7. A type of liability that can apply to both signers and nonsigners who are transferors is called ________ ________.

8. The type of liability that (1) makers of promissory notes and certificates of deposit and (2) acceptors of drafts and checks have is called ________ ________.

9. A person who acts an accommodation party who signs on behalf of a maker is called an ________ ________.

10. To be ________ is to be relieved from liability on negotiable instruments; this can be accomplished by (1) payment, cancellation, or reacquisition of an instrument, (2) impairment or recourse of collateral, or (3) alteration of the instrument.

True/False:

1. ____ A holder in due course can never acquire better rights then the transferor had.

2. ____ A holder does not necessarily have to own the instrument to transfer it, negotiate it, enforce payment of it in his own name, or discharge it.

3. _____ The UCC concept of value and the contract concept of consideration are the same and can be used interchangeably.

4. _____ The defense of fraud in the inducement is a real defense and can be raised against a holder in due course.

5. _____ Implied warranties are made when a negotiable instrument is originally issued.

6. _____ Transfer warranties are not made when an instrument is transferred for no value.

7. _____ Notice of presentment and notice of dishonor are effective when sent, not when received.

8. _____ Both an authorized agent and her principal are liable if the agent signs a negotiable instrument indicating that she is an agent, but failing to specifically identify her principal.

9. _____ When a party other than the indorser pays a negotiable instrument, that party is discharged from liability but none of the subsequent parties are discharged.

10. _____ The Federal Consumer Product Safety Commission has eliminated the HDC status for the sale of certain consumer goods.

Multiple Choice:

1. To the extent that a holder of a negotiable promissory note is a holder in due course, the holder takes the note free from which of the following defenses?
   A. Nonperformance of a condition precedent.
   B. Discharge of the maker in bankruptcy.
   C. Minority of the maker where it is a defense to enforcement of a contract.
   D. Forgery of the maker's signature.

2. Wealthy William gives Busy Betsy a check for $100,000 as payment for Betsy's business. Betsy indorses the check and negotiates it to Contractor Cathy who has just built a house for Betsy. Which of the following is correct?
   A. Cathy is a holder in due course because building the house satisfies the contract concept of consideration, and that is all that is needed under the UCC.
   B. Cathy is a holder in due course because she performed the agreed upon consideration which satisfies the UCC concept of value.
   C. Cathy is not a holder in due course because she has not satisfied the UCC concept of value.
   D. None of the above.

3. Borrowing Bob borrows $500 from Loaner Linda. Bob signs a promissory note promising to repay Linda the $500 plus interest at the end of the year. Linda sells the note to Holder Henry for $300 in cash and $200 due in 30 days. Henry is:
   A. A limited holder in due course to the amount of $300.
   B. A limited holder in due course to the amount of $200.
   C. A holder in due course to the amount of $500.
   D. None of the above.

4. Assume the same facts as in question 3 except Linda sells the note to Henry at a discount. Henry pays only $250 for the $500 note. Henry is:
   A. A limited holder in due course to the amount of $250.
   B. A holder in due course to the amount of $250.
   C. A holder in due course to the amount of $500.
   D. Not a holder in due course or a limited holder in due course since he did not pay the full $500.

5. Naive Nancy acquires an instrument from Dealing Diedre. The instrument has a face value of $1,000. However, Nancy, who has never dealt with such negotiable instruments before, is ecstatic because she believes she got a great deal since she paid only $250. Which of the following is correct?
   A. Nancy did not take the instrument in good faith since under an objective test for honesty no reasonable person could have believed this was a legitimate negotiation.
   A. Nancy may be a holder in due course if, under the subjective test, she believed the negotiation was legitimate.
   B. Nancy did not take the instrument in good faith unless Diedre actually believed the negotiation was legitimate.
   D. Nancy did not take the instrument in good faith since she purchased it at a deep discount.

6. Trusting Tina writes a check to Sneaky Sam for $50. Sam alters the check by adding a zero and rewriting the amount to read $500. Sam then negotiates the check to Creditor Carol as payment for a debt he owed. Which of the following is correct?
   A. Tina must pay Carol $500 since Carol took the check free of any defenses.
   B. Tina must pay Carol $500 since material alteration is a personal defense that Tina cannot assert.
   C. Tina does not have to pay $500 since material alteration is a real defense that she may assert.
   D. None of the above.

7. Blare bought a house and provided the required funds in the form of a certified check from a bank. Which of the following statements correctly describes the legal liability of Blare and the bank?
   A. The bank has accepted; therefore, Blare is without liability.
   B. The bank has not accepted; therefore, Blare has primary liability.
   C. The bank has accepted, but Blare has secondary liability.
   D. The bank has not accepted, but Blare has secondary liability.

8. Edward writes a check to Fred for $500. Fred indorses the check "without recourse" and negotiates the check to Grace. Which of the following is correct?
   A. Fred has disclaimed the transfer warranty that no defenses of any party are good against the transferor.
   B. Fred has disclaimed the transfer warranty that all signatures are genuine or authorized.
   C. Fred has disclaimed any presentment warranties.
   D. None of the above. Fred cannot disclaim any implied warranties.

9. Amy delivers bearer paper to Brian. Brian then negotiates the check to Carmen. Which of the following is correct?
   A. Amy is liable to both Brian and Carmen for transfer warranties.
   B. Carmen can recover on transfer warranties from both Amy and Brian.
   C. Carmen can recover on transfer warranties from only Brian.
   D. There are no transfer warranties on bearer paper.

10. Unstable Edwin writes a check to Middle Mildred for $300. Mildred endorses the check and negotiates it to Suspicious Susan. Susan believes Edwin's checking account is overdrawn and the check may bounce. Which of the following is correct?
    A. Susan can hold Mildred liable since she is afraid Edwin's checking account is overdrawn.
    B. Susan can hold Edwin liable since she is afraid Edwin's checking account is overdrawn.
    C. Susan cannot hold anyone liable until the check is actually dishonored.
    D. None of the above.

V. Answers to Objective Questions

Terms:

1. *Performance of agreed upon consideration.* The second basic building block in the construction of HDC status is to show that the holder has given value for the negotiable instrument. Under Section 3-303 of the UCC, both scenarios listed in the question qualify as giving value for purposes of this section.
2. *Limited holder in due course.* HDC status will be extended to this person only to the extent of his or her interest purchased in the instrument.
3. *Red light.* This doctrine says that a holder cannot qualify as an HDC if he or she has notice of unauthorized signatures or alternations to the instrument, or of any adverse claim against it, or of any defenses to its payment.
4. *Shelter principle.* This rule applies to anyone in the chain of title who can trace his or her title back to a holder in due course. In order to prevent laundering of negotiable instruments, this rule cannot be used for the benefit of persons who were parties to fraud or illegalities affecting the instrument. It also applies to prior holders who had notice of a defense or claim against the instrument.
5. *Real defense.* These defenses are considered too important as a matter of public policy to be cut off in favor of HDCs. They include minority, mental incompetence, extreme duress, fraud in the inception, bankruptcy, forgeries, and material alterations (only to the extent of the alteration).
6. *Personal defense.* These defenses ordinarily arise out of the initial contract relationship that gave rise to the instrument, such as breach of contract or fraud in the inducement. Because of possible abuses of having these defenses cut off by the HDC status, the Federal Trade Commission rules on credit transactions have modified these defense cut-offs vis-à-vis certain transactions involving consumer goods.
7. *Warranty liability.* The law implies certain warranties on transferors of negotiable instruments. These warranties not only apply to signers but also to transferors. The theory behind these warranties is to shift the risk of loss to the party who was in the best position to prevent the loss, i.e., the person who dealt face to face with the wrongdoer.
8. *Primary liability.* This is an absolute liability to pay a negotiable instrument, subject only to certain real defenses. Secondary liability is similar to that of a guarantor of a simple contract. It arises when the party primarily liable on the instrument defaults and fails to pay the instrument when due.
9. *Accommodation maker.* A person who signs an instrument guaranteeing payment is primarily liable on the instrument. That is, the other party can seek payment on the instrument from the accommodation maker without first seeking payment from the maker.
10. *Discharged.* Discharge is defined as an action or event that relieves certain parties from liability on negotiable instruments. All the items listed in this question qualify as a basis for discharge from liability on a negotiable instrument under the UCC.

True/False:

1. False. A holder in due course, by definition, can acquire better rights than the transferor had. A holder in due course takes the instrument free from all claims and all personal defenses that can be asserted against the transferor.
2. True. A holder need not necessarily own the instrument to negotiate it.
3. False. Certain things may constitute consideration but not meet the UCC definition of value.
4. False. The defense of fraud in the inducement is a personal defense and, therefore, cannot be raised against a holder in due course.
5. False. Implied warranties are only made when the instrument is transferred, not when it is originally issued.
6. True. An instrument must be transferred for value to have transfer warranties.
7. False. Notice of presentment is effective when it is received. Notice of dishonor is effective when it is sent.
8. False. If an agent signs a negotiable instrument indicating that she is an agent but does not identify her principal, only the agent is liable. The principal is not liable on such an instrument. However, the agent can recover from the principal if she is made to pay.
9. False. When a party other than the indorser pays a negotiable instrument, that party and all subsequent parties are discharged from liability.
10. False. The Federal Trade Commission has eliminated the HDC rules with regard to certain consumer credit transactions.

Multiple Choice:

1. C. A holder in due course takes an instrument free of personal defenses but is subject to real defenses. Breach of contract or nonperformance of a condition precedent describes a personal defense. B is incorrect because bankruptcy is a real defense. C is incorrect because when a minor disaffirms a contract, it is treated as a real defense. D is incorrect because a forgery of a maker's or drawer's signature is a real defense.
2. B. Performance of the agreed upon consideration satisfies the UCC concept of value. A is incorrect because the holder must satisfy the UCC concept of value, not the contract concept of consideration. Though here they are similar, what qualifies as consideration may not necessarily qualify as value. C is incorrect because Cathy has satisfied the UCC concept of value.
3. A. Henry is a limited holder in due course to the amount for which he has paid--$300. In 30 days, when Henry pays the remaining $200, he will be a holder in due course to the full amount of $500. B is incorrect because Henry is a holder in due course to the amount that he has paid, not the amount he has not paid. C is incorrect because Henry is not a holder in due course to the full amount of $500 until he pays the remaining $200 due in 30 days.
4. C. Although Henry bought the note at a $250 discount, he is a holder in due course to the full amount of the note--$500. A is incorrect because Henry is not a limited holder in due course, but a holder in due course to the full amount of the face value of the note. B is incorrect because Henry is a holder in due course to the full amount of $500, not merely the discounted price he paid for the note--$250. D is incorrect because buying a negotiable instrument at a discount does not destroy a holder's ability to be a holder in due course. In fact, if all other requirements are met, the holder is a holder in due course to the full amount of the face value of the instrument.
5. B. In determining good faith, courts use a subjective test in deciding the honesty of the holder. Here, the circumstances suggest that Nancy was acting in good faith since she was inexperienced in such transactions. A is incorrect because the test for honesty is not an objective, but a subjective, test. The test is based on the person in question, not the reasonable person. C is incorrect because the

subjective test is applied only to the holder, not to the transferor. D is incorrect because purchasing an instrument at a deep discount does not necessarily destroy the holder's good faith.

6. C. Material alteration is a real defense and may be raised against Carol in this situation. Carol's only recourse is against Sam. A is incorrect because Carol takes the instrument free from any personal defenses, but not real defenses. Material alteration is a real defense and can be raised in this situation. B is incorrect because material alteration is a real defense, not a personal defense. Therefore, it can be raised in this situation.
7. C. A certified check is a check in which the payer bank has accepted the check and has agreed in advance to pay. Thus, the bank is primarily liable. The drawer, Blare, has secondary liability on the check. A is incorrect because the drawer retains secondary liability. B is incorrect because the bank accepts the check when it certifies it by accepting the check. The bank has primary liability, and the drawer has secondary liability. D is incorrect because although Blare has secondary liability, the bank has accepted.
8. A. Generally, transfer warranties cannot be disclaimed. However, here Fred indorsed the check "without recourse" which is a disclaimer to the transfer warranty that no defenses of any party are good against the transferor. Such an indorsement warrants that the transferor has no knowledge of defenses. B is incorrect because the transfer warranty that all signatures are genuine or authorized cannot generally be disclaimed. C is incorrect because such an indorsement does not affect presentment warranties. D is incorrect because Fred can disclaim at least one implied transfer warranty.
9. C. When bearer paper is delivered, the transferor makes the transfer warranties only to the immediate transferee. Therefore, here, Carmen can only recover against Brian, but not Amy. Amy does have transfer warranty liability to Brian. A is incorrect because Amy is not liable to Carmen for transfer warranties. B is incorrect because Carmen cannot recover for transfer warranties from Amy. D is incorrect because there are transfer warranties on bearer paper although they are somewhat limited.
10. C. Both Edwin and Mildred are secondarily liable on the check. However, such liability does not come into effect until (1) the instrument is properly presented for payment, (2) the instrument is dishonored and (3) notice of the dishonor is timely given to the person to be held secondarily liable on the instrument. Here, Susan only suspects Edward's checking account is overdrawn, but the check has not been dishonored yet. Susan cannot hold anyone liable until the check is actually dishonored. A and B are incorrect because Susan cannot hold anyone liable until the check has been dishonored.

VI. Answers to Essay Question:

It appears that the UCC rules of negotiation were complied with up to the point of Acme having acquired these checks from third parties. These checks were payable to Acme, and Acme qualified as a holder by way of having physical possession of the checks. It may also be assumed that the checks were obtained by Acme in exchange for value by way of the company having provided goods or services for them. In addition, prior to Tippee's involvement with the check indorsement process, there was no evidence of any bad faith on the part of Acme, or of any notice, actual or constructive, of any problems, including forgeries alterations or irregularities. Thus, Acme qualifies as an HDC on these checks.

Because Acme is an HDC, that status will allow it to take these checks free of personal defenses that may have arisen in the negotiation process prior to Acme getting the checks. This HDC status will not protect Acme vis-à-vis any real defenses that may have arisen such as minority, adjudicated mental incapacity, extreme duress, illegality, fraud in the inception, or bankruptcy discharge. In addition, if the checks were related to the sale of consumer goods, then the personal defenses may not have been cut off due to the preemption of UCC rules by the FTC consumer credit laws. In this case, it appears that the products in question were most likely to be used by the buyers as farm equipment, in spite of a trend of having pot-bellied pigs as personal pets.

The Hogsbreath Bank may be liable under a theory of conversion. The bank can successfully argue that no forgery occurred. Had there been an out and out forgery, the bank would be liable for improper payment of the checks on presentation for not knowing its indorser or for failing to handle the document in the proper manner.

The signatures on the checks by the original drawer and subsequent indorsers prior to the checks arriving at Acme are not in dispute here. Primary liability on the checks remains with the acceptor of the check. The checks themselves were not dishonored at the point of presentment, so no secondary liability is at issue.

In addition, the original parties lived up to the transfer and presentment warranties. The real problem here was the breach of warranties by the indorser, Tippee, acting in violation of his employment contract. Tippee was engaged in forgery when he proceeded to deposit the checks made to Acme's order into his own personal account at Hogsbreath Bank.

The respective liabilities of the various parties to these instruments are as follows:

1. The drawees of the original instruments upon presentment are discharged from any further liability.
2. The original drawers of the instrument, the people who wrote the checks, are also discharged because of the payments made by the drawer.

Here, there was a principal/agency relationship established between Acme and Tippee Tonic. The scope of the agency was limited, and that limitation was duly communicated to Hogsbreath Bank. Because the acts of Tippee constituted forgery, they were beyond the scope of authority granted to Tippee by Acme. Hogsbreath Bank should be charged with that knowledge. Tippee was authorized to indorse Acme's checks for purposes of deposit, not into his personal account. The bank is therefore liable to Acme by reason of allowing forged and unauthorized indorsements to go through the negotiation process. The bank must now credit Acme's account based on its mistakes and seek reimbursement from Tippee if he can be found. They should try either Myorca or Yourorca.

Chapter 21
CHECKS AND DIGITAL BANKING

There is time for work and time for love. That leaves no other time.
Coco Chanel

I. Overview

As a target of derision, the banking profession has always kept pace with the members of the bar. When the Bank of America world headquarters building was dedicated in San Francisco, a large amorphous stone sculpture was placed near the portico. It was quickly nicknamed the "Bankers Heart" by local pundits. This resentment may be based on some of the same sources of contempt for any number of professions. The larger society resents its dependence on them. It objects to code language used by the members of these groups and wishes life would go back to a simpler time when one did not need to formally contract for everything or make a federal case out of every dispute. Would that it were so! The realities of today's financial services marketplace simply have put paid that sort of wishful longing. We live in an age of Spaceship Earth that is quickly rendering the use of paper money obsolete. The transfer of money substitutes is more complicated than ever because it is, at least in part, more efficient than ever. Money sitting in Tokyo can be transferred to New York City in seconds. As with any fast moving train, a few people get left at the station and resent it. In the meantime, welcome aboard for the ride.

The rules of rights and duties involving checks are really not all that new to the layperson. Most students have an ongoing contractual relationship with a bank and are already familiar with the basic rules involving negotiable instruments. The best way to approach issues involving checks is to first examine the underlying contractual obligations between the respective parties. Then determine what public policy based limitations are imposed on the parties by the UCC and the courts. The duties imposed on banks can be very strict and difficult to avoid. Finally, look to see if the transaction falls into the sphere of hi-tech transfers of money substitutes.

Many futurists predict that the year 2000 A.D. will herald the age of the paperless society. Given the proclivity of law to sometimes inadequately protect those of us in an unequal bargaining position, equity and good conscience should dictate the rules of the electronic road as well as the paper road. Recent revisions to the UCC vis-à-vis wire transfers and the like reflect attempts by both scholars and practitioners of banking to have the code better reflect this inexorable move into the next millennium. In all events, the law of checks, wire transfers, and money substitute transfer systems cannot afford to disserve its constituents by being a lager.

II. Hypothetical Multi-issue Essay Question

Patric Vanman worked long and hard at Bailout Bank as a paper-pushing, computer-punching bankocrat. He needed a vacation. He bought a brand new sports car and decided to drive it across Canada from Vancouver to Nova Scotia on scenic Highway One. He was to leave on the 15th of the month and come back four weeks later. Because he was to be paid in the interim, he filled out a direct deposit form asking Bailout to credit his paycheck directly to his personal account while he was gone. Bailout sent his check to his house instead, and Patric bounced checks all across Canada without knowing it. The Royal Canadian Mounted Police were just about to catch him as he crossed the border. What results?

III. Outline

Parties to a Check
- Drawer
  - Customer who maintains the checking account and writes checks against the account
- Drawee
  - The bank on which the check is drawn
- Payee
  - The party to whom the check is written

Electronic Funds Transfer Systems
- Automated teller machines
- Point-of-sale terminals
- Direct deposits and withdrawals
- Pay-by-telephone systems
- Pay-by-Internet

Special Types of Checks
- Certified checks
- Cashier's checks
- Traveler's checks

Honoring Checks
- Stale checks
  - A check that has been outstanding for more than six months
- Incomplete checks
- Death or incompetence of the drawer
- Stop payment orders
  - An order by a drawer of a check to the payor bank not to pay or certify a check
- Overdrafts
  - The amount of money a drawer owes a bank after it has paid a check despite insufficient funds in the drawer's account
- Wrongful dishonor
  - Occurs when there are sufficient funds in a drawer's account to pay a properly payable check, but the bank does not pay the check

Forged Signatures and Altered Checks
- Forged instrument
  - Is wholly inoperative as the signature of the drawer
  - Is not properly payable

Altered check
A check that has been altered without authorization that modifies the legal obligation of a party

Check Collection Process

Final Settlement
Occurs when the payor bank
Pays the check in cash
Settles for the check without having a right to revoke the settlement
Fails to dishonor the check within certain statutory time periods

"On Us" v. "On Them" Checks
"On us"
A check that is presented for payment where the depository bank is also the payor bank, i.e., the drawer and payee or holder have accounts at the same bank
"On them"
A check presented for payment by the payee or holder where the depository bank and payor bank are not the same bank

"Four Legals" that Prevent Payment of a Check
Receipt of a notice affecting the account
Receipt of service or a court order of other legal process that freezes the customer's account
Receipt of a stop-payment order from the drawer
Payor bank's exercise of its right of setoff against the customer's account

IV. Objective Questions

Terms:

1. A certified check, cashier's check, or traveler's check, the payment for which the bank is solely or primarily liable for are all examples of a _______________ ________________.

2. When there are sufficient funds in a customer's account to pay a check, but the bank does not do so, this is a _______________ ________________.

3. The rules for bank deposits and collections for checking accounts offered by commercial banks, NOW accounts, and other check-like accounts are set out in _______________ _________________ of the UCC.

4. A bank in the collection process, which is not the depository or payor bank, is called a _______________ ________________.

5. The term for crediting a payee's account against the drawer's check prior to actual clearance is known as a _______________ ________________.

6. The rule that allows banks to fix an afternoon hour of 2:00 pm or later as a cutoff hour for the purpose of processing items is called the _______________ _______________ _______________.

7. A check that is presented for payment by the payee or holder where the depository bank is also the payor bank is known as an _______________ _______________ _______________.

8. A check presented for payment by the payee or holder where the depository bank and the payor bank are not the same bank is called a ______________ ______________ ______________.

9. There are four notices or actions that affect the payment of a check that has been presented for collection and is in the process of being posted. These are called the ______________ ______________.

10. An order by a drawer of a check to the payor not to pay or certify a check is called a ______________ ______________ ______________.

True/False:

1. ____ There are three parties to an ordinary check: the drawer, the drawee, and the payee.
2. ____ Certified checks become stale six months after they are issued.
3. ____ If the holder of a check requests certification on the check, the drawer and any party who indorsed the check before certification are discharged from liability.
4. ____ A drawer can stop payment on a certified check just as he can on an ordinary check.
5. ____ A traveler's check must be signed twice to be considered a negotiable instrument.
6. ____ Most states have legislation that a bank is liable if it pays a stale check unless it received the customer's permission to do so.
7. ____ Under the UCC, if the drawer of a check omits certain information, the risk of loss on such a check is on the bank if the incorrect amount is charged to the customer's account.
8. ____ Drawers, payees, holders, and indorsers can all stop payment on a check.
9. ____ Generally, loss on a forged signature check falls on the bank, not on the customer's account since the forger has usually disappeared.
10. ____ Under the right to setoff, a bank can apply funds from a customer's account to pay any overdue loans the customer may have.

Multiple Choice:

1. Customer Chris opens a checking account at Busy Bank. Chris writes a check to Landlord Len for rent. Chris receives a paycheck from Boss Beatrice and takes it to the bank for deposit. Which of the following is correct?
   A. Busy Bank must accept the check from Beatrice as a deposit.
   B. Busy Bank must collect on the check from Beatrice.
   C. Busy Bank must honor the check written to Len.
   D. All of the above.

2. Assume the same facts as in question 1. Which of the following is correct regarding Chris' various transactions?
   A. A creditor/debtor relationship has been formed.
   B. A principal/agent relationship has been formed.
   C. Both of the above.
   D. None of the above. There is no special relationship between a customer and his bank.

3. Customer Cassandra has a checking account with Bakerville Bank. Cassandra presents a check for certification by Bakerville Bank. The Bank certifies the check. Cassandra then gives the check to Dealer Dan as down payment on a new car she bought. Dan indorses the check and delivers it to Boater Bob for a new boat Dan bought. Which of the following is correct?
   A. Cassandra is secondarily liable in the check.
   B. The check is payable only for six months, but not beyond that since it will be stale.
   C. Bakerville Bank can revoke certification of the check at its option.
   D. All of the above are correct.

4. Assume the same facts as in question 3 except that Dan, not Cassandra, has the check certified before delivering it to Bob. Which of the following is correct?
   A. Cassandra is secondarily liable on the check.
   B. The check is payable only for six months, but not beyond that since it will be stale.
   C. Bob cannot hold Cassandra liable on the check.
   D. All of the above are correct.

5. Traveler Tina has planned a world tour and decides to get traveler's checks from her bank. Which of the following is incorrect regarding Tina's traveler's checks?
   A. The traveler's checks are issued without a named payee.
   B. The traveler's checks are three-party instruments.
   C. Tina can stop payment on the traveler's checks if they are lost or stolen, but not negotiated.
   D. All of the above are correct regarding Tina's traveler's checks.

6. Bumbling Betsy never balances her checkbook and is never quite sure how much money she has. Unfortunately, Betsy only has $100 in her account when she writes a $200 check to Kitchen Komfort for a new set of dishes. Kitchen Komfort seeks payment on the check. Which of the following is correct?
   A. Betsy's bank can dishonor the check.
   B. Betsy's bank can pay the check for Betsy although she has insufficient funds, but is not required to do so.
   C. Both of the above.
   D. None of the above. Betsy's bank must pay the check since dishonoring the check would be unfair to Kitchen Komfort.

7. Assume the same facts as in question 6 except Betsy has $500 in her account when she writes the check. Also, assume Betsy's bank wrongfully dishonors the check to Kitchen Komfort. Which of the following is correct?
   A. The bank is liable to Betsy for wrongfully dishonoring the check.
   B. Kitchen Komfort can sue the bank for wrongfully dishonoring the check.
   C. Kitchen Komfort has no recourse and must bear the loss.
   D. Betsy has no recourse and must bear the loss.

8. Juggling John knows his checking account is low, and he would be unable to pay a check at present. However, in a week, John will have more money in his account. John writes a check to Understanding Ulga, his landlord, on June 1 and postdates the check for June 10. Ulga forgets to hold the check and takes it to the bank on June 2. Which of the following is correct?
   A. If John's bank pays the check, it is liable for any actual damages suffered by John.
   B. Ulga is liable for any actual damages suffered by John since she did not hold the check.
   C. The postdated check is not negotiable.
   D. John must bear the cost of any actual damages suffered since he should not have postdated the check.

9. Preoccupied Patty owes Shady Sherman $50. Patty writes a check to Sherman, but in her haste, does not fill in the amount. Patty tells Sherman to fill in $50 and gives him the signed check. Sherman, though, fills in the amount for $1,000. Sherman then takes the check to the bank for payment. Which of the following is correct?
   A. If the bank pays the check, it is liable for $950.
   B. If the bank pays the check, it is liable for $1,000.
   C. If the bank pays the check, it is not liable at all since Patty bears the risk of loss in this situation.
   D. None of the above.

10. Which of the following is not one of the four legals?
   A. Receipt of a stop payment order from the drawer.
   B. The payor's bank's exercise of its rights of setoff against the customer's account.
   C. Receipt of service of a court order or legal process that "freezes" the customer's account, such as a writ of garnishment.
   D. All of the above are one of the four legals.

V. Answers to Objective Questions

Terms:

1. *Bank check.* These checks are often used when a payee is concerned that a drawer's normal check may not be honored by the drawee due to insufficient funds. With these checks, the payee has more comfort because they are considered much closer to a cash equivalent.
2. *Wrongful dishonor.* When a customer opens a checking account at a bank, the customer is expected to keep sufficient funds in the account to pay any checks written against it. Thus, when the drawer bank receives a properly drawn and payable check, the bank is under a duty per Section 4-401(a) of the UCC to honor the check and change the drawer's account in the amount of the check.
3. *Article 4.* Article 4 controls if Article 3 and 4 conflict per UCC 4-102(a). In addition, Article 4 was substantially amended in 1990 to better incorporate the hi-tech changes that have taken place in the banking industry.
4. *Intermediary bank.* This bank is used to help facilitate the collection process on negotiable instruments. This bank, by definition, is not the depository or payor bank.
5. *Provisional credit.* Under Section 4-201(a) a bank may provisionally credit the customer's account for a deposit of a check into his or her account. Even though the depository bank does not have to pay the customer the amount of the check until final settlement, this section allows payment in the meantime. If the payor bank dishonors the check, the provisional credit is reversed.
6. *Deferred posting rule.* This rule is set out under Section 4-108 and applies to all banks in the collection process. Any check or deposit of money received after the 2:00 p.m. cutoff hour is treated as received on the next banking day.

7. *On us item.* If both the drawer and payee or holder have accounts of the same bank, the depository bank is also the payor bank. The check is then called an "on us" item when the payee or holder presents it for payment.
8. *On them item.* If the drawer and the payee or holder have accounts at different banks, the payor and depository bank are not the same bank. A depositor has the option of physically presenting the check for payment at the payer bank if he or she so chooses.
9. *Four legals.* These notices or actions can prevent the payment of a check if the payor bank receives them before it has finished its process of posting the check for payment.
10. *Stop payment order.* An oral stop-payment order is binding on the bank for fourteen days; a written stop-payment order is binding for six months and may be renewed for additional six-month periods.

True/False:

1. True. There is a debtor/creditor relationship between the bank and its customer. When a check is drawn on a checking account, the drawer (creditor) is ordering the drawee (bank) to transfer his or her money to a third party, the payee.
2. False. Certified checks do not become stale after six months. They are payable even six months after they are issued.
3. True. The bank has assumed primarily liability for the negotiable instrument.
4. False. A drawer cannot stop payment on a certified check. In addition, the certifying bank can only revoke its certification under certain circumstances.
5. True. A traveler's check must have the second signature to be considered a negotiable instrument.
6. True. Remember, the UCC is not a federal statute but rather a uniform act that the individual states can adopt either in part or entirely. Most states have adopted this provision of the UCC.
7. False. Under the UCC, if the drawer of a check omits certain information, the risk of loss on such a check is on the drawer, not the bank if the incorrect amount is charged to the customer's account.
8. False. Only drawers can stop payment on a check since only they have a contractual arrangement with the bank.
9. True. Although the bank could try to recover on a forgery from the forger, the forger is usually not available or is judgment proof. Therefore, the bank bears the loss.
10. True. The banker's right of setoff is derived from common law principles which were in place prior to the enactment of the UCC. It is considered one of the "four legals" which can prevent the payment of a check if received by the payor bank before it has finished its process of posting.

Multiple Choice:

1. D. A, B, and C are all examples of Busy Bank's duties in servicing its customer, Chris.
2. C. A is correct because a creditor/debtor relationship is formed when Chris deposits money in the account. B is correct because a principal/agent relationship is formed when Chris deposits his paycheck and when he writes a check to Len. D is incorrect because these special relationships are formed.
3. A. Cassandra, as the drawer of the check, remains secondarily liable on the check although it has been certified. B is incorrect because a certified check does not become stale after six months. C is incorrect because the bank is not permitted to revoke certification of the check at its option.
4. C. Once Dan has certified the check, Cassandra, as the drawer of the original check, is released from liability. Therefore, Bob cannot hold Cassandra liable on the check. A is incorrect because Cassandra is released from liability on the check as discussed above. B is incorrect because a certified check does not become stale after six months.

5. B. Traveler's checks are two-party checks because the bank acts as both the drawer and the drawee. A is a true statement. C is a true statement.
6. C. A is correct because the bank can dishonor a check if the customer's account has insufficient funds. B is correct because the bank can pay the check and charge Betsy's account for the overdraft later. D is incorrect because the bank is not required to honor a check if the customer's account has insufficient funds.
7. A. If the bank wrongfully dishonors the check, it is liable to its customer, Betsy. B is incorrect because the bank is not liable to Kitchen Komfort since Kitchen Komfort is not its customer and therefore has no contractual relationship with the bank. C is incorrect because Kitchen Komfort has recourse against Betsy. D is incorrect because Betsy can hold the bank liable and therefore, need not bear the loss.
8. A. If the bank pays the postdated check before the date on the check (June 10), the bank is liable for any actual damages suffered by John. B is incorrect because Ulga is not liable. The bank is responsible for honoring the post-date. C is incorrect because a postdated check is a negotiable instrument. D is incorrect because John is permitted to postdate a check. In addition, the bank must honor the post-date.
9. C. If Patty, the drawer, leaves part of the check blank, the bank may pay the check as it is filled out by the payee (Sherman) in good faith. Patty, not the bank, bears the risk of loss in such a situation. A and B are incorrect because the bank is not liable if it paid the check in good faith. Patty must bear this loss.
10. D. A, B, and C are each one of the four legals. In addition, the fourth is receipt of a notice affecting the account, such as a notice of the customer's death, judgment of incompetence, or bankruptcy.

VI. Answers to Essay Question

Poor Patric is lucky not to be in jail, and his travels certainly did not help foster better U.S.-Canadian relations. The basic relationship that needs to be examined first is the one between Patric and Bailout Bank. Even though Patric may be an employee of the bank, and for some purposes, an agent, his key role here is customer. Because he has a personal account at the bank, a creditor/debtor contract relationship has been established between him and the bank. When he writes checks on his account, he is the drawer and the bank is the drawee. A debtor/creditor relationship has been created. The bank is the debtor, and Patric is the creditor.

The checks involved here would be considered normal drafts on Patric's personal checking account at Bailout Bank. There is no evidence of the bank having undertaken any extra or special duties related to certification or cashiers checks. Maybe Patric should have used traveler's checks. They would have provided him with protection in case of theft. But even more important, the payees would have also been better protected than they were in accepting Patric's personal checks.

The point of negotiation in question is dishonor of Patric's checks at presentation for payment. The dishonor was proper because of insufficient funds in Patric's account at the time. The payor bank may accept and process a deposit per instructions it had agreed upon with Patric before he left for his trip. Patric can argue, that even though there were insufficient funds in the account, the insufficiently was created by Bailout's failure to live up to its contractual obligations. That failure created a wrongful dishonor of Patric's checks when they were presented for payment. Because of their negligence, Patric can now claim damages proximately caused by the wrongful dishonor. In addition, he should be compensated for any consequential damages or liabilities caused by criminal prosecution. At the bare minimum, Bailout should now honor the checks, pay any bounce fees, and write letters of apology to all parties involved in the negotiation of Patric's checks. Patric loved his trip and doesn't want to be *persona non-grata* in Canada.

Chapter 22
CREDIT AND SECURED TRANSACTIONS

A trifling debt makes a man your debtor, a large one makes him your enemy.
Seneca

I. Overview

Credit is really a shibboleth for risk. Inherent in our business enterprise system is the principle that risk will be rewarded if well placed and punished if not. The riches of wise credit extension are the stuff of financial dynasties, and it is no accident that most economic measures of a nation's growth are tied to the success of its financial institutions. Conversely, failure to wisely administer and manage these key elements in our society leads to costly debacles such as the current savings and loan bailout. The law has long reflected this win/lose dichotomy of the credit marketplace. Many of the protections and remedies accorded to creditors have long historical tracings in the common law. In today's highly codified scheme of things, most of the rights and duties of the both debtor and creditor are found in the statutes, but the principles hearken back to an earlier age of debtor's prisons and the like. The possibilities for fraud, chevisance, and conspiracy by charlatans in this area of the law have provided both the challenge and opportunity to seek a balance between the competing rights and duties of debtors and creditors. If the law takes an uneven hand to either side, the long-term interest of the entire economic system is disserved.

In your examination of credit relationships, remember that most disputes have more than one level of controversy. On the surface, the most apparent problem revolves around resolution of the differences between the debtor and creditor. At the sub strata, a second dispute is often found: that of competing creditors fighting over priority rights to the debtor's property. A proper examination of any credit case answers both aspects.

There are three basic relationships, which you must establish before going into the details set out in the key question checklist. How was the debtor/creditor relationship established? Which body of case law, statute, or combination thereof is to be used in defining the respective rights and duties of the debtor and creditor? What about the rights and duties of third parties as they relate to this issue?

Earlier chapters of this book compared the relationship of the UCC to the common law of contracts as being akin to an eclipse. The UCC may cover a segment of the law heretofore resolved by contract, but the penumbra of common law still shines on underneath. Article 9 of the UCC is a particularly good example of this interaction between the old and new. The original sources of the law of secured transactions involving personal property can be found in a combination of common law contracts, real property law, debtor and creditor law, and the law of liens. The UCC has sought to interpolate the best elements of all these areas into a cogent and organized structured system of facilitating secured credit transactions for the sale of personal property.

This system is premised on the legal realism that merchants doing business with each other are expected to live up to a higher standard of both knowledge and behavior of the law. In addition, that same realism attempts to protect the innocent third parties' good faith reliance on the legitimacy of the marketplace whenever possible. These ends are mainly sought through the use of the recordation principles long established in real property law transactions. These recordation rules are designed to give creditors notice not only of the debtor's obligation, but even more important, to give notice to other creditors that this security-based transaction has taken place which will establish an order of priority or pecking order among the creditors.

Several practical factors make the law of secured transactions in personal property particular troublesome. First, compared to real estate, personal property is portable. In many legal systems the distinction between real and personal property is described as *portables* versus *nonportables*. The same is true for many multilateral tax and trade treaties. Being more moveable, the role of making sure that the security interest attaches and stays with portable goods is of critical importance. Second, in a society based on credit, there is a strong likelihood that most large personal goods, such as automobiles and the like, may have more than one creditor looking to that good for security. This reliance on that good may come from either the original acquisition of the good or from subsequent transactions after the good is acquired. The ordering of priorities between multiple creditors having claims to the same goods becomes a critical issue.

Finally, just as there are likely to be multiple creditors, so are there likely to be multiple users. The vertical chain of distribution starts with supplies of raw materials to manufacturers, distributors, and retailers. On the horizontal level, the users start with the consumer, his or her family, and go on to third party users or acquirers. With each shift, there lies the probability of having to recognize new duties and obligations vis-à-vis all parties having a legal relationship to those goods. All in all, it can very quickly become a tangled web, but the UCC and the common law rules with regard to liens, surety, guaranty, and collection provide some very good rules for both debtor and creditor.

II. Hypothetical Multi-Issue Essay Question:

Ms. Zoie Zonna lived and commuted for many years in Los Angeles working as a zygostates for the L.A. Mint. Her job was golden, but she hated putting up with the tailgaters who were always nearly running into her. She invented and patented the "Z.Z. Tail-Gator." This device was attached to an auto tailpipe and emitted an odious methane-based flatulence when someone got too close for comfort. Zoie was on her way to making a fortune because so many drivers saw her invention as a dream come true.

Based on her expected high income, Zoie moved to Colorado and bought the famous "How's Your Aspen? Ranch." The financing was arranged with $2 million down, a first mortgage deed of trust payable to Snow Bank for $10 million and a second mortgage deed of trust payable to Downhill Bank for $2 million. Snow Bank's loan officer was high on the Rockies and failed to record the mortgage before the Downhill Bank's recordation.

Unfortunately for Zoie, both the California Air Resources Board and the EPA have now banned her product for being in violation of tailpipe emissions standards. She is nearly broke and doesn't want to go back to her old lifestyle. What results?

III. Outline

Types of Credit
- Unsecured credit
  - Credit that does not require any security (collateral) to protect the payment of the debt
- Secured credit
  - Credit that requires security (collateral) to secure payment of the loan

Security Interests in Personal Property
- Common law lien
  - Given to artisans, laborers, innkeepers, common carriers, and other service providers on personal property of customers to secure reasonable payment for services rendered
- Statutory lien
  - A lien that codifies by statute a service provider's common law lien

Parties to a Mortgage

Parties to a Note and Deed of Trust

Right of Redemption
- The mortgager has the right to redeem real property after default and before foreclosure

Material Person's Lien
- A contractor's and laborer's lien that makes the real property to which improvements are being made become security for the payment of the services and materials for those improvements

Surety and Guaranty Arrangements
- Surety arrangement
  - An arrangement where a third party promises to be primarily liable with the borrower for the payment of the borrower's debt
- Guaranty arrangement
  - An arrangement where a third party promises to be secondarily liable for the payment of another's debt

Collection Remedies
- Attachment
  - Seizure by the creditor of property in the debtor's possession in order to collect on a debt while their lawsuit is pending
- Execution
  - Postjudgment seizure and sale of the debtor's property to satisfy a creditor's judgment against the debtor
- Garnishment
  - Postjudgment remedy that is directed against property of the debtor that is in the possession of third persons

Assignment for the Benefit of Creditors
- An assignment that allows debtors voluntarily to assign title to their property to a trustee or an assignee for the benefit of their creditors

IV. Objective Questions

Terms:

1. An arrangement by which a property owner borrows money from a creditor and uses a deed as collateral for repayment of the loan is called a _______________.

2. A situation in which a third party purchases real property that secures a debt in the form of a mortgage or deed of trust and agrees to make future payments on the related debt is known as _______________.

3. A party who acts as a surety in a transaction without receiving compensation for acting in that capacity is an _______________ _______________.

4. A postjudgment remedy directed against the property of a debtor in the possession of a third party is known as _______________.

5. The security interest of a creditor in property of a debtor that minimizes the creditor's risk of extending unsecured credit is called _______________.

6. In a secured transaction, the debtor and the secured party enter into a _______________ _______________ to create or provide for a security interest.

7. Those items of tangible personal property that have been affixed to real estate so as to become a part thereof are called _______________.

8. A writing that evidences both a monetary obligation and a security interest is known as _______________ _______________.

9. The order in which competing claims may be satisfied from the same collateral is known as _______________.

10. Money lent to the same debtor by the same creditor secured by the same collateral is known as a _______________ _______________.

True/False:

1. ____ In a transaction involving unsecured credit, the creditor may look to collateral pledged by the debtor to satisfy the debt obligation in the event the debtor fails to make payment when due.

2. ____ A creditor (lienholder) can have the sheriff seize property subject to a lien and sell it at a judicial sale immediately after the debtor fails to make a payment when due.

3. ____ Failure to record a mortgage or deed of trust in accordance with the applicable recording statute renders the instrument invalid and legally unenforceable between the parties to the instrument.

4. ____ If a mortgagor defaults on a mortgage, the mortgagee can declare the entire debt due and payable immediately.

5. ____ To exercise the right of redemption with respect to real property, a mortgagor is required to pay only those amounts then in default.

6. ____ To perfect a security interest in fixtures, the secured party must file a financing statement in the office where mortgages on real estate are filed or recorded.

7. ____ Typically, the lender or seller in a secured transaction grants a security interest to the borrower or buyer.

8. ____ When collateral is disposed of by a secured party, the disposition discharges the security interest under which it is made, but does not discharge any subordinate security interests or liens.

9. ____ If collateral subject to a perfected security interest increases in value, the secured party may hold the increased value or profits as collateral.

10. ____ Repossession tactics allow for some breach of the peace as a price for protecting creditors from errant debtors.

Multiple Choice:

1. Ike Investor sells 200 acres of prime farmland to Frank Farmer. Ike and Frank enter into a five-year land sale contract pursuant to which Frank agrees to pay Ike principal and interest in five equal annual installments. Which of the following is correct?
   A. If Frank defaults in his payment obligation, Ike must wait until the end of the five-year cure period before he may declare a forfeiture and retake possession of the property.
   B. Ike retains title to the land until Frank has satisfied all of his payment obligations.
   C. During the five-year contract period, Frank has the right to possession and use of the property.
   D. B and C above.

2. Charlie Creditrisk borrowed $10,000 from Aloof Bank. In investigating Charlie's financial status after Charlie defaulted on his loan payments, the Bank discovered that Charlie owns a speed-boat worth $15,000, a stereo and other entertainment equipment worth $7,500, and miscellaneous sporting equipment worth $6,000. Charlie supports himself on the $2,000 his parents give him each month. The Bank can obtain:
   A. A writ of garnishment immediately for a portion of the $2,000 Charlie's parents give him each month.
   B. A lien on Charlie's stereo equipment by recording its lien and notifying Charlie.
   C. A writ of execution for the seizure of Charlie's speed-boat after obtaining a judgment against Charlie.
   D. A writ of execution for the immediate seizure of the speed-boat.

3. Assume the same facts as in question 2. If the Bank obtains a judgment and a court order:
   A. Upon the seizure and sale of Charlie's property, a certain amount must be set aside so that Charlie can provide shelter for himself and his family.
   B. The Bank will be unsuccessful in recovering anything from Charlie's assets because of statutory exemptions.
   C. The gifts of money received from Charlie's parents cannot be reached through garnishment.
   D. The Bank will have liens on Charlie's property that may then be foreclosed.

4. The lien granted to contractors and laborers who make improvements to real property is known as a(n):
   A. General lien.
   B. Mechanic's lien.
   C. Artisans lien.
   D. Chattel lien.

5. Danny Digithead borrowed $15,000 from George Generous to provide startup capital for his new accounting firm. George required that Danny find a cosignor and that he pledge certain corporate securities as collateral. Fran Friend agreed to cosign. When unrest in Eastern Europe caused Danny to fear a drop in the stock market, George allowed him to sell the corporate securities. As a result of the sale of the securities, Fran:
   A. Remains liable in the event that Danny defaults.
   B. Is completely discharged by the sale of the collateral.
   C. Is partially discharged to the extent of the value of the collateral sold.
   D. None of the above.

6. Huge Company owned a series of variety stores across the country. Huge Company borrowed $30,000 from Giant Company and gave Giant Company bonds to hold as security for repayment of the loan. Which of the following is correct?
   A. Giant Company may only perfect its security interest by properly filing a financing statement.
   B. This arrangement would not be classified as a secured transaction under the UCC because it is not possible to grant a security interest in bonds.
   C. Giant Company has a perfected security interest with respect to the bonds.
   D. If the proceeds of the bonds were originally used to build a warehouse, Giant Company would need to file a financing statement in the real estate records to perfect its security interest.

7. Lynnette borrowed $5,000 from Jeannette and signed a written agreement. The agreement stated that Lynnette agreed to repay the loan on January 15 and that Jeannette could have possession of an antique automobile she was to inherit from her elderly Aunt Annette if Lynnette failed to repay the loan. Which of the following is correct?
   A. Jeannette is the secured party and has a security interest in the automobile.
   B. There was no security interest created because Jeannette did not sign the writing.
   C. There was no security interest created because Lynnette has no possessory or ownership interest in the automobile.
   D. A security interest was created only if the proceeds of the loan were used to purchase the automobile.

8. Mary purchased new furniture in the amount of $10,000 from Joe's Furniture in Sacramento, California and agreed to make payments at $100 per month for 100 months. In addition to its purchase money security interest, Joe's Furniture filed a valid financing statement to protect its security interest (assume that financing statements are valid for five years in California). After 4 years, Mary moved to New York and took the furniture with her. Which of the following is correct?
   A. A secured creditor in New York who files a valid financing statement with respect to the furniture will have priority over Joe's Furniture in any event.
   B Joe's Furniture has four months to perfect its security interest by filing a new financing statement in New York.
   C. Joe's Furniture has one year to perfect its security interest by filing a new financing statement in New York.
   D. None of the above.

9. Under Sec. 90504 of the UCC, where a creditor chooses not to retain collateral, but rather to dispose of it, he or she must first apply the proceeds to:
   A. His or her personal bank account.
   B. Satisfy subordinate security interests.
   C. Pay reasonable expenses related to the disposition costs.
   D. The creditor may keep all proceeds in any event.

10. GI-GO (Garbage In/ Garbage Out) is a high-technology oriented firm that develops computer hardware. It has secured a patent for its latest hardware system. To finance the marketing for this system, GI-GO took out a loan from Easy Money Finance Company and gave its rights under the patent and chattel paper as security. If GI-GO defaults:
    A. Easy Money is entitled to whatever it can collect from the collateral after proper notification to the debtors and obligors.
    B. Easy Money has no recourse against GI-GO beyond the collateral.
    C. GI-GO is entitled to any surplus of collections made by Easy Money after deduction of expenses.
    D. Easy Money has no collection rights with respect to licensing of the patent since patents are not subject to the UCC.

VI. Answers to Objective Questions:

Terms:

1. *Mortgage.* The vast majority of real estate in this country, and throughout the world for that matter, is purchased through the use of this or some similar sort of collateralized credit arrangement.
2. *Assumption.* This sort of provision has become less commonly used in the past twenty years with the advent of large interest rate swings in the 1970's. Most financial institutions now insist on a "fresh start" with each new buyer and therefore limit the use of these sorts of provisions.
3. *Accommodation surety.* Most people who take on this role are either family members or close friends of the principal debtors.
4. *Garnishment.* This action is usually directed at wages due the debtor, banks in possession of funds belonging to the debtor or against third parsons in possession of the property of the debtor.
5. *Collateral.* The two main sources of law with regard to the rights and duties of the parties to collateral are common law vis-à-vis real property and Article 9 of the UCC vis-à-vis personal property.
6. *Security agreement.* Under Sec. 9-105(1), this can be any agreement between the debtor and the secured party that secures payment on performance of an obligation.
7. *Fixtures.* These personal properties can become part of real property by way of attachment, weight, or most important, by reason of the intent of the parties.
8. *Chattel paper.* This is defined by UCC Sec. 9-105(1)(b) and, in most cases, consists of a negotiable instrument coupled with a security agreement.
9. *Priority.* As much of the law of Article 9 is directed at establishing rights between competing creditors as it is towards defining creditors versus debtors rights.
10. *Future advance.* These are used in conjunction with revolving lines of credit and the like where certain personal property of the debtor is designated as collateral for future loans.

True/False:

1. False. Unsecured credit does not require any security (collateral) to protect the payment of the debt.
2. False. If no statutory procedure exists, the creditor must bring a lawsuit and obtain a judgment against the debtor. Once a judgment has been obtained, the sheriff can seize the property and sell it at a judicial sale.
3. False. Failure to record a mortgage or deed of trust does not affect either the legality of the instrument or the rights and obligations of the parties to the instrument. Failure to record may affect the priority of the holder's rights vis-à-vis others who have recorded.
4. True. This is a standard provision found in most mortgage instruments subject to state law limitations.
5. False. To exercise the right of redemption with respect to real property, a mortgagor must pay the full amount of the debt including principal, interest, and other costs incurred by the mortgagee because of the mortgagor's default.
6. True. The secured party must file a financing statement.
7. False. In the typical secured transaction, the borrower or buyer grants a security interest in his property to the seller or lender.
8. False. The disposition discharges the security interest under which it is made as well as any subordinate security interests or liens.
9. True. However, if the collateral is money, any increases or profits must either be remitted to the debtor or applied to reduce the balance of the secured debt.

10. False. Under Sec. 9-503, the creditor may not create a breach of the peace to repossess collateral upon a debtor's default. Breach of the peace is defined locally by each state and not by the UCC.

Multiple Choice:

1. D. Answers B and C are both correct. Answer A is incorrect because under a land sale contract, Ike may declare a forfeiture and retake possession of the property at the point in time that Frank defaults in his payment obligation.
2. C. Execution is a post-judgment remedy that may only be given effect after a judgment has been obtained.
3. C. Garnishment is directed against property of the debtor in the possession of another. These gifts are not Charlie's property until made, and therefore garnishment is not available with respect to these gifts. Answer A is incorrect because the homestead exemption relates to real property, not those items of personal property listed. Answer B is incorrect because most states do not provide an exemption for the items of personal property listed. Answer D is incorrect because a lien does not arise merely as a result of obtaining a judgment. Some further action such as attachment or execution is required.
4. B. These titles vary from state to state depending on local statutes.
5. B. A surety is completely discharged from liability if the creditor enters into an agreement with the principal debtor to alter the debtor's obligation without the consent of the surety.
6. C. Generally, possession is the only method available to perfect a security interest in investment securities such as bonds.
7. C. To create a valid security interest, the debtor must have rights in the collateral. Here, it is merely speculation that Lynnette will ultimately inherit the automobile, and thus no security interest was created. Answer B is incorrect because it is not necessary that the secured party sign the security agreement to create a valid security interest. Answer D is incorrect because there are other types of security interests than purchase money security interests.
8. B. Joe's Furniture has the shorter of four months or the period of time remaining under the original California perfection (here, one year) to file in New York. Answer A is an incorrect statement because the original perfection continues if the proper steps are taken to perfect in the new state. Answer C is incorrect because the secured party must file the new financing statement in the shorter of the two periods discussed above.
9. C. Under Sec. 9-504(1)(a), reasonable expenses of retaking, holding, and preparing the collateral for sale, lease, or other disposition are paid first.
10. C. The secured party must account to the debtor for any surplus from collection, less reasonable expenses incurred. Answer A is incorrect because Easy Money is liable for any surplus to GI-GO. Answer B is incorrect because GI-GO will be liable for any deficiency. Answer D is an incorrect statement.

VI. Answers to Essay Question

The nature of these transactions centers around a series of bilateral contracts entered into between Ms. Zonna and two creditors. In both cases, credit was extended by the banks in exchange for a promise to repay. The transactions called for a writing because contracts involving interest in real estate are covered by the Statute of Frauds. The creditors provided a service to the debtor which benefited her ability to acquire her interest in real property.

From the facts as presented, Ms. Zonna signed two mortgaged deeds of trust to Snow and Downhill banks respectively. In both cases, she borrowed the money and used it towards the purchase price of the ranch. Also, in both cases, the creditors secured the debt owed to them by having a mortgage note signed to their favor by Ms. Zonna. In addition, under the deed of trust method of financing the purchase of real property, the legal title to the property is handed over to a trustee during the payment period covered by the note. In the interim period, Ms. Zonna is allowed to retain legal possession of the property as owner/trustor. In addition, she will also be responsible for taxes and the like relating to the property. It appears that not all the proper filing procedures were followed in that the first bank to issue credit (Snow Bank) failed to file its mortgage in a timely manner. This will not relieve Ms. Zonna of any liability on her note to Snow Bank, but it may affect the priority of Snow Bank's claim on the property in case of foreclosure.

The most likely method of enforcement of a debt secured by real property is foreclosure and sale. The parties may have modified this likelihood by contract but such contract provisions are narrowly construed and may not be as valid against public policy. In this case, both the banks and Ms. Zonna have an interest in the ranch. She has made a substantial down payment plus any equity value based on appreciation in value since the time of the purchase. The banks, in turn, have their respective security interests per the unpaid balances of the mortgage notes owed to them by Ms. Zonna. If Ms. Zonna finds herself in default because of her inability to pay, the banks may seek foreclosure on the ranch, subject to any redemption rights provided by statute. They may then seek to have the court conduct a judicially supervised sale. Most states, including Colorado, will allow a deficiency judgment against the debtor if the sale does not raise enough revenue to pay the debt. If the state of Colorado had an antideficiency statute similar to the one enacted in California, Ms. Zonna could not be held liable for more than the value of the property at the time of the court-ordered sale. In case the property has appreciated in value, Ms. Zonna would be entitled to the net surplus from the proceeds of the sale.

There was no involvement of any surety or guarantor in this transaction. There was, however, the aforementioned failure by Snow Bank to file its mortgage note in a timely manner. Even though that delay does not directly affect the amount owed to them by Ms. Zonna, it may have an adverse effect on their ability to collect. If the ranch had depreciated in value, and Ms. Zonna was judgment proof, Snow Bank's rights to proceeds from the judicial sale would be subrogated to the rights of Downhill Bank. For example, if the ranch were worth $5 million, Downhill would get its $2 million first. Snow Bank would get the remaining $3 million but would probably lose out on the remaining $7 million owed to it due to Ms. Zonna's insolvency. Snow Bank's loan officer will probably have to find a new high.

It's too bad that the California Air Resources Board and the EPA did what they did. The world really needs a good "Z.Z. Tail-Gator," and Ms. Zonna could be enjoying the fruits of her labor on the ranch.

Chapter 23
BANKRUPTCY AND ORGANIZATION

Bankruptcy: life after debt.
D. Robert White

I. Overview

The basic underlying premise of bankruptcy law is founded on a simple reality: bad things happen to good people. How many of us can really provide ourselves with a safe haven from financial disasters brought on by bad health, economic downturns, financial institution failures, and the like? The early bankruptcy laws of England first recognized that businesses can and do fail in spite of the best good faith efforts of their proprietors. That failure should not, in effect, act as a life sentence in keeping that business or its proprietor from reentering the marketplace. Bankruptcy is really one of the earliest forms of recycling, a recycling of economic opportunity for good faith debtors who deserve a second chance.

As with any legal favor, there are people and business entities that get too greedy in asking for the benefit of the law. Bankruptcy is built on a cornerstone of good faith. Where debtors' actions are motivated by bad faith attempts to avoid legitimate obligations, both the law and the larger societal public policies are subverted. The history of the law of bankruptcy is riddled with cases of clear abuse and creditor victimization that have created a dilemma for legislators who must draft our bankruptcy statutes.

The recent history of federal bankruptcy reforms in the U.S. illustrates Congress's attempts to deal with this dilemma of trying to make the law more humane while trying to curb abuses. The Bankruptcy Act of 1978 provided for sweeping reforms that sought to destigmatize bankruptcy in an economy that had grown too dependent on the manna of credit. Unfortunately, with this liberalization came a horde of abuses of the law. Graduates of long and expensive professional studies began their lucrative careers with a bankruptcy discharge of school loans. Many consumers loaded up on all sorts of trinkets on credit and kept them debt free under bankruptcy. Corporations began to use the reorganization provisions of the law as a management wedge to get out from under otherwise binding executory contracts, or even worse, tort judgments. Congress responded with the Bankruptcy Amendments and Federal Judgeship Act of 1984 and other subsequent revisions. This legislation sought to pull in the reins on many of these abuses as outlined in the text.

There is still a great need to further balance the basic dichotomy between relief and prevention of abuse. On one side, the mechanics of bankruptcy law must be made more cost efficient. The law should provide a larger slice of the economic pie for creditors than is presently available while retaining the fresh start opportunities presently provided for good faith debtors. At the same time, present loopholes in the law must be closed. There are still far too many opportunities for charlatans, deadbeats, and unscrupulous managers to abuse the system. The losses generated by these people add a colossal weight to the cost of credit borne by all of us. Witness the mockery of law made by the key protagonists of the junk bond or savings and loan scandals.

A famous commentator on American history, Allistair Cooke, ended his TV series "Allistair Cooke's America" on PBS with a quote from an immigrant from Italy who had lived in the U.S. for four decades. He said: "There is no free lunch." These words are as prophetic as ever.

II. Hypothetical Multi-issue Essay Question

Mr. Pinky Flash is not a modest fellow. His personal credo has always been: "If you've got it, flaunt it; and if you flaunt it, be sure to telemarket it!" Pinky invented and patented the original computer-voice telemarketing machine. This is the very same machine that seems to instinctively know exactly when you are about to have dinner or take a shower. Evangelists, phone companies, and penny stock promoters favor it because it allows them to disturb you in spite of having an unlisted phone number.

As Pinky's business prospered, he proceeded to indulge in his favorite pastime--buying expensive jewelry on credit. On every one of his fingers he wore a precious pink gem of at least five carats. His supplier, Rocks R Us, extended a $1 million line of unsecured credit to Pinky, and he used it fully. In addition, he owed $5 million each to two other unsecured creditors, Phakky Peripherals Inc. and Magnetic Voices, Inc. Both of them supplied parts for his telemarketing machines. In addition, he owed $1 million to Bailout Bank for his home mortgage. His home's value has gone down to $500,000 since he had it plastered in a shade of luminescent pink to match his rings.

Last week Pinky's world fell in on him when Congress passed a constitutional amendment that banned the manufacture, use, or sale of Pinky's telemarketing machines. The amendment was ratified in record time through the use of e-mail. Pinky has no other assets other than his house, jewelry, and half-finished machines. In addition, he still owes $100,000 from last year's income taxes to the IRS. What are Pinky's prospects in Bankruptcy Court?

III. Outline

Fresh Start in Bankruptcy
- The goal of federal bankruptcy law is to discharge the debtor from the burdensome debts and allow him or her to begin again

Chapter 7 Bankruptcy Liquidation
- Debtor's nonexempt property is sold for cash
- Cash is distributed to creditors
- Unpaid debts are discharged
- Any person, including individuals, partnerships, and corporations may be debtors in a Chapter 7 proceeding
- Debtor gets a fresh start

Federal Exemptions from the Bankruptcy Estate
- Exemptions include:
  - Homestead exemption—interest up to $15,000 in property used as a residence and burial plots
  - Interest up to $2,400 for one motor vehicle
  - Interest up to $400 per item for household goods

Voidable Transfers

The bankruptcy code prevents debtors from making unusual payments or transfers of property on the eve of bankruptcy that would unfairly benefit the debtor or some creditors at the expense of others

- Preferential transfers within 90 days before bankruptcy
- Preferential liens
- Preferential transfers to insiders
- Fraudulent transfers

Priority of Creditors

Secured creditors have priority over unsecured creditors

Unsecured creditors claims are satisfied out of the bankruptcy estate in order of their statutory priority

Nondischargeable Debts

Nondischargeable debts include

- Claims for taxes accrued within three years prior to bankruptcy filing
- Certain files and penalties payable to federal, state, and local government units
- Alimony, maintenance, and child support
- Claims based on the consumer-debtor's purchase of luxury goods of more than $1000 from a single creditor
- School loans

Chapter 11 Bankruptcy Reorganization

Reorganizes the debtor's financial affairs under the supervision of the Bankruptcy Court to give the debtor a new capital structure so that it will emerge from bankruptcy as a viable concern

Available to individuals, partnerships, corporations, unincorporated associations, and railroads

May be filed voluntarily by the debtor or involuntarily by its creditors

Rejection of Collective Bargaining Agreements

Companies that file for Chapter 11 reorganization sometimes argue that agreements with labor unions may be rejected in bankruptcy

The U.S. Supreme court has upheld the right of companies to reject contracts in bankruptcy if necessary for the successful rehabilitation of the debtor

Confirmation

Acceptance method

The bankruptcy court must approve a plan of reorganization if

The plan is in the best interests of each class of claims and interests

The plan is feasible

At least one class of claims votes to accept the plan

Each class of claims and interests in nonimpaired

Cram-down method

A method of confirmation of a plan of reorganization where the court forces an impaired class to participate in the plan of reorganization

Chapter 13 Bankruptcy Consumer Debt Adjustment

Permits the courts to supervise the debtor's plan for the payment of unpaid debts by installments

Debtor retains more property than is exempt under Chapter 7

Less expensive and complicated than Chapter 7

Creditors may recover a greater percentage of debts than under Chapter 7

IV. Objective Questions

Terms:

1. By discharging burdensome debts of the debtor, the federal bankruptcy law gives the debtor a _______________ _______________.

2. A stay that results from the filing of a bankruptcy petition, whether voluntary or involuntary, and suspends certain actions by creditors against the debtor or the debtor's property is called _______________ _______________.

3. An unsecured creditor is required to timely file a _______________ _______________ _______________ before he will be entitled to share in the distribution of the bankruptcy estate.

4. Those unusual transfers or payments made by the debtor on the eve of bankruptcy that can be avoided by the trustee because they benefit some creditors at the expense of others are called _______________.

5. A situation in which a debtor transfers her property with the actual intent to hinder, delay, or defraud her creditors is known as a _______________ transfer.

6. If all of the property in the bankruptcy estate is exhausted in satisfaction of allowed claims, remaining unpaid claims will be ___________________.

7. An arrangement by which the debtor voluntarily agrees to pay an unsatisfied debt that is dischargeable in bankruptcy is called a _______________ _______________.

8. In a Chapter 11 case, the debtor who is authorized to continue to operate her business is known as a _______________ _______________ _______________.

9. The document by which a bankruptcy proceeding is commenced is known as a _______________.

10. The legal representative of the bankrupt debtor's estate is the _______________.

True/False:

1. ____ Only partnerships and corporations are entitled to have their debts discharged in bankruptcy proceedings.

2. ____ Alimony, but not child support, is dischargeable in a bankruptcy proceeding.

3. ____ In a Chapter 7 proceeding, the debtor's assets (except for exempt property) are sold for cash and the cash is distributed to the creditors.

4. ____ A voluntary petition must state that the debtor is unable to pay his debts as they become due.

5. ____ In a bankruptcy proceeding, secured and unsecured creditors are given the same priority with respect to the distribution of the assets in the bankruptcy estate.

6. ____ If the value of collateral securing the debt of a particular creditor is less than the claim of that creditor, that person becomes an unsecured creditor to the extent of the difference.

7. ____ The bankruptcy trustee is never permitted to liquidate exempt property of the debtor.

8. ____ A debtor who was granted a discharge in bankruptcy five years ago will be denied a discharge of debts.

9. ____ The transfer of assets by an insolvent debtor to her parents within one year is not permissisble.

10. ____ The creditor commences voluntary bankruptcy proceedings.

Multiple Choice:

1. A downturn in the economy caused Charlie, a construction contractor, to suffer many financial losses. As a result, he was forced to declare bankruptcy. Charlie owns a home that he built valued at $85,000 on which he still owes $77,000, tools and other minor construction equipment worth $700, and a gold watch that he inherited valued at $600. Which of Charlie's assets will not be fully exempted under the Bankruptcy Act?
   A. The house.
   B. The tools.
   C. The watch.
   D. All of his assets will be fully exempted.

2. Paul Playboy was forced into bankruptcy involuntarily. The value of all his nonexempt assets amounted to $75,000. After paying priority claims, $30,000 remained. There is $100,000 of provable unsecured claims. These unsecured creditors will each receive:
   A. Nothing.
   B. One dollar for each dollar of debt owed them.
   C. Thirty cents for each dollar of debt owed them.
   D. None of the above.

3. On March 1, Jack loaned Janet $3,000 to purchase a sailboat. However, Jack insisted that he have possession of Janet's $4,000 diamond ring until she repaid the $3,000. On May 15, Janet filed a voluntary petition in bankruptcy. Fearing he may not be repaid, Jack sold the ring to Hillary Honest on May 20. Which of the following is correct?
   A. The transfer from Jack to Hillary constitutes a voidable preference.
   B. Jack is protected because he sold the ring to a bona fide purchaser for value.
   C. Hillary is entitled to keep the ring unless she knew that Janet had filed a petition in bankruptcy.
   D. None of the above.

4. Which of the following actions is not stayed in a bankruptcy proceeding?
   A. The enforcement of judgments obtained against the debtor.
   B. The institution of legal actions to collect debts against the debtor.
   C. An action to recover alimony from the debtor.
   D. All of the above.

5. Which of the following is not considered an asset of the bankruptcy estate?
   A. A software program created by the debtor.
   B. A time-share apartment in Mexico in which the debtor has an interest.
   C. $11 million the debtor won from a lottery ticket purchased with a dollar bill found on the street four months after filing the voluntary petition in bankruptcy.
   D $40,000 the debtor inherited from a distant relative three months after filing the voluntary petition in bankruptcy.

6. Sally Slowpay operates a printing business. Her business has been declining slowly for several years. Sally has been unable to pay any of her ten creditors for several months. George Greedy, to whom Sally owes $4,000, is frustrated and has lost his patience with Sally. He signs and files an involuntary bankruptcy petition against Sally. Which of the following is correct?
   A. The petition is invalid because at least three creditors must sign an involuntary petition.
   A. The petition is invalid because unsecured creditors with claims aggregating at least $5,000 must sign the petition.
   C. The petition is invalid if it fails to list Sally's current income and expenses.
   D. None of the above.

7. Which of the following is not a responsibility of the trustee in a Chapter 7 proceeding?
   A. Employ disinterested accountants, attorneys, and appraisers to serve as legal representatives of the estate.
   B. Take immediate possession of the debtor's property.
   C. Make reports to the court, creditors, and the debtor regarding the administration of the estate.
   D. Investigate the debtor's financial affairs.

8. Which of the following is not a primary purpose of the federal bankruptcy laws?
   A. To protect creditors from actions of the debtor that would diminish the value of the bankruptcy estate.
   B. To protect debtors from abusive activities by creditors in collecting debts.
   C. To give the debtor a fresh start by freeing her from legal responsibility for future debts.
   D. To preserve existing business relations of the debtor.

9. CompuTech, a manufacturer of computer components, was beginning to feel the pinch of the slow economy. Several months ago, Bully's Supply, the only supplier of certain components used in CompuTech's manufacturing process, began requiring CompuTech to pay for components upon delivery. In addition, Bully's required CompuTech to give it a note secured by a mortgage on certain properties owned by CompuTech to evidence outstanding balances due Bully's for components delivered in the past. The note and mortgage were executed on June 2, 2001 at which time the mortgage was filed. On August 8, 2001, CompuTech filed a voluntary petition in bankruptcy. Which of the following is correct?
   A. Any payments made to CompuTech for components delivered in June, July, or August are voidable preferences.
   B. The mortgage given to Bully's was a voidable preference.
   C. The mortgage was not a voidable preference because it was executed and filed on the same day.
   D. None of the above.

10. Which of the following acts by a debtor would not bar discharge of the debts of the debtor in a bankruptcy proceeding?
    A. An unusual payment or transfer of property on the eve of bankruptcy that would unfairly benefit some creditors at the expense of others.
    B. Failure to appear at a hearing on an objection to discharge.
    C. Falsifying, destroying, or concealing records of his financial condition.
    D. The transfer, concealment, or removal of property from the estate with the intent to hinder, delay, or defraud creditors.

V. Answers to Objective Questions

Terms:

1. *Fresh start.* This element represents the recycling aspect of bankruptcy law.
2. *Automatic stay.* This is a procedural "time out."
3. *Proof of claim.* Most proofs are relatively straightforward if they are based on a contract debt.
4. *Preferences.* Note that some of these transfers are presumed to have been made in bad faith.
5. *Fraudulent.* Many of these transfers subvert the intent of the law.
6. *Discharged.* The debt has now been legally ended.
7. *Reaffirmation agreement.* The code is concerned with the possible abuse of marketplace leverage by these creditors and wants to be sure that the debt be reinstated as a matter of free will.
8. *Debtor-in-possession.* Most cases will allow this unless there is a showing of some sort of bad faith on the part of management.
9. *Petition.* This is the opening procedural step.
10. *Trustee.* The trustee is not considered the sole agent of the creditors nor the debtor but is, in fact, responsible to the court.

True/False:

1. False. Only individuals may have their debts discharged in bankruptcy. Partnerships and corporations are required to liquidate.
2. False. Neither alimony nor child support is dischargeable in a bankruptcy proceeding.
3. True. This is why it is called liquidation.
4. False. A voluntary petition is only required to state that the debtor has debts.
5. False. A secured creditor's claim to the debtor's property has priority over the claims of unsecured creditors.
6. True. Many creditors find themselves in this two-track position.
7. False. If the debtor's equity in certain exempt property exceeds the exemption limit, the trustee may liquidate the property.
8. True. A debtor will be denied discharge if discharge was granted within six years of the commencement of the present case.
9. True. Relatives are considered insiders, thus making the one-year avoidance period applicable.
10. False. The debtor commences voluntary proceedings.

Multiple Choice:

1. C. An interest in jewelry up to $500 that is held for personal use is exempt under the Bankruptcy Act.
2. D. The Bankruptcy Code provides that unsecured claims are to be satisfied out of the bankruptcy estate in the order of their statutory priority. In this situation, there is not enough information to determine the priority in which the unsecured claims will be paid.
3. C. A creditor who has received a preferential transfer may convey good title to an innocent third party purchaser.
4. C. Congress did not want these sorts of obligations delayed.
5. C. Property acquired after the petition is not property of the bankruptcy estate. However, inheritance is an exception to this rule, and therefore answer D is incorrect. Intangible property, and any other property in which the debtor has a legal or equitable interest, wherever located, is considered property of the estate, and therefore answers A and B are incorrect.
6. B. If a debtor has fewer than twelve creditors, an involuntary petition may only be filed by a creditor, or creditors, with claims aggregating $5,000.
7. A. The trustee has the responsibility to hire disinterested persons to assist in the administration of the estate. However, the trustee is the legal representative of the estate.
8. C. The purpose is to give a fresh start by freeing the debtor from existing debts.
9. B. CompuTech gave Bully's a preferential lien within 90 days of filing the bankruptcy petition; the lien was given for amounts past due (antecedent debt); and it is presumed that Bully's would receive more because of the lien than as an unsecured creditor in liquidation. Answer A is incorrect because Bully's required payment simultaneously with delivery, and therefore such payments are not made in respect of antecedent debts.
10. A. A preferential transfer can be avoided in bankruptcy. It does not, however, bar discharge of debts. B, C, and D are all items that would bar discharge of debts.

VI. Answers to Essay Question

Things look blue for Pinky. Assuming he has not incorporated his business, he may be able to eventually have most of his debts discharged if he follows the steps outlined under the Bankruptcy Code. His business does not fall into any of the specific activities covered by separate statutes such as banks, railroads, or savings and loan associations, so he does have standing to petition the court as an individual.

It appears from the case as presented, Pinky's best strategy would lie in a Chapter 7 liquidation of his assets. Chapter 11 reorganization appears unlikely given Congress's decisive (and rare) action in getting rid of these truly obnoxious devices. There is very little likelihood that Pinky's business could be reorganized unless the machines could be reconfigured for some more socially productive use.

Assuming that a Chapter 7 liquidation is in order, the first issue that needs to be resolved is the filing of the petition for bankruptcy. Pinky may chose to file voluntarily, but if he does not, he appears to have little control over this issue. His creditors can force an involuntary filing upon him once he has stopped paying his credit bills on time and is in a position of liquid insolvency, i.e., not being able to pay his bills as they become due. Once the petition is filed, an order for relief will be issued. This order will suspend more creditors' actions against Pinky for the moment. The key exception to this stay would involve Bailout Bank's claims against Pinky's house. The bank is a secured creditor to the extent of the reduced value of the home and an unsecured creditor as to amounts owed above the collateral value. Bailout may therefore seek to get cash payments to cover depreciation costs or indubitable equivalents, but both remedies seem unlikely in this case.

Once the stay has been ordered, Pinky must meet with all his creditors and fully disclose his financial affairs to them. Soon thereafter, the Bankruptcy Court will appoint a trustee who will take control of Pinky's property. The trustee will seek to garner any assets properly included in Pinky's estate. This would include any accounts receivable for sales of his machines and the like. In addition, the trustee would look to see if there have been any voidable preferences, preferential transfers, or possible fraudulent transactions. The trustee will also examine the claims filed by the creditors.

When all the assets have been brought into the estate, the trustee will proceed, with court approval, to allocate those assets according to priorities set out the Bankruptcy Code. Pinky will be able to keep some exempted property. He will be able to claim a homestead based on either federal or state law. In addition, he will be able to keep one very small pinky ring valued at no more than $500. He may also keep any other property qualified under the listed exemptions set out in the code.

Assuming the home is sold to pay off the secured creditor, Bailout Bank (net of Pinky's homestead claim amount), Bailout Bank's remaining deficiency amount owed will then be classified as a general unsecured debt. As such those creditors will be paid in the priority of payment allowed by the code. The general unsecured creditors will be paid only after the fees and expenses of the administration of the estate are paid along with numerous other preferred creditors. In all likelihood, the computer suppliers and Rocks R Us are not going to fare well in bankruptcy.

When the remaining assets are sold off and the proceeds distributed, the trustee will file his or her report with all parties concerned. The Bankruptcy Court will then discharge Pinky from any further obligations on all his debts except the $100,000 owed to the IRS. Because that debt occurred within three years prior to the filing of the petition, it is not discharged. In addition, if Pinky for some reason chooses to reaffirm any of these debts, he must do so with the consent of the Bankruptcy Court.

All in all, Pinky might consider some other line of work once he gets his fresh start. He should get into a line of work that will not foster invasive intrusions into people's homes in the middle of the dinner hour.

Chapter 24
AGENCY

Those who consent to the act and those who do it shall be equally punished.
Sir Edward Coke

I. Overview

Qui facit per alium facit per se. He who acts through another acts himself. This simple Latin phrase provides the keystone upon which the mutual obligations of agency law rest. Agency is defined by Section 1 of the *Second Restatement of Agency* as "the fiduciary relation which results from the manifestation of consent by one person to another that the other shall act on his behalf and subject to his control, and consent by the other so to act."

Agency is a legally recognized relationship that allows an attribution of one person's behavior to another. This carryover process is two-sided in that both benefit and burden inure to parties involved in the agency relationship.

Under the basic doctrine of agency, the principal is allowed to reap the beneficial harvest of the agent's actions made in his or her behalf. For example, assume an agent has agreed to be paid a set salary of $100 for selling certain kinds of goods. The principal gets to keep the net profits from that agent's selling activities, be they $100 or $1,000,000. This net gain is what allows the use of agency theory to maximize one's profits through the actions of another. Exponential growth of most any sort of enterprise is almost always directly tied to effective use of the talent of others through agency law. There are some limits on this ability to designate others to act on one's behalf based on uniqueness of personal services or on public policy grounds that forbid use of agents, such as voting or serving a criminal sentence. As a practical matter, business as we know it today simply could not be conducted on any scale beyond sole proprietorship without extensive use of agency relationships.

The benefits of agency are not without counterbalancing detriment. The Latin maxim *respondeat superior* may be familiar to you. It means that, in certain instances, a principal is liable to third parties for the acts of his or her agent. Just as the benefits of agency can be great, so can the burdens. One of the fastest growing areas of management specialization in today's business environment is risk management. This area generally concerns business financial responsibility for exposure to specified contingencies or perils. Included in these perils are the acts of the agents for which the principal may be liable. The ironic aspect of all this is that the very same people who help a business grow can lead that same enterprise to financial ruin.

Every agency liability question having an involvement with third parties has three subquestions that must be answered in order to come to a final resolution of the issues at hand. They are:

1. What are the responsibilities of the principal and agent vis-à-vis each other?

2. What are the responsibilities of the agent vis-à-vis the third party?
3. What are the responsibilities of the principal vis-à-vis the third party?

Invariably a certain pattern of facts emerges. First there is some sort of principal/agent relationship established. This relationship may be based on actual, implied, apparent, or ratified authority. In all events, once that authority line has been drawn, the question of the legal consequences to the principal and agent vis-à-vis each other must be answered. These consequences would include their respective rights and duties to each other as outlined in the key question checklist.

Once the first subquestion is resolved, the rights, duties, and obligations of the agent and principal respectively, must be examined vis-à-vis the third party. Often there will be some sort of wrongful and unauthorized act committed by the agent. That act will result in probable liability for both the agent and the principal to the third party who was harmed by the act. Think of the three subquestions as a loop that must be closed in order for the whole case to be resolved. The loop is similar to a legal description used to outline the surveyed ownership of real property. A proper legal description always closes at the point where it began. So must an agency issue. It starts with the establishment of the agency relationship. It goes through the rights and duties of third parties. It terminates back where it began, with a determination of the ultimate responsibilities of the principal and agent vis-à-vis each other.

II. Hypothetical Multi-Issue Essay Question

Jacko Frinton is the quintessence of opportunistic entrepreneurship. No silly rules on employment taxes and social security were going to stop him from creating the largest door-to-door sales force in used toothbrushes ever assembled. He specifically hired all his salespeople as independent contractors. Under the terms of his agreement with them, he controlled the time, routes, and all working conditions on their sales routes. They, in turn, were to be paid only on commission. In addition, because they were designated as independent contractors, Jacko saved millions on FICA and other employment taxes. The Internal Revenue Service got wind of Jacko's operation from a disgruntled brush salesman who turned him in for an informant's fee. Does Jacko owe any employment taxes to the IRS?

III. Outline

Nature of Agency
- Agency relationships are formed by the mutual consent of a principal and an agent
  - Principal
    - The party who employs another person to act on his or her behalf
  - Agent
    - The party who agrees to act on behalf of another
  - Agency
    - The principal-agent relationship

Employment Relationships
- Employer-employee
  - A relationship that results when an employer hires an employee to perform some form of physical service
- Principal-agent
  - A relationship that results when an employer hires an employee and gives that employee authority to act and enter into contracts on his or her behalf
- Principal-independent contractor
  - A relationship that results when a person or business that is not an employee is employed by a principal to perform a certain task on his or her behalf

Forming the Agency Relationship
- Express agency
  - An agency that occurs when a principal and an agent expressly agree to enter into an agency agreement with each other
- Implied agency
  - An agency that occurs when a principal and an agent do not expressly create an agency, but it is inferred from the conduct of the parties
- Apparent agency
  - Agency that arises when a principal creates the appearance of an agency that in actuality does not exist
- Agency by ratification
  - Agency that occurs when
    - A person misrepresents himself or herself as another's agent when in fact he or she is not
    - The purported principal ratifies the unauthorized act

Termination of an Agency by Acts of the Parties
- Mutual agreement
- Lapse of time
- Purpose achieved
- Occurrence of a specified event

Termination of an Agency by Operation of Law
- Death of the principal or agent
- Insanity of the principal or agent
- Bankruptcy of the principal
- Impossibility of performance
- Change in circumstances
- War between the principal's and the agent's countries

Wrongful Termination of an Agency Contract
- Wrongful termination is the termination of an agency contract in violation of the terms of the agency contract
- The nonbreaching party may recover damages from the breaching party

Irrevocable Agency
- An agency coupled with an interest is a special type of agency relationship that is created by the agent's benefits
- It is irrevocable by the principal

IV. Objective Questions

Terms:

1. An agency that occurs when a principal and agent agree to enter into an agency agreement with each other is called an _______________ agency.

2. An agency that occurs when a principal and agent do not expressly create an agency, but it is inferred from the conduct of the parties is called an _______________ agency.

3. An agency that arises when a principal creates the appearance of an agency that in actuality does not exist is known as an _______________ agency.

4. An agency that occurs when (1) a person misrepresents herself as another's agent when in fact she is not and (2) the purported principal ratifies the unauthorized act is called an _______________ _______________ _______________.

5. The party who employs another person to act on her behalf is the _______________.

6. An agency results if the third party entering into the contract knows (1) that the agent is acting as an agent for a principal and (2) the actual identity of the principal. This is called a _______________ _______________ agency.

7. A rule that says an employer is liable for the tortious conduct of its employees or agents while they are acting within the scope of its authority is called _______________ _______________.

8. The rule that says a principal is generally not liable for injuries caused by its agents and employees while they are on their way to or from work is called the _______________ _______________ _______________ _______________.

9. The test to determine the liability of the principal: if the agent's motivation in committing the intentional tort is to promote the principal's business, then the principal is liable for any injury caused by the tort is known as the _______________ test.

10. The test to determine the liability of a principal: if an agent commits an intentional tort within a work-related time or space, the principal is liable for any injury caused by the agent's intentional tort. This test is called the _______________ _______________ test.

True/False:

1. ____ An agency relationship must be based on consideration. That is, no agency relationship is created if the agent is not paid for his services.

2. ____ A power of attorney, which is an express agency agreement, may be established by either an oral or written agreement.

3. ____ A principal's ratification of an unauthorized contract must be by express approval. Implied consent is not sufficient to bind a principal under ratification principles.

4. ____ Constructive notice is sufficient to inform parties who dealt with an agent that the agency has been terminated.

5. ____ An employee in an employer/employee relationship is not an agent since she cannot enter into contracts for her principal.

6. ____ In a partially disclosed agency, both the agent and the principal are liable on any contract that the agent makes.

7. ____ An agent of a fully disclosed principal is liable on a contract if he does not specify the principal's name in signing the contract.

8. ____ Both agents and principals are liable for an agent's breach of the implied warranty of authority.

9. ____ A principal is not liable for the criminal conduct of his agent since the principal did not have the criminal intent to commit a crime.

10. ____ An agent is prohibited from using his agency for personal benefit even if the principal is aware of and consents to such actions.

Multiple Choice:

1. Roman, an agent for Tom, has the express authority to sell Christine's goods. Roman also has the express authority to grant discounts of up to five percent of list price. Roman sold Sidney goods with a list price of $1,000 and granted Sidney a ten percent discount. Sidney had not previously dealt with either Roman or Tom. Which of the following courses of action may Christine properly take?
   A. Seek to void the sale to Sidney.
   B. Seek recovery of $50 from Sidney only.
   C. Seek recovery of $50 from Roman only.
   D. Seek recovery of $50 from either Sidney or Roman.

2. Mean Master wanted to hire Billy Butler as his personal "man." Generally, under the laws of agency, which of the following acts may Mean Master not hire Billy Butler to do?
   A. Wash, clean, and hand press his dirty jogging clothes.
   B. Clean out his backed up sewer line with a toothbrush.
   C. Brush Mean Master's teeth bare handed.
   D. Have Billy serve Mean Master's time in federal prison for employment tax evasion.

3. Kathleen Company dismissed Sabrina, its purchasing agent. It published a notice in the appropriate trade journals that stated that the Kathleen Company no longer employed Sabrina. Sabrina called on several of Kathleen's suppliers with whom she had previously dealt, and when she found one who was unaware of her dismissal, she would place a substantial order for merchandise to be delivered to a warehouse in which she had rental space. Sabrina also called on several suppliers with whom Kathleen had never dealt; she would present one of her old business cards to the secretary and then make purchases on open account in the name of Kathleen. Sabrina then sold all the merchandise and absconded with the money. In this situation:
   A. Sabrina had continuing express authority to make contracts on Kathleen's behalf with suppliers with whom she had previously dealt as Kathleen's agent if they were unaware of her dismissal.
   B. The suppliers who previously had no dealing with Kathleen cannot enforce the contracts against Kathleen even if the suppliers were unaware of Sabrina's lack of authority.
   C. Kathleen is liable on Sabrina's contracts to all suppliers who had dealt with Sabrina in the past as Kathleen's agent.
   D. Constructive notice via publication in the appropriate trade journals is an effective notice to all third parties regardless of whether they had dealt with Sabrina or read the notice.

4. Worker Wendy is an employee in an employer/employee relationship with her employer, Car Parts Company (CPC). In her job, Wendy loads various car parts on a truck for delivery. CPC comes to you, its attorney, and asks about its liability with respect to Wendy. Please advise them.
   A. CPC is liable for any torts committed by Wendy and any contracts she enters into.
   B. CPC is liable for any contracts Wendy enters into, but never for any torts she commits.
   C. CPC is liable for any torts committed by Wendy while acting on the company's behalf.
   D. CPC is not liable for any torts committed by Wendy or any contracts she enters into since she is only an employee.

5. Principal Peter hires Agent Anne to act as a sales agent for his company, Complete Computers. Anne and Peter have a written agreement that states that Anne may negotiate sales contracts and receive customer payments. Although the agreement is silent on the issue, Anne also delivers the computers to her customers. Which of the following is correct regarding Anne's authority?
   A. Anne is an expressly appointed agent with some implied authority.
   B. Anne is an implied agent.
   C. Anne is an agent by ratification.
   D. None of the above.

6. Which of the following is not a duty of an agent towards the principal?
   A. Duty of performance.
   B. Duty to provide safe working conditions.
   C. Duty of loyalty.
   D. Duty of obedience.

7. Agent Amy has negotiated a contract for the sale of computer equipment to Merchant Mike. Amy's principal, Principal Pat believes that the equipment is not to be shipped until October 10. However, Mike told Amy he needed the equipment shipped on October 2. Amy does not notify Pat of this change. Consequently, the equipment is not shipped until October 10 and Mike incurs damages. Which of the following is correct?
   A. Amy is liable to Pat for any damages her failure to notify may have caused.
   B. Pat is not liable to Mike since he never knew of the changed date.
   C. If Amy is a gratuitous agent, she does not have a duty to notify Pat.
   D. All of the above.

8. Scientist Sam works for Medical Miracles (MM). During his work for MM, Sam acquires knowledge of the company's greatest technological innovation, a single dose pill that immediately replaces hair loss. The pill is known as Single Sensation. Sam leaves MM soon after acquiring this knowledge to start his own medical supplies company. Which of the following is correct?
   A. Sam can use his knowledge of Single Sensation for personal benefit after he leaves MM.
   B. Sam's use of his knowledge of Single Sensation would be a violation of Sam's duty of notification.
   C. Sam cannot use his knowledge of Single Sensation even after he leaves MM.
   D. None of the above.

9. Assume the same facts as in question 8. Also, assume Sam does use the information after leaving MM. What are MM's remedies?
   A. Obtain an injunction against Sam?
   B. Sue Sam for damages?
   C. Both of the above?
   D. None of the above? MM has no recourse once Sam is no longer its employee.

10. Agent Arnold negotiates a contract with Theo on behalf of Principal Peter. Theo is aware that Arnold is working for a principal. However, Theo does not know that Peter is Arnold's principal. Theo and Arnold reach agreement on the contract and sign it. Arnold signs as follows: Agent Arnold, agent. Regarding liability on the contract, which of the following is correct?
    A. Arnold is liable on the contract.
    B. Peter is liable on the contract.
    C. Both of the above.
    D. None of the above. Arnold is not liable since Theo knew he was only an agent, and Peter is not liable since his name is not on the contract.

V. Answers to Objective Questions

Terms:

1. *Express.* This is the most common form of agency. They can be either oral or written unless the Statute of Frauds stipulates that they must be written.
2. *Implied.* The extent of the agent's authority is determined from the facts and circumstances. These can include industry custom, prior dealings between the parties, or any other factors the court may deem relevant.
3. *Apparent.* This agency is also sometimes called an agency by estoppel. If the agency is established, the principal is estopped from denying the relationship. It is the principal's actions (or lack thereof) that can create an apparent agency.
4. *Agency by ratification.* Note, at the time of the purported agency, there was, in fact, no agency in existence. It is the act of ratification (acceptance) by the principal that renders him or her now liable for the acts of the agent.
5. *Principal.* Most principals are known as employers although that has its own set of special characteristics.
6. *Fully disclosed.* In a fully disclosed agency, the contract is between the principal and the third party and the agent is not generally liable.
7. *Respondent superior.* This term is derived from the Latin term meaning: "Let the master answer." The doctrine represents the burden that comes to the principal where he or she can also reap the benefits of the agent's act.
8. *Coming and going rule.* As the labor market adjusts to more off-site work situations through the use of electronic networking, the definition of this employer's defense theory will become more difficult.
9. *Motivation.* If the agent's motivation in committing the intentional tort was personal, then the principal is not liable.
10. *Work-related.* This doctrine rejects the motivation test and will hold the principal liable, regardless of motivation, if the act was committed within a work-related time or space.

True/False:

1. False. An agency relationship may be created even without consideration. Such an agency is known as a gratuitous agency.
2. False. A power of attorney, which is an express agency agreement, must be established by a written agreement. An oral agreement is not sufficient to establish a power of attorney.
3. False. Implied consent may be shown by the principal's conduct and is sufficient to show the principal's intent to ratify an unauthorized contract.
4. False. Parties who have dealt with an agent must be given direct notice of termination of the agency. Constructive notice is sufficient for parties who have no knowledge of the agency.

5. True. An agent is empowered to enter into contracts for her principal. An employee in an employer/employee relationship does not have this power.
6. True. Since the actual identity of the principal is not known to the third party in a partially disclosed agency. Both the agent and the principal are liable on such a contract.
7. True. Even though agents of fully disclosed principals are not usually liable on contracts for the principals, if the agent does not specify the principal's name in signing the contract, he may be held liable on it.
8. False. Only an agent is liable for an agent's breach of the implied warranty of authority.
9. True. The principal cannot have the requisite *mens rea* to be convicted of a crime that the agent committed and the principal was not involved in.
10. False. An agent may personally benefit from his agency if the principal is aware of such conduct and consents to it.

Multiple Choice:

1. C. Roman exceeded his express authority when he granted the ten percent price discount to Sidney; therefore, Roman would be liable to Christine for the amount of damage caused by Roman's breach, $50. A, B, and D are incorrect because Sidney had not previously dealt with Roman or Tom; therefore, Roman was acting with apparent authority when he granted the ten percent price discount to Sidney. Consequently, Christine would not be able to void the sale or recover from Sidney.
2. D. Public policy prevents Mean Master from trying to have Billy Serve his sentence. He can ask Billy to do A, B, and C, assuming no labor or minimum wage laws are broken. Billy might want to consider another line of work.
3. B. When the agency is terminated, the principal must notify customers of the agent's termination. Publication of the notice in the trade journal (constructive notice) precluded Sabrina's apparent authority to those suppliers who previously had no dealings with Kathleen. A is incorrect because Sabrina did not have continuing express authority. However Sabrina did have continuing apparent authority on behalf of suppliers with whom she had previously dealt if they did not know of her dismissal. C is incorrect because Kathleen is not liable on the contracts with those suppliers who read the published notice and had actual knowledge of the discharge that terminated Sabrina's apparent authority. D is incorrect because constructive notice by publication is sufficient notice only for the third parties who had not previously dealt with Sabrina.
4. C. An employer in an employee/employer relationship is liable for the torts of its employees if an employee acting on the principal's behalf committed the tort. A is incorrect because CPC is not liable for any contracts Wendy enters into since she has no authority to do so. B is incorrect because CPC is not liable for contracts Wendy enters into for the reason stated above. In addition, CPC is liable for torts Wendy commits while acting on its behalf. D is incorrect because CPC is liable for some torts even though she is only an employee in an employer/employee relationship.
5. A. Since Anne and Peter have an express agreement regarding Anne's agency for Peter, Anne is an expressly appointed agent. In addition, expressly appointed agents may also have implied or apparent authority to carry out certain activities which are not expressly stated. Here, Anne's authority to deliver computers is implied, not expressed. B is incorrect because the agreement between Anne and Peter is express, not implied. C is incorrect because there is no indication of ratification here. Also, Anne clearly has an express agency agreement.
6. B. The duty to provide safe working conditions is owed by the principal to his or her agent(s). A, C, and D are duties owed by the agent to his or her principal.
7. A. Amy has breached her duty to notify Pat. Therefore, she is liable for any damages resulting from this breach. B is incorrect because Pat is liable to Mike since Amy's knowledge of the changed date is imputed to Pat. C is incorrect because even a gratuitous agent must perform the necessary duties owed by an agent to a principal.

8. C. Sam owes a duty of loyalty to MM not to use its trade secrets or technological innovations even after Sam leaves the company. Such a use would be a misuse of confidential information. A is incorrect because this is a breach of the duty of loyalty. B is incorrect because use of the knowledge is a breach of the duty of loyalty, not the duty of notification.
9. C. MM can seek either of these remedies.
10. C. Both A and B are correct because both an agent and a principal are liable on a contract in a partially disclosed agency situation. D is incorrect because an agent may be held liable in spite of his agency capacity, and a principal is always liable for the authorized contracts of his agents.

VI. Answers to Essay Question

In this case, it appears that there is an express agreement entered into between Jacko and his salespeople. Under this agreement, the salespeople are given specific authority to act on behalf of Jacko with regard to entering into sales contracts for his products. Implied in this actual authority would be reasonably related activities involved in the sale of the product.

What will happen to this agency relationship revolves around a very common scenario currently being played out in many of the U.S. District and Tax Courts throughout the U.S. A number of business enterprises have sought to avoid responsibility for increasingly higher employment taxes by seeking to create independent contractor status for their employees. The IRS and local taxing authorities are, in turn, challenging these arrangements as being specious and in contravention to the intended coverage of the tax code. What is particularly unfortunate about these cases is that the "employee" winds up footing the bill one way or the other. If employees are real independent contractors, they must pay all of their own social security taxes. If they are not, often employers' failure to pay into those funds leads to no credit for the time worked in the employee's ultimate computation of retirement or disability benefits. Either way the worker pays.

Because of the large amount of control that Jacko has chosen to exercise over his salespeople, it would appear that the IRS might want to argue that they are, in fact, employees rather than independent contractors. The tax rules on this issue are extensive and go beyond the intended scope of this answer. Assuming that the IRS did challenge Jacko's arrangement, his best rebuttal would be the fact that the salespeople's compensation was based entirely on commissions. That fact alone, however, would not likely shield Jacko from liability for employment taxes. It looks as if Jacko will have to find a better way to do business. Perhaps refilling used toothpaste tubes would be more fulfilling to him?

Chapter 25
EMPLOYMENT AND LABOR LAWS

Adam ate the apple, and our teeth still ache.
Hungarian Proverb

I. Overview

Organized labor has suffered from a long and steady decline in membership, power, and influence over the past forty years. Much of this slide has been of its own doing, traceable to poor union management, a fat cat image, and sometimes-silly work rules that have no economic justification in the face of changed technology. In spite of all this bad news for unions, consider the working conditions that existed before them. It is a hallmark of advanced industrialized economies that the work force is highly organized and has a strong bargaining power over its affairs. The immediate post-Civil War era of industrialization saw the possibilities for abuse of the work force not only become reality, but also a tragedy, when it came to workers' safety. Most modern social legislation, ranging from the minimum wage, to child labor laws, to workplace and anti-discrimination statutes are traceable not to the largess of employers but rather to hard fought collective bargaining agreements. It is no accident that union representation is low in areas of the world still noted for the exploitation of their labor force. Corollaries of low wages are low levels of worker protection, environmental callousness, and an overall diminished standard of living.

The basic employer/employee relationship is a contractual one. As with any contract, both parties are expected to enter into the relationship with their own best interests at heart. The quaint notion of a paternalistic employer who cares for his workers over and above his own interests is simply unrealistic in today's age of cutthroat economic competition. What is realistic is enlightened self-interest. Each side of the labor management relationship still looks out for itself. But in looking out for number one, both must realize their mutual interdependence on each other. Labor must realize that it cannot sustain its own survival on the backs of failed companies brought down by union imposed inefficiencies. Labor must adjust to "Real World 101" and make concessions to both the technological and economic realities of trying to compete in a global economy. Management, in turn, cannot forever continue to erode our economic consumer base at home by running overseas at every opportunity for lower wages and less restrictive hospitalities for doing business.

What good will all those do in the long run if we have allowed our home economy to become depleted of workers who earn real living wages? We cannot let our economy degenerate into one of only two classes: the very rich and minimum wage service workers. Our industrial base was built on a working partnership between management and labor. Like any marriage this partnership was not always easy to live with. But it did thrive on a mutual respect for the other's role in the larger scheme of things. What we have had instead for the past forty years is a willingness on the part of both sides to forgo the long-term societal benefits that can inure from good faith bargaining. The end result is that our economy is in danger of not only being nonunion, but also more noncompetitive, nonproductive, and nongrowth than ever. Unions and management both need to wake up and smell the coffee before we become a fallen third-rate economic power.

A phase commonly used in the area of pension and employee benefits law is "golden ball and chain." This description is meant to illustrate the two-sided nature of many of the perquisites associated with the employment relationship. On one side, the gold comes from the financial security afforded by a steady income and the various employment-related insurance protections designed to buffer one from the vicissitudes of life. The ball and chain aspect comes from the price paid for those perks. This cost not only includes the obvious time and energy commitments involved in doing one's job, but also the lost opportunity costs associated with having chosen one employer over others. For many the employment relationship becomes one of love-hate or at least a standoff based on economic necessity. As one demoralized employee recently stated: "My pension is simply a form of compensatory damages for having endured terminal boredom."

In the past, the dominant contractual form this relationship has taken has been found in the *employment at will* doctrine. This doctrine assumes that given equal bargaining power, and absent express or implied agreement to the contrary, either the employee or employer may end the relationship. This termination can come about at any time, for any reason, bilaterally or unilaterally. The doctrine has long been under fire because of being myopic on two main scores: it presumes equality of bargaining power between the employer and employee and that the employment relationship is a totally private contractual matter between the contracting parties.

Recent cases and legislative enactments have greatly eroded the doctrine. On the legislative front, a number of states have decided that public policy interests in favor of certain kinds of activities must take precedent over the employer/employee relationship. Examples would include voting rights, antidiscrimination measures, whistleblower protections, and employee health and safety protections. In addition, a number of courts have seen fit to interpret employer handbooks, written and oral job policies, and other acts as indicia of an implied contract between the employer and employee. So what may have appeared originally to be the employee's economic ball and chain has also become the employer's.

The simple truth is that there is a growing involvement of government at every step of the employer/employee relationship. It sets the ground rules for hiring, working conditions, paying for harm, termination, and ultimate payment of pensions and or death benefits. It all has come a long way from the simplistic and archaic notion that the employment contract is only the business of the immediate parties involved.

II. Hypothetical Multi-Issue Essay Question

Mr. Quincy Harpie has a problem. As a matter of fact, he has a lot of problems. He is chronically complaining about all sort of illnesses, both actual and perceived. His mother was never proud of him and kicked him out of the house at age thirty-two. Finally Mr. Harpie had to face reality and get a job.

He found his first job with Mega Widgets Corporation in Backwater as a crystal widget assembler. Two weeks into the job, he complained of severe backaches and sought worker compensation for his back problems. He was awarded his claim of partial disability and proceeded to stay at home while seeking other employment.

He did find other employment at Mini Widgets Corporation in the adjoining state of Massazona. He did not tell Mini Widgets about his employment history with Mega Widgets. Two weeks on the job as an assembler, guess what happened? His back flared up again. He successfully filed a worker compensation claim against Mini Widgets. He continued to also collect from Mega Widgets claiming that he was still disabled.

Mr. Harpie decided to try once more and moved to East Moncalla where he found employment with Micro Widgets Corporation. Micro had a strict employment at will policy and informed Mr. Harpie of that policy in writing. After two weeks of work on Micro's assembly line, Mr. Harpie again claimed a back injury. He filed yet a third claim under workers compensation. As in the prior incidents, he did not inform Micro of any of his prior job history.

Micro fought Mr. Harpie's claim and also fired him under the employment at will doctrine. Mr. Harpie claims that the doctrine is being wrongfully used against him. He is arguing that he was being fired only because he asserted his rights under the workers compensation law. At the time Micro did not know about Mr. Harpie's history with Mega and Mini Widgets Corporations. Should Mr. Harpie be reinstated at Micro Corporation?

III. Outline

Major Federal Labor Law Statutes
- Norris-LaGuardia Act
- National Labor Relations Act
- Labor-Management Relations Act
- Labor-Management Reporting and Disclosure Act
- Railway Labor Act

Organizing a Union
- Section 7 of the NLRA gives employees the right to join together and form a union
- The group that is seeking representation is called an appropriate bargaining unit
- Employers may restrict solicitation activities to the employees' free time and to non-working areas

Collective Bargaining
- Collective bargaining
  - The act of negotiating contract terms between an employer and the members of a union
- Collective bargaining agreement
  - The resulting contract from a collective bargaining procedure

Union Security Agreements
- Union shop
  - An employee must join the union within a certain number of days after being hired
- Agency shop
  - Employees do not have to be union members but must pay an agency fee (an amount equal to union dues) to the union
- Closed shop
  - An establishment where union membership is a condition of employment
  - Closed shops are illegal under the Taft-Hartley Act

Crossover and Replacement Workers
- Crossover worker
  - A person who does not honor a strike by either
    - Choosing not to strike

Returning to work after joining the strikers for a time

Replacement worker

A worker who is hired either on a temporary or on a permanent basis to take the place of a striking worker

Illegal Types of Strikes

- Violent strikes
- Sit-down strikes
- Partial or intermittent strikes
- Wildcat strikes
- Strikes during the 60-day cooling off period
- Strikes in violation of a no-strike clause

Secondary Boycott Picketing

A type of picketing where unions try to bring pressure against an employer by picketing his or her suppliers or customers

Employment at Will

- At-will employees do not have employment contracts
- Under common law, at-will employees could be discharged by an employer at any time for any reason
- Today, there are many exceptions to the at-will doctrine
  - Statutes
  - Contracts
  - Public policy
  - Tort

Wrongful Discharge

Wrongful discharge

The discharge of an employee in violation of a statute, an employment contract, or public policy, or tortiously.

The employee can recover damages or other remedies

Statutes that Protect Workers

- Workers' compensation acts
- Occupational Safety and Health Act (OSHA)
- Fair Labor Standards Act
- Employee Retirement Income Security Act (ERISA)
- Consolidated Omnibus Budget Reconciliation Act (COBRA)
- Immigration Reform and Control Act (IRCA)
- Unemployment compensation
- Social security system
- Family and Medical Leave Act

Worker's Compensation Acts

- Compensates workers and their families if workers are injured in connection with their jobs
- Worker's compensation is an exclusive remedy

Occupational Safety and Health Act
- Promotes safety in the workplace
- OSHA has regulations to enforce the safety standards established by the act
  - Specific duty standards
  - General duty standards
- OSHA is empowered to inspect places of employment for health hazards and safety violations

Fair Labor Standards Act
- Prohibits child labor
- Establishes overtime pay
- Establishes minimum wage

Employee Retirement Income Security Act
- Employers are not required to establish pension plans for their employees
- This statute is designed to prevent fraud and other abuses associated with private pension funds

Consolidated Omnibus Budget Reconciliation Act
- Permits employees and their beneficiaries to continue their group health insurance after an employee's employment has ended

Immigration Reform and Control Act
- Makes it unlawful for employers to hire illegal immigrants
- All U.S. employers must fill out an INS Form I-9 for each employee stating that the employer has inspected the employee's legal qualifications to work

Social Security
- Provides limited retirement and death benefits to covered employees and their dependents
- Employers and employees must make contributions into the Social Security fund
- Self-employed persons must pay Social Security taxes equal to the combined employer-employee amount

IV. Objective Questions

Terms:

1. The administrative procedure created so that workers receive compensation for injuries that occur on the job is called _______________ _______________.

2. The 1970 Occupational Safety and Health Act created the _______________ _______________ _______________ _______________, a federal administrative agency empowered to administer the Act and adopt rules and regulations to interpret and enforce the Act.

3. In 1986, the Immigration Reform and Control Act granted _______________ to illegal immigrants who were living in the U.S.

4. Since 1935, most workers who are temporarily unemployed are entitled to _______________ _______________.

5. Under the Fair Labor Standards Act, employers are required to pay a _______________ wage to employees.

6. The act of negotiating is called _______________ _______________.

7. A union may operate a _______________ _______________ that acts as a job referral service and clearinghouse between employees seeking work and employers seeking workers.

8. In a _______________ shop, an employee must join the union within a certain number of days after being hired.

9. When striking employees walk in front of an employer's premises carrying signs announcing their strike, this is called _______________.

10. A contract where an employer agrees with a union not to use, handle, transport, deal in, or purchase the products made by nonunion workers of other employers is a _______________ _______________ contract.

True/False:

1. ____ An at-will employee could be discharged by an employer at any time and for any reason under common law.

2. ____ The Employee Polygraph Protection Act is administered by the Department of Justice.

3. ____ Under the Fair Labor Standards Act, an employer cannot require nonexempt employees to work more than 40 hours per week unless they receive overtime pay.

4. ____ Payments under workers' compensation programs include hospital and medical benefits, disability benefits, and death benefits to dependents, but not recovery for certain body parts.

5. ____ Employers are required to establish pension plans for their employees.

6. ____ The American Federation of Labor was formed in 1886 under the leadership of John L. Lewis.

7. ____ Once a union is elected and certified, covered employees may negotiate individually with the employer regarding employment benefits.

8. ____ Employees with a bona fide religious objection to joining or financially supporting a union do not have to join.

9. ____ If a collective bargaining agreement cannot be reached, the union may call a strike.

10. ____ Unions cannot spend union dues on political activities.

Multiple Choice:

1. Under the Fair Labor Standards Act, which of the following does not define lawful child labor?
   A. Children aged 14 and 15 may work limited hours in nonhazardous jobs approved by the Department of Labor.
   B. Children aged 16 and 17 may work unlimited hours in nonhazardous jobs approved by the Department of Labor.
   C. Child actors and performers are exempt from child labor law restrictions.
   D. Children under the age of 14 cannot work as newspaper deliverers.

2. Edwin Employee worked for ABC Creditcard, an employer legally permitted to use polygraph testing. Which of the following statements is not true with regard to Edwin?
   A. ABC must give notice to Edwin.
   B. ABC must use licensed examiners.
   C. The polygraph examiner may ask Edwin any questions he wants to ask.
   D. The polygraph examiner may not ask questions relating to Edwin's religion or sexual behavior.

3. Edna Employee is employed as a secretary by Megacorp. Since this is a slow week for dictation, Megacorp asked Edna to work only 30 hours this week and 50 hours next week. When Edna received her check, she received no overtime pay for the extra 10 hours during the 50 hour week. Edna feels she is entitled to the overtime pay. What result?
   A. Under the Fair Labor Standards Act, an employer can require nonexempt employees to work more than 40 hours per week without overtime compensation.
   B. For purposes of determining overtime compensation, weeks are not treated separately. Therefore, Edna is not entitled to overtime pay for 10 hours of the 50-hour week.
   C. For purposes of determining overtime compensation, each week is treated separately. Therefore, Edna is entitled to overtime pay for 10 hours of the 50-hour week.
   D. None of the above.

4. Lunching Linda was on her personal lunch hour at the Et Yet Cafe. Distracted Doris spilled boiling hot split pea soup on Linda. Because Linda does not think Doris has any money, she decides to seek recovery from her employer under workers' compensation. Which of the following must Linda prove?
   A. That she is an employee and not an independent contractor.
   B. That the injury arose out of and in the course of her employment.
   C. That she was at Et Yet on an errand for her employer.
   D. All of the above.

5. Which of the following statements about Social Security is untrue?
   A. Self-employed persons do not have to pay Social Security taxes.
   B. Social Security includes retirement benefits.
   C. Survivors' benefits to family members of deceased workers are included in Social Security.
   D. Social Security is not like a savings account.

6. Which act created the National Labor Relations Board?
   A. Labor-Management Reporting and Disclosure Act.
   B. Norris-LaGuardia Act.
   C. National Labor Relations Act.
   D. Labor-Management Relations Act.

7. Where a union solicitation is being conducted by fellow employees, which of the following may an employer not do?
   A. Restrict solicitation to the employees' free time.
   B. Limit solicitation to nonworking areas such as the cafeteria, etc..
   C. Dismiss employees who do not violate solicitation rules.
   D. Bar off-duty employees from union solicitation on company premises.

8. Which of the following is not a permissive subject of collective bargaining?
   A. Size and composition of supervisory force.
   B. Corporate reorganization.
   C. Work discrimination.
   D. Location of plants.

9. Sally Striker was on strike with her union protesting unsafe working conditions. Her employer hired a replacement worker to replace her during the strike. Now the strike is over and Sally wants her job back, but her employer says she is discharged. What results?
   A. Economic strikers who reapply for their jobs are entitled to a nondiscriminatory review of their job applications prior to new applicants.
   B. Employees who are striking against an unfair labor practice are entitled to reinstatement even if the employer has to discharge replacements hired during the strike.
   C. Illegal strikers may be discharged by the employer and have no right to reinstatement.
   D. Strikers do not have to be paid while they are on strike.

10. Picketing is lawful when it:
   A. Is accompanied by violence.
   B. Prevents nonstriking employees from entering the employer's premises.
   C. Does not obstruct customers from entering the employer's place of business.
   D. Prevents pickups and deliveries at the employer's place of business.

V. Answers to Objective Questions

Terms:

1. *Worker's compensation.* The basic statutory scheme is designed to avoid the waste and expense of seeking compensation for work-related injuries through the courts. It is one of the earliest forms of contract-based strict liability.
2. *Occupational Safety and Health Administration.* This agency has power to promote and regulate safety in virtually all workplaces except federal, state, and local government workplaces. It is part of the Department of Labor, and it imposes a number of record-keeping and reporting requirements on employers.
3. *Amnesty.* This act continues to be controversial in that members of certain ethnic groups allege that it is being enforced against them in a discriminatory manner. People on the opposite side of the political fence are saying that the act is not being enforced enough to be effective. You decide.
4. *Unemployment compensation.* The federal government sets the guidelines for these benefits, but the actual administration of these employer-contributed taxes is handled by the individual states.
5. *Minimum.* This is one of the provisions of the Fair Labor Standards Act of 1938 and it is occasionally amended to reflect the higher costs of living due to inflation. Much political debate revolves around this provision due to its effects on the job market at the lowest end of the employment spectrum.
6. *Collective bargaining.* Once a union has been elected, the employer and union will engage in this process in an attempt to negotiate a contract.

7. *Hiring hall.* Most job assignments made through these halls are made on the basis of seniority.
8. *Union.* Union shops are lawful. Closed shops are not lawful.
9. *Picketing.* This practice is lawful unless it is accompanied by violence or some other improper interference with the employer's business.
10. *Cross-over workers.* These are individual members of a union who chose not to honor a strike.

True/False:

1. True. This common law doctrine is still very important in the study of employment law in spite of all the statutory and public policy based case law exceptions written into it in today's legal environment.
2. False. The Employee Polygraph Protection Act is administered by the Department of Labor. As new detection technologies emerge, we can expect to see more legislation of this sort to help protect employees from abuses of that technology.
3. True. This provision of the act was designed to combat some of the sweatshop abuses commonly used in the early shift of our society from an agricultural base to the industrial era.
4. False. Worker's compensation legislation includes recovery for certain body parts. These schedules are either listed by statute or agreed upon between the contracting parties.
5. False. Private pension plans are not required *per se*. But if they are established, they must conform to all applicable federal and state laws.
6. False. The A.F.L. was formed by Samuel Gompers.
7. False. The union has now assumed that role.
8. False. Such employees must contribute an amount of money equal to union dues to an approved charity.
9. True. The strike must still be conducted legally in order to be protected by federal law.
10. False. Unions are very active in the political process although their influence has decreased dramatically in recent years.

Multiple Choice:

1. D. Newspaper delivery work has always been allowed as an exception to the general rules of child labor. It is interesting to note, however, that hard times are forcing more and more adults into these jobs.
2. C. There were a number of abuses by employers, coupled with the scientific shortcomings of polygraph inaccuracies, which led Congress to pass the Employee Polygraph Protection Act in 1988.
3. C. This is one of the requirements of the Fair Labor Standards Act of 1938, as amended.
4. D. All three are elements required in order to show that Linda is qualified for coverage and payment under typical worker's compensation acts.
5. A. Under the Self-Employment Contributions Act, self-employed individuals must also pay into Social Security. The amount of taxes self-employed individuals must pay is equal to the combined employer/employee amount.
6. C. This Act remains the cornerstone of U.S. labor relations law.
7. C. If the employees comply with the solicitation rules, the employer may not dismiss them for such activities.
8. C. Civil rights laws cannot be violated in the process of collective bargaining.
9. B. If the employer has engaged in an illegal labor practice, he or she must reinstate wrongfully discharged workers who struck against the illegal practice.
10. C. A, B, and D are all examples of illegal picketing practices.

VI. Answers to Essay Question

The employer/employee relationship between Mr. Harpie and Micro Widgets Corporation is clearly one of employment at will. The parties have agreed to this in writing. Under that doctrine, either party may terminate the relationship at any time for any reason, subject to limitations imposed as a matter of public policy, by statute or by court precedent.

There may be limitations imposed by the State of East Moncalla on the employment at will doctrine. Examples might include protection of whistleblowers or protection of persons who refuse to break the law at the behest of their employers. But none of these limitations are designed to protect the employee to the extent of shielding his or her own wrongdoing.

Worker's compensation fraud is one of the fastest growing and costliest crimes that today's employers face. States have not acted in concert to prevent this fraud as well as they should have. That failure to enforce against this sort of fraudulent behavior has made cases like Mr. Harpie's very common and very costly. Under the employment at will doctrine, unless Mr. Harpie can show that one of the exceptions applies based or public policy, case law, or statute, Micro Widgets can fire him. If they know haw bad an apple he really is, they should seek to have criminal charges filed also. Then maybe he can develop a bad back while punching out license plates.

Chapter 26
EQUAL OPPORTUNITY IN EMPLOYMENT

I am free of prejudice. I hate everybody equally.
W. C. Fields

I. Overview

No single American legal issue is inflamed with more controversy than discrimination in the workplace. The genesis of our nation's heritage is rooted in a diversity of peoples who immigrated to the New World in order to flee the royalist, class, or caste systems that so often predestined their opportunities for social and economic advancement. The U.S. Declaration of Independence, and the government founded on it, was the first major system of self-governance premised on the assumption that all persons are born equal and should be treated equally in the eyes of the law. As we all know, that equality has often been a hope rather than a reality for many.

The same diversity that has been a source of national pride has also been the basis of disparate treatment of persons in the workplace for many years. The term *discriminate* has within it two distinct and opposite meanings. On the positive side, discrimination is simply a fact of life. We are not all equal in all ways. We have different talents, strengths, levels of training, and abilities. Employers, in turn, should be allowed and expected to seek utilization of these divergent talents and strengths in their own best interests. To discriminate in the positive sense is to reward ability and merit. The positive aspect of discrimination really says that uniqueness should be discerned, differentiated, distinguished, and rewarded in the workplace. In the end, economic marketplace factors are blind to any other factors but job performance. Like it or not, positive discrimination is a simple economic necessity which is no different than the laws of nature and cannot be ignored.

The negative side of discrimination is found in a wrongful process of selection. For a society founded on a premise of equality, we have certainly had more than our share of unequal treatment in the workplace. The negative side of discrimination is inequality of treatment based on wrongful motive, justifications, or rationalizations. Each choice not based on talent, ability, and merit is a step away from the inherent basis of equality before the law. Wrongful discrimination is like cancer. If it is allowed to grow unchecked, it will kill. None of us can afford to look the other way and say: "It's not my problem." Wrongful discrimination against any group is a wrong upon the society at large. Most everyone appreciates that fact intuitively, if not intellectually.

Law is an end product of many larger movements in our society. It is a combination of morality, religious doctrines, social sciences, social engineering, and philosophy. In the end, law reflects, rather than dictates, choices made by society as to how it will help or hurt its members. The evolution of justice has never been painless. With each growing awareness of the pain incurred by wrongful discrimination, the cure will become even more difficult. Law is more readily administered when we have drawn a clear moral line. It is easy to justify the jailing of a rapist in order to protect society. But when there are competing, socially justifiable "goods" on both sides of the issue, the job of judgment is far more difficult.

Consider the dilemma posed by affirmative action. Past history provides vivid proof that patterns of discrimination have deep roots and will continue unless some sort of corrective social engineering takes hold in the workplace. Yet, is the cure worse than the disease? Can reverse discrimination really be justified? Is it any wonder that every day this quandary seems to get deeper? It is like kicking in quicksand—the more you try to move, the faster you sink.

One element that provides hope for positive change is goodwill. Where people of goodwill cling to the basic rightness of equity before the law, that equity will eventually result in a changed culture. Until then, law and our courts will continue to be the testing grounds for this monumental change in the social order. It may sound corny, but we all must do our part to live and let live.

II. Hypothetical Multi-issue Essay Question

An item recently appeared on the news wires:

"Major department store says no to Santa!"

A person who had been working for a major New York City department store was seeking to be reinstated into his seasonal job assignment as a Santa for the holidays. He had performed this job satisfactorily for a number of years.

He was denied reinstatement because he tested positive for HIV since the last holiday season. The department store was concerned that the health and safety of children sitting on Santa's lap would be placed in jeopardy. Even if the risk to health was minimal, the department store argued that the fear of parents to bring their children to a person known to have been tested positive for HIV would cost the store much lost business and damage its goodwill.

"Santa" decides to file a complaint with the EEOC. What will result?

III. Outline

- Title VII of the Civil Rights Act of 1964
  - Protected classes
    - Race
    - Color
    - National origin
    - Sex
      - Includes pregnancy
      - Includes sexual harassment
      - Does not include sexual preference
    - Religion

Title VII of the Civil Rights Act of 1964
- Covered Employers and Employment Decisions
  - Employers who
    - Have 15 or more employees for 20 weeks in the current or preceding year
    - All employment agencies
    - Labor unions with 15 or more members
    - State and local governments
    - Most federal agencies
  - Decisions regarding
    - Hiring
    - Promotion
    - Demotion
    - Payment of salaries, wages, and fringe benefits
    - Dismissal
    - Job training and apprenticeships
    - Work rules
    - Any other term, condition, or privilege of employment

Title VII of the Civil Rights Act of 1964—Types of Discrimination
- Disparate treatment discrimination
  - Discrimination against a specific individual because that person belongs to a protected class
- Disparate impact discrimination
  - Discrimination that occurs where a neutral-looking employment rule causes discrimination against a protected class

Title VII of the Civil Rights Act of 1964—Defenses and Remedies
- Defenses
  - Merit
  - Seniority
  - Bona fide occupational qualification (BFOQ)
- Remedies
  - Court may order payment of 2 years back pay, issue an injunction, and award reinstatement, fictional seniority, or other equitable remedy

Equal Pay Act of 1963
- Criteria that justify a differential in wages
  - Seniority
  - Merit
  - Quantity or quality of product
  - Any factor other than sex

Other Statutes Affecting Equal Opportunity in Employment
- Age Discrimination in Employment Act of 1967
  - Prohibits age discrimination practices against employees who are 40 and older
- Rehabilitation Act of 1973
  - Prohibits discrimination against handicapped persons by an employer who receives federal contracts or assistance
- Civil Rights Act of 1866
  - An act enacted by the Civil War that states that all persons "have the same right to make and enforce contracts … as is enjoyed by white persons"
  - Prohibits racial and national origin employment discrimination

Americans with Disabilities Act of 1990

Imposes obligations on employers and providers of public transportation, telecommunications, and public accommodations to accommodate individuals with disabilities

Affirmative Action

Policy that provides that certain job preferences will be given to minority or other protected class applicants when an employer makes an employment decision

IV. Objective Questions

Terms:

1. The right of all employees and job applicants (1) to be treated without discrimination and (2) to be able to prosecute employers if they are discriminated against is generally known as ______________ ______________.

2. The five-member federal agency responsible for enforcing most federal antidiscrimination laws is called the ______________ ______________ ______________ ______________.

3. A document given to an employee by the EEOC in the event that the EEOC does not want to act on behalf of an employee in a civil rights suit that gives the employee the right to sue her employer in the appropriate federal district court is called the ______________ ______________ ______________ letter.

4. The treaty that allows "nationals and companies of Japan" to hire specified employees of their choice and creates an exemption from Title VII coverage is called the Treaty of ______________ ______________ ______________ ______________ with Japan.

5. The practice of refusing to hire or promote someone unless she has sex with the manager or supervisor is illegal and is called ______________ ______________ ______________ discrimination.

6. The duty that employers owe employees to accommodate (1) the occurrence of religious practices, observances, or beliefs if it does not cause undue hardship to the employer or (2) a handicapped person's special requirements in the workplace is known as the duty of ______________ ______________.

7. The selection or promotion qualification that is based on work, educational experience, and professionally developed ability tests is called ______________ selection or promotion.

8. Discrimination based on sex, religion, and national origin is permitted if there is a legitimate reason for such discrimination and if it can be shown to be based on ______________ ______________.

9. A hiring policy that provides that certain job preferences will be given to minority or other protected class applicants when an employer makes an employment decision is called ______________ ______________.

10. Discrimination against a group that is usually thought of as a majority by a group that is usually thought of as a minority is generally referred to as ______________ ______________.

True/False:

1. ____ Federal workers are the only class that is exempt from the Equal Pay Act of 1963.

2. ____ Under the Equal Pay Act, employers can either increase the wages of the discriminated-against employee, or decrease the wages of the other employees.

3. ____ Back pay, liquidated damages, and overtime pay are all remedies available under the Equal Pay Act.

4. ____ Under Title VII, a private complainant must first file a complaint with the EEOC before bringing suit against the employer.

5. ____ When the EEOC decides to bring suit on behalf of a complainant, it issues a right to sue letter.

6. ____ Both monetary damages and extensive equitable remedies are available to prevailing complainants in Title VII cases.

7. ____ Only women are protected from sex discrimination under Title VII.

8. ____ Title VII, like many local ordinances, prohibits discrimination based on a person's sexual preference.

9. ____ Because of the Age Discrimination In Employment Act of 1967, most nonfederal employers cannot establish mandatory retirement ages for their employees.

10. ____ If the EEOC brings a suit under the Age Discrimination In Employment Act, a private suit arising from the same violation cannot be brought.

Multiple Choice:

1. Employee Edward has been with Super Sales Company (SSC) for ten years. Employee Ellen was hired by SSC two years ago. Edward and Ellen are both managers, perform many of the same tasks, and have similar responsibilities. However, Edward is paid more than Ellen. Is this permissible under the Equal Pay Act?
   A. No, because two people in essentially the same job can never have different pay levels.
   B. Yes, because the Equal Pay Act does not cover managers.
   C. Yes, because Edward can be rewarded for seniority.
   D. Yes, because SSC can set whatever pay level it chooses.

2. Disadvantaged David wants to bring a suit under the Equal Pay Act against his employer, Discriminating Daniel, for pay discrimination. Daniel hired Successful Sarah at the same time David was hired for essentially the same position. However, Sarah is paid almost twice what David is paid. Which of the following is correct?
   A. David cannot sue Daniel since the Equal Pay Act only protects women.
   B. David must submit a complaint with the EEOC before filing suit.
   C. Both of the above.
   D. None of the above.

3. Prevailing Pam has won a judgment against her employer under the Equal Pay Act. Which of the following is correct?
   A. Pam can receive back pay as damages.
   B. The other employees' wages may be lowered to eliminate disparity.
   C. Pam can receive overtime pay as damages.
   D. All of the above.

4. Infuriated Irene, a well-qualified computer programmer, applied for a position with Computer Corp. (CC). However, CC denied her the position and sought other applicants with Irene's qualifications to fill the position. Irene wants to bring a suit against CC under Title VII. What type of discrimination should Irene base her case on?
   A. Disparate impact discrimination.
   B. Pattern or practice discrimination.
   C. Disparate treatment discrimination.
   D. Any of the above.

5. Assume the same facts as in question 4. Also assume Irene prevails in her suit against CC. What remedies may Irene receive?
   A. Reasonable attorney fees.
   B. An injunction to compel hiring.
   C. Both of the above.
   D. None of the above. Irene can only receive punitive damages.

6. Discouraged Doug believes he has been fired from his job with Narrow-Minded Manufacturers (NMM) because of his homosexuality. Doug wants to sue NMM. Under what authority should Doug bring his case?
   A. Title VII of the Civil Rights Act.
   B. The Equal Pay Act.
   C. A local ordinance forbidding such discrimination.
   D. None of the above.

7. Employee Edgar believes Youthful Ventures, Inc. (YVI) has discriminatorily fired him because of his age. If Edgar wants to bring suit against YVI under the Age Discrimination in Employment Act, which of the following requirements must be met?
   A. YVI must have at least 20 nonseasonal employees.
   B. Edgar must be at least 40 years old.
   C. YVI must be a nonfederal employer.
   D. All of the above.

8. Assume the same facts as in question 7. Also, assume Edgar prevails in his suit against YVI. What remedies may Edgar receive?
   A. Unpaid back wages.
   B. Attorney's fees.
   C. Reinstatement.
   D. All of the above.

9. Regarding the Rehabilitation Act of 1973, which of the following is incorrect?
   A. People with a history of alcoholism have been held to be handicapped under the Act.
   B. The Labor Department administers the nonfederal aspects of the Act.
   C. Private plaintiffs cannot recover damages.
   D. People with diabetes are considered handicapped under the Act.

10. Regarding the Americans with Disabilities Act of 1990, which of the following is correct?
    A. Employers must make reasonable accommodations to accommodate people with disabilities.
    B. Preemployment medical examinations are forbidden before a job offer.
    C. Title I of the ADA is administered by the EEOC.
    D. All of the above.

V. Answers to Objective Questions

Terms:

1. *Equal opportunity.* This simple sounding phrase raises some of the most vexing legal issues of our day.
2. *Equal Employment Opportunity Commission.* Justice Thomas was once the head of this agency.
3. *Right to sue.* Most of the cases are farmed out through the use of these letters.
4. *Treaty of Friendship, Commerce and Navigation.* This treaty illustrates the bargaining power that comes with being a major economic trading partner.
5. *Quid pro quo.* In Hollywood parlance, this is known as the casting couch.
6. *Reasonable accommodation.* There are limits on how much the employer can be expected to do based on undue hardship.
7. *Merit.* This is the best qualification of all.
8. *Bona fide occupation qualification.* The burden of proof is on the employer.
9. *Affirmative action.* Debate on this issue will not end in the foreseeable future.
10. *Reverse discrimination.* Two wrongs do not make a right.

True/False:

1. True. They are covered under their own federal work rules.
2. False. Under the EPA, employers may increase the wages of the discriminated-against employee, but may not lower the wages of the other employees.
3. False. While back pay and liquidated damages are available remedies under the EPA, overtime pay is not.
4. True. Under Title VII, a private complainant must first file a complaint with the EEOC. If the EEOC decides not to bring suit on behalf of the complainant, it will issue a right to sue letter.
5. False. A right to sue letter is issued by the EEOC when it decides not to bring suit on behalf of the complainant.
6. True. The remedies granted can be one or both.
7. False. Both men and women are protected from sex discrimination under Title VII. However, women bring the vast majority of the cases.
8. False. Though many local ordinances do prohibit discrimination based on sexual preference, Title VII does not.
9. True. There are still some exceptions such as airline pilots.
10. True. This is outlined in the Age Discrimination in Employment Act.

Multiple Choice:

1. C. Under the EPA, employees can receive different wages if the payment system is based on seniority, merit, quantity or quality of product, or any factor other than sex. Here, Edward has been employed for eight years longer than Ellen. Therefore, the wage differential is probably justified. A is incorrect because two people in essentially the same job can have different pay scales if it is based on a system described above. B is incorrect because the EPA does cover managers. D is incorrect because SSC must follow the guidelines of the EPA and cannot arbitrarily or discriminatorily set pay scales.
2. D. A is incorrect because the EPA protects both men and women. B is incorrect because a private complainant does not have to submit a complaint to the EEOC before filing suit. C is incorrect because both A and B are incorrect.
3. A. Under the EPA, a prevailing employee may be awarded back pay and liquidated damages. B is incorrect because the employer must raise Pam's wages; he cannot lower the other employees' wages. C is incorrect because overtime pay is not awarded as damages under the EPA.
4. C. Disparate treatment discrimination occurs when an employer treats a person less favorably than another because of that person's race, color, national origin, sex, or religion. A and B are incorrect because they are types of discrimination that affect a whole class, not just one specific person.
5. C. Under Title VII, a successful plaintiff can recover up to two years' back pay, reasonable attorney's fees and, where appropriate, compensatory damages. In addition, several equitable remedies are available including an injunction to compel hiring. D is incorrect because punitive damages are seldom awarded and are certainly not the only remedy available.
6. C. At present, none of the acts discussed in this chapter protect employees from discrimination because of sexual preference. An employee's only option is to sue under a local ordinance, if there is one. A and B are incorrect because these acts do not protect employees from discrimination based on sexual preference.
7. D. These are all threshold tests for bringing a lawsuit under the ADEA.
8. D. All three remedies are provided under the ADEA.
9. C. Private plaintiffs may recover damages in a suit under the Rehabilitation Act. However, the EEOC and the Labor Department administer provisions of the act. A, B, and D are all correct statements.
10. D. Answers A, B, and C are found in the provisions of the ADA.

VI. Answers to Essay Question

The passions involved in many of these antidiscrimination issues run deep. They are often founded on claims by both sides of being morally, economically, or politically correct. Here the alleged wrongful discrimination is based on a physical characteristic or condition, i.e., testing positive for HIV. The outbreak of AIDS has raised a raucous debate in this country about what body of law is applicable to persons afflicted with this physical condition.

On one side, many argue that it should be treated solely as a health problem. As such, laws with regard to communicable diseases should control. These measures would be based on the theory that certain prophylactic measures are necessary where there is a real and present danger to oneself and others from the disease.

On the other side, a number of commentators have argued that fear of HIV and AIDS-related afflictions have generated the worst sort of civil rights abuses against persons afflicted with those conditions. Discrimination in housing, jobs, education, health care, and insurance have all surfaced as a result of the fears raised by the disease. Where there is no provable risk of communication of the virus, civil rights, including job rights, should not be compromised.

The real answer probably lies somewhere in between. Society does have the right and responsibility to take reasonable measures to protect itself from deadly threats to public health. But, when measures are taken, society should also meet the burden of proof that the measures are reasonable and proper. It cannot and should not trample on civil rights based on panic and ill-founded fear.

In this case, Congress has sought to balance the needs of both society and persons afflicted with HIV through the Americans with Disabilities Act of 1990. Under that Act persons who have tested positive for HIV are classified as disabled persons. As such they are entitled to seek protection under the Act from discriminatory practices in the workplace.

Conversely, the employer is allowed to use the defenses listed in the statute. One of those defenses is the health and safety of others. Another is undue hardship to the employer. In both cases, the burden of proof is on the employer to show that the defenses are applicable. If the department store can show that a real health and safety hazard exists for the persons with whom Santa will have contact, it will prevail. In addition, if the department store can show that using this person in this setting would create an undue hardship, it again would prevail. In either case, it must meet the burden of proof. Failure to do so will allow Santa to prevail under the Act. Santa's best strategy is to seek protection under the Americans with Disabilities Act of 1990.

Chapter 27
ENTREPRNEURSHIP, FRANCHISING, AND LICENSING

In all companies, there are more fools than wise men.
Francois Rabelais

I. Overview

This chapter has two main objectives. The first is to introduce students to the law of business entity choices. One of the key roles of attorneys engaged in the practice of modern business law is advising their clients on the selection of the best venue for doing business. What seems like a relatively limited set of options is, in fact, quite extensive. These choices can run the gamut from the simplest lemonade stand set up for a youngster to a multinational publicly-traded corporation.

With each choice, the law provides a list of pros and cons. For example, if a person seeks maximum privacy in his or her financial affairs, a private form of sole proprietorship may be best. Compare, however, the business person who wants to leverage the maximum utilization of other people's money while limiting her personal financial exposure. That person may find the corporate form best suited for her needs.

The law literally has something for everyone. The real issue is first finding out what options are legally available and then choosing the best fit. That fit should be tailored by sound advice from a number of quarters including law, accounting, finance, and business management strategy. It is this constant interdependent equation that makes the practice of business law so difficult yet so interesting. The vast majority of the users of this book may never go to law school. Yet that same majority will be influenced every working day by the business entity law choices made in whatever business pursuits they chose.

Take, for example, sole proprietorships. It remains the most widely used form of business entity even though it may no longer be the most important in sheer economic terms. With the advent of the so-called "information highway" and more emphasis on entrepreneurial niche marketing of goods and services, this form of business may enjoy a renaissance in the Twenty-First century. The most recent U.S. census report indicates that the self-employed individual represents one of the fastest growing forms of business entity choice. This trend is confirmed by statistics reported by the IRS that show a dramatic increase in the use of Schedule C (for self-employed) on Form 1040 reported by U.S. taxpayers.

The second main objective of this chapter is to introduce students to several of the many forms of conducting international business. In many respects, these methods are really variations on the themes established in the first part of this chapter. Yet, the governing rules of business entities will be derived from a combination of two or more sovereign entities. International business law planning is one of the fastest growing areas of law practice today.

The world of franchising combines concepts of marketing, management, finance, and many other diverse skills into a special kind of cooperative venture that is facilitated through the law. Like any business marriage, it can lead to the best of all worlds or the worst. Careful planning, skilled research, and, most of all, a good faith willingness to let each participant do what he does best appears to be the key to today's most successful franchise operations. Under the franchise system, there is ample evidence of the positive effects of good franchise planning. The original basic technology, patent, process, or other trademarked service or product is allowed to reach far more users or consumers through the franchise system. With intelligent planning and quality control, the original franchisor of the product or service can see phenomenal growth through the use of equity-sharing participants in that growth. Witness the fast food industry, convenience stores, food product production, all sorts of consumer good retailing systems, or even professional sports teams. All of these industries rely heavily on the franchise concept to further their businesses.

Another interesting aspect of franchising is its tie to basic capitalism for the little guy. With sufficient start up capital and a willingness to provide a lot of personal effort, the franchising concept allows the small businessperson to ride the coattails of the goodwill, advertising, and technology development of large multinational enterprises.

There are some down sides to franchising. The "get rich quick" mentality of franchising has led to a number of abuses on the part of would be franchisors. Many people have lost substantial sums of money trying to invest in pie-in-the-sky sales of bogus franchises. In addition, the franchise industry has seen more than its share of pyramid schemes, shallow capitalization, adhesion contracts, and behavior in violation of the antitrust laws. It is interesting to note that many of the rules promulgated by the Federal Trade Commission are designed to protect persons about to enter into franchise agreements rather than the ultimate consumer of the franchise's goods or services.

In addition, the franchise device has not always served the third party well. Because a franchisee is an independent contractor, the franchisor is not normally responsible to third parties for torts or contracts that the franchisee has been involved with. That may all sound well and good in legal terms, but does it always make equitable sense? If the consumer of the goods or services thought he or she was dealing with a megacorporation, why not hold a megacorporation responsible rather than just its franchise? Law works best when the benefit/burden balance is fairly struck. Where large business entities grow through the utilization of its franchise efforts, they should also be willing to take responsibility for their harm.

II. Hypothetical Multi-Issue Essay Question:

Mr. Dorgis Dimmie was a student at IOU trying to figure out a way to pay for his school expenses. He was majoring in Artificially Intelligent Gamma Rays, and the cost of his lab equipment and supplies was out of sight. He noticed that people really ate and drank a lot at the weekend IOU football games in spite of the high concessionaire fees. He decided he could offer them cheaper food if it could be pre-sold before the fans came into the stadium.

He contacted a local eatery, the Et Yet? Cafe and made a deal with them. He would take orders for, sell, and deliver sandwiches prepared by Et Yet? and they would split the proceeds down the middle. During the first five games of the home season, Dorgis was doing well and sold over 5,000 Et Yet Mystery Meat sandwiches to his fellow students and assorted football fans. Last week, things turned ugly, however, when numerous fans that ate these sandwiches became ill due to meat poisoning. Many of them are now suing Dorgis and Et Yet?

Dorgis claims he was only acting as Et Yet's? disclosed agent. Et Yet? in turn, is saying that it sold the sandwiches to Dorgis, and that he resold them directly to the fans as an independent sole proprietor. What is the result?

III. Outline

Forms of Conducting Domestic Business
- Sole proprietorship
- General partnership
- Limited partnership
- Limited liability partnership (LLP)
- Limited liability company (LLC)

Forms of Conducting Domestic Business
- Corporation
- Close corporation
- Public corporation
- S corporation
- C corporation
- Franchise
- Joint venture
- Syndicate or investment group
- Business trust
- Cooperative

Advantages of Sole Proprietorship
- Owner is the business; no separate legal entity
- Easy and inexpensive to form
- Owner has right to make all management decisions concerning the business
- Sole proprietor owns all of the business and has the right to receive all of the business's profits
- Sole proprietorship can be easily transferred and sold at owner's discretion

Personal Liability of Sole Proprietors
- Sole proprietor bears the entire risk of loss of the business
  - He or she will lose his or her entire capital contribution if the business fails
- Sole proprietor has unlimited personal liability
  - Creditors may recover claims against the business from the sole proprietor's personal assets

Conducting International Business Using a Branch Office

Conducting International Business Using a Subsidiary Corporation

Forms of Conducting International Business
- Direct export and import sales
- Sales agents and representatives
- Distributorship
- Joint venture
- Branch
- Subsidiary
- Licensing
- Franchising

What Is a Franchise?
A franchise is established when one party licenses another party to use the franchisor's trade name, trademarks, commercial symbols, patents, copyrights, and other property in the distribution and selling of goods and services

Parties to a Franchise
Franchisor
The party who does the licensing in a franchise situation
Franchisee
The party who is licensed by the franchisor in a franchise situation

Parties to a Typical Franchise Arrangement

Common Terms of a Franchise Agreement
Quality control standards
Training requirements
Covenant not to compete
Arbitration clause

Common Franchise Fees
Initial license fee
Royalty fee
Assessment fee
Lease fee
Cost of supplies

Trademarks and Service Marks
A distinctive mark, symbol, name, word, motto or device that identifies the goods of services of a particular franchisor, e.g.,™
Anyone who uses a mark without authorization may be sued for trademark infringement

Contract and Tort Liability of Franchisors and Franchisees
Franchisors and franchisees are liable for their own contracts and torts
Typically, the franchisor deals with the franchisee as an independent contractor

IV. Objective Questions

Terms:

1. A business entity that is noncorporate in nature, and that is not considered a separate taxpaying entity for income tax purposes is known as a ________________ ________________.

2. Most states require all businesses that operate under a trade name to file a ________________ ________________ ________________.

3. If a sole proprietorship fails in business, the owner is subject to ________________ ________________ liability for claims against the business.

4. A corporation that has a small number of shareholders and is treated like a partnership for tax purposes is called a ________________ corporation.

5. An arrangement whereby an owner of a patent, trademark, trade secret, or product grants a license to another to sell products or services using the name of such owner is generally known as a _______________.

6. In a franchise arrangement, the franchisee may be allowed to use the trademarks and service marks of the franchisor. In addition, the franchisee may be allowed to use _______________ _______________, which are certain ideas that make a franchise successful but which do not qualify for a trademark, copyright, or patent.

7. Several franchisees may enter into an informal agreement whereby they establish a price at which they will sell the franchisor's products or services. Such an arrangement is known as _______________ _______________ and constitutes a violation of the Sherman Antitrust Act.

8. Many franchise agreements permit the franchisor to terminate the relationship for cause. _______________ _______________ occurs if the franchise is terminated without cause.

9. In a franchise setting, one party, known as the _______________ licenses another, known as the ___________ _______________ to use a trade name, trademark, commercial symbol, or the like in the distribution and selling of goods and services.

10. One who misappropriates a trade secret may be sued for _______________ _______________ and be responsible for damages or have an injunction issued against him to prevent further unauthorized use.

True/False:

1. ____ The general partners of a limited partnership have limited personal liability for the debts and obligations of the limited partnership.

2. ____ The personal liability of a sole proprietor is limited to the extent of her capital contribution to the business.

3. ____ Sole proprietorships are prohibited by law from operating under a fictitious name.

4. ____ A corporation is not a separate taxpaying entity for federal income tax purposes.

5. ____ Dot.com businesses are generally successful.

6. ____ The remedies for breach of a lawful franchise agreement are based on contract theories for breach of contract.

7. ____ In a franchise arrangement where the franchisee uses the same name as the franchisor, the franchisee will be deemed to be the agent of the franchisor.

8. ____ A franchise agreement cannot limit the geographical territory in which the franchisee can operate.

9. ____ The license fee in a franchise arrangement generally does not cover the cost of buildings.

10. ____ Generally, a license fee is the only compensation that a franchisor can legally collect.

Multiple Choice:

1. There are several advantages to the sole proprietorship form of operating a business. Which of the following is not one of its advantages?
   A. Formation is simple and low in cost.
   B. The sole proprietor retains the right to make all management decisions concerning the business.
   C. The sole proprietor is legally responsible for the debts and obligations of the business only to the extent of his capital contribution.
   D. A sole proprietorship can be transferred or sold quite easily.

2. Which of the following organizations is a separate taxpaying entity for federal income tax purposes?
   A. Corporation.
   B. General partnership.
   C. Limited partnership.
   D. Sole proprietorship.

3. Which of the following statements about the liability of general partners for the debts and obligations of a partnership is most accurate?
   A. The liability of a general partner is like that of a corporate shareholder.
   B. The liability of a general partner is like that of a sole proprietor.
   C. The liability of a general partner is like that of a beneficiary of a business trust.
   D. None of the above.

4. Which of the following statements about limited partnerships is correct?
   A. Limited partnerships are separate taxpaying entities for federal income tax purposes.
   B. Every limited partnership must have at least one general partner.
   C. Limited partners contribute capital to the business in exchange for a stated rate of return on their capital contribution.
   D. None of the above.

5. Tom Taxbreak is hoping to create a general partnership. Which of the following persons or organizations could not be partners with Tom?
   A. A limited partnership in which Tom is a partner.
   B. Tom's wife.
   C. Tom's best friend who has been adjudicated insane.
   D. Tom's 12-year-old son.

6. Magic Burger Corporation has discovered a new special sauce that has sent its sales off the top of the charts. For a modest fee, Magic Burger Corporation grants franchisees the right to use the special sauce and to use the Magic Burger trade name for marketing purposes. The arrangement between Magic Burger Corporation and its franchisees is best described as which of the following?
   A. A chain-style franchise.
   B. An area franchise.
   C. A processing plant franchise.
   D. A distributorship franchise.

7. Assume the same facts as in question 6. In addition, assume that a franchisee served a burger with special sauce that had spoiled causing serious injuries to the customer. If the customer sues Magic Burger Corporation, which of the following is correct?
   A. A franchisee is an independent contractor, and therefore Magic Burger Corporation is not liable.
   B. A franchisor and franchisee relationship constitutes a principal and agent relationship, and therefore the franchisor is liable.
   C. By serving spoiled sauce, the franchisee acted outside the scope of agency and therefore Magic Burger Corporation is not liable.
   D. None of the above.

8. Assume the same facts as in question 7. With respect to termination of the relationship, which of the following is correct?
   A. Unless stated otherwise in the agreement, the franchise will automatically terminate after one year.
   B. Unless stated otherwise in the agreement, the grant of a franchise is irrevocable.
   C. Serving tainted burgers would probably be enough to constitute "cause" to terminate the relationship.
   D. None of the above.

9. Which of the following is not an advantage of franchising?
   A. Franchising allows the franchisor to reach new markets with its product or services.
   B. Both the franchisee and the franchisor are protected from competition in the marketplace.
   C. Consumers are generally assured of uniform product quality.
   D. The franchisee benefits from the knowledge and resources of the franchisor.

10. Which of the following best describes the legal relation of a franchisor and franchisee where there is a high degree of independence between the two?
    A. Principal/agent.
    B. Limited partners.
    C. Employer/independent contractor.
    D. Partners.

V. Answers to Objective Questions

Terms:

1. *Sole proprietorship*. Under the tax laws, business income and losses are reported on the individual tax returns of the sole proprietor. Although this has many disadvantages, one major benefit is that the problem of corporate double taxation is avoided.
2. *Business name statement (or certificate of trade name)*. These statements are designed to give the authorities information about the principals of the business such as names, addresses, etc. In addition, most of these statutes required publication of this information in a newspaper of general publication.
3. *Unlimited personal.* Because there is no legal distinction between the sole proprietor and his or her business, the liabilities of the business and the proprietor are considered one and the same.
4. *"S."* This form of favorable corporate tax treatment is limited to those entities that qualify under the "S" subchapter of the corporate tax provisions of the Internal Revenue Code, beginning at Section 1361. These rules were substantially overhauled in 1982, but have been part of the tax code since 1958.
5. *Franchise.* This method of doing business has become one of the most important of all. Consider its worldwide implications in such key service industries as food, tourism, and the sale of essential

commodities. A growing component of international law involves the interpretation and application of long-term multinational franchise contract agreements.

6. *Trade secrets*. The interesting aspect of trade secrets is that they are not registered in the same way as certain kinds of other intellectual property, such as patents. Thus they are not as easy to reverse engineer and the like. But they are still considered to be a property and enjoy the protection of the law.
7. *Price fixing*. In spite of all the benefits that inure from the franchisor/franchisee relationship, the antitrust laws will not allow it to be used as a subterfuge to get around laws designed to protect competition in the marketplace.
8. *Wrongful termination*. Good faith and fair dealing are elements of franchise contracts that more and more courts are reading into franchise contracts even if the parties do not specifically list them as covenants of the agreement.
9. *Franchisor, franchisee*. These are the basic terms used in this body of law.
10. *Unfair competition*. Because the trade secret is a protected property right, wrongful use of another's such property has many legal consequences.

True/False:

1. False. A corporation is a separate taxpaying entity for federal income tax purposes. Thus, corporations are said to be subject to double taxation.
2. False. A sole proprietor bears the entire risk of loss of such a business and has unlimited personal liability.
3. False. A sole proprietorship can operate under the name of the sole proprietor or a trade name. If a trade name is used, a fictitious business name statement will probably have to be filed.
4. False. A corporation is a separate tax paying entity for federal income tax purposes. Thus, corporations are said to be subject to double taxation.
5. False. Dot.com businesses have the same or similar success as any other business. Many dot.coms have lost money trying to attract people to their website with big discounts, ect.
6. True. If the franchise agreement is reached, the injured party may sue for rescission of the agreement, restitution, and damages.
7. False. Mere use of the same name does not automatically make the franchisee the agent of the franchisor. If, however, there is no effort to inform the public of the separate legal status, the franchisee may be deemed the agent of the franchisor.
8. False. The franchise agreement is a contractual agreement. Therefore, the agreement can contain any reasonable restrictions to which parties agree.
9. True. The license fee is generally a lump sum payment for the privilege of being granted a franchise.
10. False. There are many types of franchise fees that can be collected by a franchisor including royalty fees, assessment fees, and lease fees.

Multiple Choice:

1. C. A sole proprietor has unlimited personal liability with respect to the debts and obligations of the business.
2. A. All of the others are considered conduits for tax purposes, and the taxes are attributable to the partner or proprietor.
3. B. Both general partners and sole proprietors have unlimited personal liability with respect to the debts and obligations of their respective business enterprises.

4. B. Although limited partnerships are required to file an informational return, they are not considered separate taxpaying entities, therefore A is incorrect. C is incorrect because limited partners are only entitled to a share of the profits, not a stated rate of return on their capital contributions.
5. C. One who has been adjudicated insane lacks the capacity to form a contract. A is incorrect because a limited partnership may be a partner, even if Tom is a partner in that organization. There is nothing prohibiting husband and wife from being partners, therefore answer B is incorrect. Finally, minors may be partners, but they have the ability to disaffirm such contracts prior to reaching majority.
6. A. A chain-style franchise exists when the franchisor licenses the franchisee to make and sell its products or services to the public using its name.
7. D. Although a franchisee is typically an independent contractor, there are some situations in which the franchisor and the franchisee use the same name and make no attempt to distinguish their existence. Then a principal and agent relationship will be found. However, the mere existence of a franchise setting does not automatically create a principal and agent relationship.
8. C. Failure to meet quality control standards is "just cause" to terminate for cause.
9. B. Franchises are obligated to abide by the antitrust laws the same as any other form of business organization.
10. C. If properly organized, the franchisor and franchisee are independent legal entities.

VI. Answers to Essay Question

Dorgis had a number of objectives when he set out on his entrepreneurial journey. He wanted to set up a business with as little formality as possible. He wanted to retain as much control over the management of his business as possible. He did not want to provide extensive capital contributions to the venture, and in fact, sought to have his key contribution be his “middleman" or "facilitator" services between Et Yet? and the football fans. Because he did not actively seek to create a separate business entity, such as a corporation, his personal assets will in all likelihood be exposed to possible business loss and tort liabilities.

Et Yet? appears to have been motivated by many of the same factors as Dorgis. It wanted to form a business relationship that would expand the market for its unique Mystery Meat Sandwiches. It appears that neither party contemplated the use of an actual agreement beyond simply sharing the proceeds.

There are elements of a number of possible business entities in place that allow for several interpretations. One view might argue that both Dorgis and Et Yet? are independent sole proprietors. As such, neither could be directly held liable for each other's acts as an agent. Under this theory, the injured fans would have to sue Dorgis and Et Yet? as individual business entities operating as sole proprietorships. This may allow for possible defenses in that each of the defendants might try to point to each other's acts as the source of the negligence that led to the food contamination.

Another viewpoint may argue that the elements of partnership between Dorgis and Et Yet? are in place. They are associated in carrying out a business as coowners for a profit motive. This is evidenced by the sharing of the proceeds from the sale of Mystery Meat Sandwiches. In the alternative, the plaintiffs may want to argue that even if they were not actually partners, partnership by estoppel may be applicable by reason of the appearance of partnership made to them as third parties.

A third theory might involve calling the arrangement a joint venture between Dorgis and Et Yet? Here there is an association carried on to conduct a limited purpose business--to sell Mystery Meat Sandwiches at IOU football games. The participants would be called adventurers (in more ways than one). As a joint venture that was not incorporated, most states following the Uniform Partnership Act mode would apply partnership law and hold Dorgis and Et Yet? liable as agents of each other's negligent acts.

Of the three scenarios proffered, the last seems most likely. Both defendants would likely be held responsible for the harm done. It is yet another lesson learned in short-term gain, long-term loss. Had

Dorgis and Et Yet? spent a little more time and money setting up a proper business entity, they might not be eating crow now.

Based on their track record so far, it does not appear that Dorgis or Et Yet? have to worry about overseas expansion. Had they been more successful at the outset of their domestic operations, most any overseas options would have been possible depending on the combined applicability of both domestic laws and the laws of countries into which they chose to extend their operations. Many food purveyors elect to go with franchising. Witness the expansion of American fast food operations to all corners of the globe.

Chapter 28
PARTNERSHIPS AND LIMITED LIABILITY COMPANIES

The richer your friends, the more they will cost you.
Elizabeth Marbury

I. Overview

One of the key distinctions between the partnership form of doing business and the corporate format is the corporation's ability to have an indefinite or perpetual existence. Under state laws of incorporation, a corporation is allowed to continue its juristic existence in spite of the death of its key players. This is not true with partnerships. A partnership is intrinsically tied to the continuing existence of the partners. Partnerships are literally more personal. When the partner is gone, so too is the partnership.

One of the questions students frequently ask is: how is it that multi-national business organizations, such as large accounting or law firms, stay in business as partnerships when they frequently lose partners through death or through changes in partnership associations? The technical answer is that with each of these changes, the partnership is theoretically ended and a new one is created. In fact, a well-crafted partnership agreement should have, as one of its key components, an orderly process of succession in case of death or termination. Where these circumstances are properly planned for, the transition is seamless, and the life of the partnership goes on.

Most partnerships are not, however, large and multi-national in scale. Most are created and operated by individuals who have sought to capitalize on their respective economic or talent contributions by acting together in the legal sense. These business ventures could be set for a short-term, specific goal, such as erecting a building, or extend to a full professional career such as a licensed practitioner of law, medicine, or accounting. Just like a marriage, the best intentions at the legal altar of partnership do not always work, and the rules of partnership should provide a means of graceful dissolution.

As we have seen in earlier chapters, business entity choices are strategic decisions based on a number of factors. These elements include choosing the best options for potential capital investment and financial growth, protection from personal liability, and tax planning. No one-entity format is ideal for all objectives. However, recent trends lead to the use of the limited liability company format as the best vehicle for providing the "best of both worlds." Those worlds are the single-layered conduit taxation of proprietorships and partnerships with the limited personal liability accorded to shareholders of a corporation.

The actual advent of limited liability companies does not begin with U.S. legal history. Many countries have allowed for the formation of limited liability companies for a long time. For example, many countries in Latin America provide for entities called "limitadas." Similar developments have taken place in England and Western Europe with examples such as the English "stock company" format tracing its history back to 1555. The limited liability company format was not used in the U.S. until the late 1970s.

Why is it that it took so long for this format to catch on in the U.S.? There are several key reasons. First and foremost is the tax treatment accorded to corporations subject to a "double tax," i.e., they pay a tax on corporate income, and dividends distributed to shareholders are subject to taxation on the shareholders' tax return. In 1978, the Internal Revenue Service issued a Revenue Ruling with regard to a new limited liability company statute just enacted the year before in Wyoming. Under that ruling, if certain elements of the corporate format were deemed to be missing from the limited liability company format, it would be taxed as partnership and not a corporation. This tax ruling opened the virtual floodgates. Many state legislators enacted laws similar to the Wyoming statute. They wanted to to attract new capital from overseas (where limited liability companies have long been accepted) and also provide an attractive legal infrastructure for start-up companies and the like. For example, in Colorado, local statutes provide for nine domestic forms of business: as cited by Professors Oesterle and Gazur in their article entitled, "*What's in Name? An Argument for a Small Business "Limited Liability Entity" Statute (with Three Subsets of Default Rules*," 32 Wake Forest Law Review 101 (1997)) a general partnership, a limited liability partnership (L.L.P.), a limited partnership, a limited liability limited partnership (L.L.L.P.), a limited liability company (L.L.C.), a limited partnership association (L.P.A.), a regular corporation, (C. or S. under I.R.S. rules), a professional corporation (P.C.), and a cooperative.

This myriad of business organization formats has grown in comparable fashion throughout the vast majority of states. Thus, it is not surprising that there is a call for more uniformity among the states vis-à-vis the passage of laws creating limited liability companies. The group most responsible is the National Conference of Commissioners on Uniform State Laws. In 1995, this group issued the Uniform Limited Liability Company Act (U.L.L.C.A.). This act takes the traditional precedents created by agency law, the various forms of partnerships, and corporations and marries them to a cohesive set of rules for limited liability companies. The states, in turn, are free to adopt their own variations on the basic themes set out in the U.L.L.C.A., as illustrated by the Colorado example. These statutes will be "fleshed out" over time, experience, and court decisions. All in all, it is a most interesting time in our legal history. In many ways, we are still playing catch-up with the rest of the world. But given the vitality and ingenuity of the U.S. legal system coupled with the current economic growth in so many new start-up businesses, the horizon for this type of business format looks bright indeed.

II. Hypothetical Multi-Issue Essay Question:

Bragg, Bluster, and Big was one of the oldest and most famous law firms in Backwater. Its specialty was personal injury litigation. It really went big time when it started advertising on late night television. They managed to convince hundreds of clients that their employers were actually hurting them, thus providing their clients with a chance to hit the courthouse tort lottery!

A practitioner of exceptional loquaciousness was the founder of the firm, Mr. Boskop Borzoi Bragg Esquire, affectionately known as the "Big B" to his partners and clients alike. The Big B could officiously obviate the occasional opposition with opprobrium and obfuscation into oblivion. He loved the battleground of the courts and hated to retire last year.

Big B still enjoyed hanging around the courthouse even in retirement. One day he started up a conversation and talked Ms. Felicity Pembrokershire into suing her employer for looking at her the wrong way. She had heard of the Big B's famous exploits in the courts. Felicity paid Big B a

$10,000 retainer to help his former firm prepare the case. In the meantime, Ms. Penbrokeshire realized she may have made a mistake and decided to drop the lawsuit. Big B is insulted and refuses to give back the retainer. Felicity sues Bragg, Bluster, and Big for the money. What results?

III. Outline

Sole proprietorship
A form of business where the owner is actually the business; the business is not a separate legal entity

Definition of a General Partnership
A voluntary association of two or more persons created for carrying on a business as co-owners for profit

Formation of General Partnerships
Express partnership
General partnership created by words, either verbal or written
Implied partnership
General partnership implied from the conduct of the parties
Certificate of partnership
A document that a partnership must file with the appropriate state government agency in some states to acknowledge that the partnership exists

Duration of Partnership
Partnership for a term
A partnership with a fixed duration
Partnership at will
A partnership with no fixed duration

Partnership Agreements
A written partnership agreement should contain:
The firm name
The names and addresses of the partners
The principal office of the partnership
The nature and scope of the partnership business
The duration of the partnership
The capital contributions of each partner
The division of profits and losses among the partners
The salaries, if any, to be paid to the partners
The duties of the partners regarding management of the partnership
Limitations, if any, on the authority of partners to bind the partnership
Provision for admission and withdrawal of partners from the firm
Provisions for continuing partnership upon withdrawal of a partner, death of a partner, or other dissolution of the partnership
Any other provisions deemed relevant by the partners

Property Rights of the Partnership and the Partners

Partnership property

Property that is originally brought into the partnership on account of the partnership and property that is subsequently acquired by purchase or otherwise on account of the partnership or with partnership funds

Right of survivorship

A deceased partner's right in specific partnership property vests with the remaining partners upon his or her death

Partner's interest

A partner's share of profits and surplus of the partnership

Charging order

A document that the court issues against the debtor-partner's partnership interest in order to satisfy a debt

Rights of Partners

Right to participate in management
Right to share in profits
Right to an accounting
Right to compensation
Right to indemnification
Right to return of advance
Right to return of capital
Right to information

Duties of Partners

Duty of loyalty
Duty of obedience
Duty of care
Duty to inform

Contract Liability to Third Parties

A partnership acts through its agents, i.e., partners
If partners have express, implied, or apparent authority, contracts are binding

Tort Liability to Third Parties

The partnership is liable if the tortious act of a partner or employee or agent of the partnership is committed while the person is acting within the ordinary course of partnership business or with the authority of his or her copartners

Liability of Incoming Partners

A new partner who is admitted to the partnership is liable for the existing debts and obligations of the partnership only to the extent of his or her capital contribution

The new partner is personally liable for debts and obligations incurred by the partnership after becoming a partner

Termination of a Partnership by Judicial Decree
- Partner is adjudicated insane
- Partner is incapable of performing duties
- Partner is guilty of improper conduct
- Partnership can only be carried on at a loss

Termination of a Partnership by Operation of Law
- Death of any partner
- Bankruptcy of any partner or the partnership
- Illegality

Winding-Up
- The process of liquidating the partnership's assets and distributing the proceeds to satisfy claims against the partnership

Distribution of Assets
- After the winding-up of a dissolved partnership, the assets of the partnership are distributed in the following order
- Creditors (except partners who are creditors)
- Creditor-partners
- Capital contributions
- Profits

Limited Liability Company (LLC)
- An unincorporated business that combines the most favorable attributes of general partnerships, limited partnerships, and corporations
- An LLC is a separate legal entity—an artificial person—that can own property, sue and be sued, enter into and enforce contracts, etc.
- Members are liable for the LLC's debts, obligations, and liabilities only to the extent of their capital contributions

Limited Liability Company (LLC)

Articles of Organization of an LLC

Articles of organization must set forth
- The name of the LLC
- The address of the LLC's initial office
- The name and address of the initial agent for service of process
- The name of each organizer
- Whether the LLC is a term LLC, and if so, the term specified
- Whether the LLC is to be manager-managed
- Whether one or more of the members of the LLC are to be personally liable for the LLC's debts and obligations

Three Kinds of LLCs
- Domestic
  - An LLC in the state in which it is organized
- Foreign
  - An LLC in any other state other than the one in which it is organized
- Alien
  - An LLC that is organized in another country

Liability of an LLC

An LLC is liable for any loss or injury caused to anyone as a result of a wrongful act or omission by a member, manager, agent, or employee of the LLC who commits the wrongful act while acting within the ordinary course of business of the LLC or with authority of the LLC

Member Lawsuits

Direct lawsuit

A lawsuit that a member can bring against an LLC to enforce his or her personal rights as a member

Derivative action

A lawsuit that a member may bring against an offending third party on behalf of his or her LLC when the LLC fails to bring the lawsuit or a request of the LLC to do so is excused

Fiduciary Duties Owed to an LLC

Duty of loyalty
Limited duty of care
Duty of good faith and fair dealing
No fiduciary duty owed by a nonmanager-member

Dissolution of an LLC

An LLC is dissolved upon the occurrence of any of the following
An event specified in the operating agreement
The occurrence of an event that makes it unlawful for all or substantially all of the business of the LLC to be continued
A member applies to a court to have the LLC dissolved by judicial decree
The secretary of state commences a proceeding to dissolve an LLC administratively

IV. Objective Questions

Terms:

1. When the relation of the partners changes because a partner ceases to be associated in the carrying on of the business of the partnership, a _______________ occurs.

2. In some circumstances a retiring partner may wish to be relieved of liability for certain obligations of the partnership. This can be accomplished through a _______________ agreement that in effect substitutes a third party for the retiring partner with respect to such obligations.

3. An outgoing partner may seek a release of liability for existing partnership obligations from his former partners. This can be accomplished through a(n) _______________ _______________ _______________ _______________ agreement between the outgoing partner and one or more of the continuing partners.

4. A doctrine dictating that partnership creditors may not seek satisfaction from the separate assets of individual partners until such time as the debts owed to individual creditors of such individual partners have been satisfied is called _______________ _______________ _______________.

5. Following certain dissolutions of a partnership, the remaining partners may wish to continue the operation and business of the partnership. In these situations, the remaining partners may enter

into a(n) _______________ _______________ setting forth the details of how the partnership will be operated at that point.

6. The Big Five accounting firms all operate as _______________ _______________ _______________.

7. Each member has equal rights in the management of the business, irrespective of the size of his or her capital contribution if it is a _______________ _______________ LLC.

8. An LLC must give members and their agents access to its records at the LLC's principal office or other reasonable location. This right is known as the member's _______________ _______________ _______________.

9. If a creditor of an individual member of an LLC obtains a judgment against a member, that creditor can ask the court to issue a _______________ _______________ against the debtor-member's distributional interest.

10. The process of preserving and selling the assets of the LLC and distributing the money and property to creditors and members after dissolution is known _______________ _______________.

True/False:

1. _____ A partnership that is formed for a specific time or purpose dissolves upon the affirmative vote of a majority of the existing partners following the expiration of the time or accomplishment of the objective.

2. ____ A partnership dissolves automatically by operation of law upon the death of a partner.

3. ____ A partnership will be dissolved by judicial decree of dissolution following an event that makes it unlawful for the business of the partnership to be carried on or for the partners to carry it on.

4. ____ Assignment of one's partnership interest will cause the dissolution of a partnership.

5. ____ Generally, the dissolution of the partnership terminates the partners' actual authority to enter into contracts or otherwise act on behalf of the partnership.

6. _____ Under the U.L.L.C.A. an LLC must designate an agent in the state of organization to receive service of process, notices and demands if the LLC is sued or involved in an administrative proceeding.

7. _____ An LLC organized in a foreign country and doing business in the United States is called a foreign limited liability company.

8. _____ Unless otherwise agreed, the U.L.L.C.A. mandates that a member has the right to an equal share in the LLC's profits.

9. _____ A member of a member-managed LLC and a manager of a manager-managed LLC do have a limited duty of care which does not include ordinary negligence to the LLC.

10. _____ A member of manager-managed LLC who is not a manager owes no fiduciary duty of loyalty or care, or duty of good faith and fair dealing, to the LLC or its other members.

Multiple Choice:

1. Ted and Fred entered into a written partnership agreement to operate a retail clothing store. Their agreement was silent as to the duration of the partnership. Ted wishes to dissolve the partnership. Which of the following statements is correct?
   A. Ted may dissolve the partnership at any time.
   B. Ted may dissolve the partnership only after notice of the proposed dissolution is given to all partnership creditors.
   C. Ted may not dissolve the partnership unless Fred consents.
   D. Ted must apply to a court and obtain a decree unless Fred consents.

2. Abe, George, and Thomas are partners in a bookstore specializing in books relating to history. George has loaned the partnership a significant amount of money, and the partnership has since begun the process of termination. If partnership funds are insufficient to repay the amounts owed to George, which of the following statements is correct?
   A. George will not be repaid.
   B. Abe and Thomas must contribute their proportionate share towards repayment of the loan from George.
   C. George will receive only the interest accrued on the unpaid amount.
   D. None of the above.

3. Assume the same facts as in question 2. Which of the following statements is correct?
   A. The partners will be paid their capital contributions after creditors of the partnership have been repaid.
   B. The loan made by George will be repaid before other creditors of the partnership are paid.
   C. George is not entitled to receive interest on the money advanced if not repaid when due.
   D. None of the above.

4. The partnership agreement of one of your clients provides that upon death or withdrawal, a partner shall be entitled to the book value of his or her partnership interest as of the close of the year preceding such death or withdrawal and nothing more. It also provides that the partnership shall continue. Regarding this partnership provision, which of the following is a correct statement?
   A. It is unconscionable on its face.
   B. It has the legal effect of preventing a dissolution upon the death or withdrawal of a partner.
   C. It effectively eliminates the legal necessity of a winding up of the partnership upon the death or withdrawal of a partner.
   D. It is not binding upon the spouse of a deceased partner if the book value figure is less than the fair market value figure.

5. Sandy Shorterm, who has been a partner of Shorterm, Attention, & Span for years, decides to withdraw from the partnership despite the fact that the partnership agreement prohibits such withdrawal. Which of the following is correct?
   A. Sandy's withdrawal will cause the dissolution of the partnership by operation of law.
   B. Because the withdrawal is wrongful, it will have no impact upon the partnership or the continuing partners.
   C. Sandy's withdrawal will cause the dissolution of the partnership if a court decree is sought.
   D. None of the above.

6. Which of the following is not required in order to convert an existing business to an LLC:
   A. An agreement of conversion is drafted that sets forth the terms of the conversion.
   B. The terms of the conversion are approved by all of the parties or by the number or percentage of owners required for conversion.
   C. Articles of organization of the LLC are filed with the secretary of state. The articles must state that the LLC was previously another form of business and the prior business's name.
   D. Pay a filing fee to the Internal Revenue Service.

7. Mary Sue and Billy Bob want to start an on-line music distribution company to be headquartered in Bora Bora. If they chose to organize a limited liability company and do business in Colorado, they will have to file with the Secretary of State in Colorado as:
   A. A Bora Bora S Corporation.
   B. A domestic limited liability company.
   C. An alien limited liability company.
   D. A foreign limited liability company.

8. All members of an LLC must vote on each of the following decisions except:
   A. Hiring new employees.
   B. Admitting new members.
   C. Amending the articles of organization.
   D. Amending the operating agreement.

9. A member of a member-managed LLC and a manager of a manager-managed LLC owe the following duties to the LLC:
   A. Duty of loyalty only.
   B. Duty of care and loyalty.
   C. Duty of care only.
   D. Neither duty of care nor duty of loyalty.

10. Without actual knowledge of disassociation or notice of disassociation, a member who disassociates him or herself from an LLC has apparent authority to bind the LLC to contracts in the ordinary course of business for:
    A. 1 year.
    B. 2 years.
    C. 5 years.
    D. 10 years.

V. Answers to Objective Questions

Terms:

1. Dissolution. Per section 29 of the Uniform Partnership Act, a dissolution is defined as "the change in the relationship of the partners caused by any partner ceasing to be associated in the carrying on as distinguished from the winding up of the business."
2. Novation. Novation is derived from the Latin term "novus" meaning new. In contract terms, a novation substitutes a new party and discharges one of the original parties to a contract by agreement of all the parties. In the partnership setting, novation is very often used to relieve retiring partners from certain obligations of the partnership.
3. Indemnification to hold harmless. This is a reimbursement arrangement between the old and new partners in case the old partner is somehow held liable for existing partnership obligations.
4. Marshaling of assets. This is an equitable doctrine requiring an arrangement or ranking of assets in a certain order towards the payment of debt.
5. Continuation agreement. These agreements will often be used in tandem with the creation of the new partnership agreement. These agreements are also sometimes referred to as so-called "bridge" agreements because they help facilitate the transition from the old to the new partnership entity.
6. Limited liability companies. Once the law permitted LLPs, all of the Big Five accounting firms changed their status from general partnership to LLPs.
7. *Member-managed.* Under Section 404(a) of the U.L.L.C.A. members have equal say in the management of member-managed limited liability companies.
8. Right to Information. Under Section 408 of the U.L.L.C.A. members have the right to inspect any records and the LLC's operating agreement during ordinary business hours.
9. Changing Order. This is form of judicial lien that entitles the creditor to be paid out of the debtor-member's LLC profits and distributions.
10. Winding Up. This happens if the business is not continued after a legal dissolution.

True/False:

1. False. Such a partnership dissolves automatically upon the expiration of the time or accomplishment of the objective.
2. True. Death of a partner acts as an operation of law dissolution of the partnership.
3. False. A partnership will be dissolved by operation of law following an event that makes it unlawful for the business of the partnership to be carried on or for the partners to carry it on.
4. False. Under the UPA, a partner may assign her interest without causing the dissolution of the partnership.
5. True. The partnership has technically ended.
6. True. Under Section 108 (b) of the U.L.L.C.A. an agent must be appointed for service of process, notices, demands and administrative proceedings.
7. False. An LLC organized in a foreign country and doing business in the United States is called an alien limited liability company.
8. True. Under Section 405 (a) of the U.L.L.C.A. members share profits equally unless otherwise agreed to.
9. True. If a covered member or manager commits an ordinary Negligent Act that is not grossly negligent, he or she is not liable to the LLC.
10. True. Under Section 409 (h)(1), a non-manager member of a manager-managed LLC is treated equally to a shareholder in a corporation and does not have these fiduciary duties.

Multiple Choice:

1. A. Because the partnership agreement does not state a duration, Ted may dissolve the partnership at any time without liability.
2. B. A partner who makes a loan to the partnership becomes a creditor of the partnership. Partners are personally liable for the partnership's debts and obligations.
3. A. Capital contributions are repaid only after creditors have been paid, but before profits are paid out.
4. C. Such an agreement does not prevent dissolution; it merely eliminates the necessity of the next step of winding up.
5. D. Sandy's actions will cause an automatic dissolution of the partnership, not a dissolution by operation of law or by court decree.
6. D. The first three steps are required under Section 902 of the U.L.L.C.A. Form 8832 determines the filing status with the IRS.
7. C. An alien limited liability company is an LLC organized in a foreign country and doing business in the United States.
8. A. Section 404 (c) of the U.L.L.C.A. calls for full member vote participation in all the above except A.
9. B. Both the duties of care and loyalty are owed to the LLC under Sections 409 (a) and 409 (h)(2) of the U.L.L.C.A.
10. B. Under Section 703 of the U.L.L.C.A. the apparent authority continues for two years.

VI. Answers to the Essay Question

The voluntary withdrawal of a partner from a partnership is recognized as an act of party basis for dissolution of a partnership. Here, when Big B left the firm, the existence of the old partnership ended.

The law firm should have exercised due care to provide as much notice as is legally required and prudent to both partners within the firm and third parties outside the firm. As for third parties who have had actual dealings with the partnership, they should be given actual verbal or written notice that Big B was no longer authorized to act for the firm. As for persons who knew of his association but did not deal with the firm, they should have been given some sort of media-based constructive notice. Parties who have not dealt with the partnership and do not have knowledge of it do not have to be given notice. The factual issue in this case as to how much Ms. Pembrokeshire had known about Big B and his firm prior to her dealings with him. If the firm failed to give any general media-based notice of Big B's retirement from the firm, it may be in some difficulty here. Perhaps the firm chose to trade on the "halo" effect of retaining Big B's name.

If the partnership had been technically dissolved at the time of Big B's retirement, and he is now allowed to represent himself to third parties as having apparent authority to act on their behalf, the partnership may in fact be liable under Mr. Big B's contract with Ms. Pembrokeshire. As such, if she is entitled to a refund of the retainer fee as a matter of contract law, the firm will be liable for it. The partners can, in turn, seek reimbursement from Big B if they can talk him into it. Better yet, they should talk him into true retirement out of state.

Chapter 29
DOMESTIC AND MULTINATIONAL CORPORATIONS

There is nothing new in the world, so history is the greatest lesson you can learn about what's happening now.

Garel Rhys

I. Overview

Corporations can provide many advantages for business including perpetual existence, limited liability, and numerous tax and other legal opportunities to massage the system. These advantages are not free, nor are they always easily obtained. Corporate law has always been technically intricate and demanding of legal practitioners. Corporations can also be more unforgiving to its users than sole proprietorships or partnerships if mistakes are made in its formation and financing. The stakes are simply greater because over eighty-five percent of business done in the U.S. uses the corporate format. This chapter has several objectives: to illustrate how the corporate form is established legally and to see how it is infused with its financial lifeblood, i.e., money. In addition, the process of establishing the basic ground rules for the key players will be examined.

The formation of a corporation starts with a contracting process initiated by a person called a promoter. The word promoter has an unsavory connotation to many people. In the modern vernacular, many people think of a promoter as a person seeking to get rich by doubtful means or chicanery rather than by legitimate work. Borderline sporting *cum* entertainment events have not done much to lessen this impression. In point of fact, the law of corporations has a much different expectation of the role of promoter.

A promoter of a new corporation is really the catalyst that brings together the diverse elements of law, finance, entrepreneurial talent, and technical competence that will eventually drive the fortunes of the new business entity. The role of promoter is also tied to the laws of contract, fiduciaries, and agency. He or she is expected to act for the benefit of the eventual corporation and can be expected to be personally liable for contracts entered into on its behalf in the interim.

The promoter's main duties are bifurcated towards two main audiences--the state and potential investors. He or she will be involved in contracts with both of these constituencies. With regard to the state, the actual creation of the new corporate entity is the outgrowth of a document called the charter. This document is the foundation contract between the promoter and the state. The charter takes the form of a certificate of incorporation. This certificate provides the official state-sanctioned ground rules under which the new corporate entity will be allowed to do business. Violation of these ground rules can lead to an eventual corporate death penalty, the revocation of the charter. The second critical task of the promoter is to find legal methods so that the start up of the corporation may be infused with financial lifeblood. The corporate form is unparalleled in its ability to be a fundraiser. These funds are generated by two basic methods--debt and equity financing.

Once the proper procedures for the establishment of the corporation have been complied with and adequate financing has been secured, the next step is to see what the basic ground rules will be for key players in this arena. The leading protagonists will be the board of directors, shareholders, and managers of the corporation. The distinctions between these roles are sometimes blurred when it comes to the formation and management of corporations. Yet failure to honor these distinctions can lead to disastrous consequences.

Erosion is defined as a gradual wearing away or disintegration. The wall of limited liability for corporations and their main protagonists has suffered from steady erosion over the past several decades. The floodwall is still holding, but the holes in the dike seem to get bigger every year. What is interesting about these reductions in limited liability is that outside parties often initiate them at the behest of other members of the same corporate family. Disadvantaged shareholders, disgruntled corporate officers, and quixotic directors have all done their share of diminishing the liability immunities once held so sacred in the corporate scheme of things.

There are two basic theories under which most of the limitations to corporate entity and/or corporate participants' liability are restricted or eliminated. First, public policy never has allowed the corporate form to act as a total shield from liability to third parties. From the very inception of corporate law, courts and legislators have clung to the back door option of holding someone personally liable where circumstances deemed it appropriate. A creature of the law should not be allowed to become a tool of circumvention of the law. Thus in the areas of crimes, torts, and generally undesirable behavior, a number of corporate shield-bursting mechanisms have always been in place. These have ranged from piercing the corporate veil to outright revocation of the continued existence of the corporation. In all these cases, the underlying premise is that the corporate form should not allow a person to do something that would otherwise be prohibited by public policy.

The second theory is found in the intracorporate workings of the organization itself. Just as the corporate form should not be allowed to defraud creditors and the like, so too should it not be allowed to be used by various members of corporate organizations to harm each other. The cornerstone of relationships between shareholders, directors, and officers is founded in the law of fiduciaries, loyalty, and mutual support. Even though not all aspects of these interdependent relationships could ever realistically be expected to be harmonious, some common good is expected to be interwoven into their behavior towards each other. As illustrated by the cases in this chapter, some of the most vociferous battles of all are fought in the confines of the corporate boardrooms around the country.

In both external relationships and intracorporate disagreements, the willingness of the contesting parties to go public with their grievances more frequently than ever has led to more exceptions to the rule of limited corporate liability. Time will tell whether or not this erosion bodes well for the economy. If it means a crippling of the economic engine of corporate vitality relative to comparable overseas entities, it does provide cause for alarm. If, however, it holds real wrongdoers responsible for harm, those whom would otherwise be shielded by the false front of a legal stage set, then it is time for the house of cards to come down.

The world of corporate combinations through mergers and the like has been turned upside down in recent history. Several factors have contributed to the frenzy of activity on the street, not all of it well motivated. As with so many of these sorts of fundamental changes to how society orders its affairs, there is no one underlying motive for it. As with any major shift in economic alignment, arguments pro and con can be heard through the halls of academia, in government agencies, and Wall Street.

The proponents for these changes first argue mergers, takeovers, and consolidation are simply marketplace adjustments that reflect attempts to correct inefficiencies in the marketplace. Where a company is poorly managed, running on "cruise control," and is not sufficiently lean and mean, it should be swallowed up by more efficient competitors. This view adopts the law of the jungle—only the strong survive. A really open marketplace rewards the most efficient in economic terms, and drags on that efficiency are doomed to failure and absorption in the end. The proponents of the law and economics, or "Chicago School" movement, proponents argue that mechanisms of law should not only allow this natural business evolution to take place, but that to stifle it would do harm to society in the end.

On the other side of the coin, a number of commentators have argued that runaway activity in the areas of mergers, acquisitions, buyouts, and the like breed a whole new generation of unethical, short-term thinking financial charlatans, whose only real interest is their personal bank accounts. Critics of the merger phenomenon point to the endless game-playing going on in this arena. They argue that what has been created is a big time casino where the chips are bigger than ever. Out of that gaming mentality, the larger society has had to pay for numerous financial scandals because the original players were not really motivated by the long-term good of society. In addition, long term growth, research, and development of essential business fundamentals has been cast aside in the name of greenmail, corporate spin-offs, tax manipulation, and golden parachutes. The net result of this sort of short term thinking is that while the financial market manipulators are playing high stakes poker, our industrial, financial, and transportation infrastructure has sunk into a quagmire of noncompetitiveness in the world marketplace.

The truth probably lies somewhere in between these poles when it comes to these business combinations. Time will ultimately tell if this recent history has reflected a transition cost to stay competitive as a nation, or if we are frittering away our industrial base at the gaming table. Whatever the outcome, each and every one of us will ultimately share in the processes' ultimate benefit or burden.

II. Hypothetical Multi-Issue Essay Question:

Harvey Wheat and Harry Oats have come upon an entirely new way of growing and promoting natural grain food products. By having as many farm animals as possible use their fields as a place to relieve themselves naturally, they have found their grains to be supercharged with nature's essences. They named their product Relievo Brand Inc. and decided to incorporate their new venture. They hired a recent law school graduate named Fossagium Fossa. All the common stock in the corporation was issued in the names of Harvey Wheat and Harry Oats equally. In addition, the board of directors for the corporation were Harvey Wheat, Harry Oats, John Barley, Peter Hay, and Fossagium Fossa. All of them met at the Tofu Natural Eating and Drinking Club for their board meetings.

The Secretary of State informed Harvey and Harry that their product must first meet certain health inspection requirements before they could do business in the state of Kannabras. Harvey and Harry took extreme umbrage at this insult to their farming expertise and chose simply to ignore this request. They ignored the inspection requirement and proceeded instead to grow and market their supercharged products at farmers' markets all over the country. In addition, Relievo Brand Inc. entered into numerous contracts with farm equipment suppliers, seed companies, and animal rental agencies in order to get a timely start on the growing season. The contracts totaled $1 million.

Last week Relievo Brand Inc. lost a big lawsuit to Calvin Calumny. It seems the Relievo Brand products were just too much for poor Calvin who lost his continence due to Harvey and Harry's products' unique ingredients. Relievo Brand Inc. now seeks to liquidate, and its creditors are seeking to collect against Harvey and Harry personally. Harvey, Harry, and the Board of Directors are claiming that they are not personally liable for the corporate debts of Relievo Brand Inc. What results?

III. Outline

Characteristics of Corporations
- Limited liability of shareholders
- Free transferability of shares
- Perpetual existence
- Centralized management

Classification of Corporations
- Profit and nonprofit
- Public and private
- Publicly held and closely held
- Professional

Types of Corporations
- Domestic
  - A corporation is a domestic corporation in the state in which it is incorporated
- Foreign
  - A corporation is a foreign corporation in states other than the one in which it is incorporated
- Alien
  - A corporation is an alien corporation in the United States if it is incorporated in another country

Articles of Incorporation
- Articles of incorporation must include
  - Name of the corporation
  - Number of shares the corporation is authorized to issue
  - Address of the corporation's initial registered office and the name of the initial registered agent
  - Name and address of each incorporator

Equity Securities
- Common stock
  - An equity security that represents the residual value of the corporation
- Preferred stock
  - An equity security that is given certain preferences and rights over common stock

Types of Shares

Corporate Powers
- Express
  - Powers given to a corporation by
    - The U.S. Constitution
    - State constitutions
    - Federal statutes
    - State statutes
    - Articles of incorporation
    - Bylaws
    - Resolutions of the board of directors
- Implied
  - Powers beyond express powers that allow a corporation to accomplish its corporate purpose

Methods of Dissolving a Corporation
- Voluntary dissolution
- Administrative dissolution
- Judicial dissolution
  - Dissolution by the state
  - Dissolution by creditors

IV. Objective Questions

Terms:

1. The articles of incorporation must identify the _______________ _______________ of the corporation, which is the person empowered to accept service of process on behalf of the corporation.

2. Very strict procedures must be followed when incorporating. If these procedures are not followed exactly, but there is substantial compliance with them, the resulting organization is known as a(n) _______________ _______________ _______________ under the common law.

3. Promoters frequently enter contracts on behalf of the corporation prior to incorporation. Through the process of _______________, the corporation becomes solely liable on such contracts, and the promoter is completely relieved of liability with respect to such contracts.

4. The rule, which relieves corporate officers and directors from personal liability to the corporation and the shareholders for honest mistakes of judgment, is known as the _______________ _______________ rule.

5. Directors of a corporation subject themselves to personal liability for certain decisions made when operating in their capacity as a director. To avoid personal liability for actions of the board, an individual director must either _______________ or _______________ _______________ _______________.

6. An officer or director who is personally involved in transactions with the corporation is said to have engaged in _______________ _______________.

7. The organization whose shares a purchaser seeks to buy in a tender offer is known as the _______________ corporation.

8. The payment by a target corporation of a premium over the fair market value of its shares to buy back a block of its stock previously acquired by the tendering corporation is known as _______________.

9. These documents must be filed with the secretary of state of the state of incorporation to effectuate a voluntary dissolution; they are known as the _______________ _______________ _______________.

10. In an attempt to fight off hostile takeover attempts, many corporations build provisions known as _______________ _______________ into their articles of incorporation and other relevant documents that provide for adverse consequences if corporate ownership changes hands.

True/False:

1. _____ A corporation must be incorporated in each state in which it chooses to do business.

2. _____ The purpose for which a corporation is formed must be specific and must be stated in the articles of incorporation.

3. _____ Following incorporation, a corporation automatically becomes liable for contracts entered into by the promoter on behalf of the corporation.

4. _____ A director who relied in good faith upon false or misleading information presented by a reliable employee cannot be held liable for her actions taken in reliance upon such information.

5. _____ A dissenting director may register her dissent only by entering it in the minutes of the meeting at which the action in question was taken.

6. _____ An officer may enter into a transaction with the corporation only if the transaction is fair and reasonable to the corporation.

7. _____ A consolidation requires the approval of the board of directors of both corporations and the affirmative approval of a majority of the shareholders of each corporation.

8. _____ In a tender offer, the target corporation is absorbed into the tendering corporation.

9. _____ Directors are shielded from liability for breach of their fiduciary duties if the breach occurs in an effort to defend against a tender offer.

10. _____ Under the Williams Act, a tender offer must remain open for fifteen days after commencement of the offer, and must be extended for ten days if the number of shares or the price offered changes.

Multiple Choice:

1. The Hellen-Julie Corporation filed its articles of incorporation with the state, but forgot to include one significant item of information required by the statute. Nevertheless, the state overlooked the omission and issued Hellen-Julie a corporate charter. Under common law, Hellen-Julie will be operating as a:
   A. Nonincorporated association.
   B. Partnership.
   C. *De facto* corporation.
   D. *De jure* corporation.

2. Ms. Margaux was the principal promoter of the New Wine Corporation. Prior to its incorporation, Margaux contracted to purchase $10,000 of inventory in the name of New Wine Corporation. The seller was not aware that New Wine did not yet exist. Who is liable on the contract?
   A. New Wine, unless it rescinds the contract.
   B. Margaux, unless New Wine adopts the contract.
   C. Margaux, because she had no authority to contract for New Wine.
   D. New Wine, because a promoter has authority to enter into preincorporation contracts.

3. Lord Byron entered into a contract in his own name to purchase, for $750,000, land zoned for industrial purposes. Immediately thereafter, Lord Byron formed the Tort Mining Corporation, retaining 50 percent of the stock and a seat on the board. At the first board of directors meeting, Lord Byron influenced the members to purchase the land for $1 million, its appraised fair market value, without disclosing his gain. With respect to the $250,000 of profit:
   A. Tort Mining Corporation may return title to the land and receive its money back.
   B. Lord Byron is liable to Tort Mining Corporation for the amount of profit.
   C. Lord Byron is not liable since he did not own more than 50 percent of the stock and the board voted on the purchase.
   D. Tort Mining Corporation has no recourse because the judgment of the board of directors was conclusive under the circumstances.

4. Ms. Kathleen Beacon is a director of both Court, Inc., and Pearl Street Corporation. Which fact requires Ms. Kathleen Beacon to make a disclosure to Court's other directors?
   A. Ms. Beacon is on Pearl Street Corporation's board of directors even if Court and Pearl Street Corporation are not in competition and do no business together.
   B. Pearl Street Corporation is contemplating a major purchase from Court at Court's preset prices.
   C. Ms. Beacon is a limited partner in a partnership that owns a cruise ship that Court is contemplating purchasing.
   D. Court is purchasing a nominal amount of supplies from Pearl Street Corporation.

5. Johnny Mephisto travels the country looking for shareholder derivative actions that can and should be brought for the betterment of mankind. Which of the following statements is correct with respect to Johnny's right to bring a derivative action?
   A. Johnny must persuade at least one other shareholder to participate before he can bring a derivative action.
   B. Johnny can acquire stock of a corporation in order to file a derivative action.
   C. Johnny can bring a derivative action without trying to persuade the officers and directors to take action to correct the matter.
   D. Johnny must have been the owner of stock of the corporation at the time of the alleged offensive action.

6. Assume the same facts as in question 5. Johnny acquires a small amount of stock in Dawgdoo Corporation and comes to you for advice concerning his rights and duties as a shareholder. Which of the following is not a right or duty of Johnny as a shareholder of Dawgdoo?
   A. Johnny owes fiduciary duties to the corporation and the other shareholders when exercising his rights as a shareholder.
   B. Johnny may be liable to creditors of the corporation if a court disregards the corporate entity.
   C. Johnny has the right to vote in elections of directors and other elections involving essential corporate changes.
   D. Johnny has the right to bring a derivative suit.

7. Boulder Bikes, Inc. united with Backpacks and Bikes, Inc. to form the Outdoor Shop, Inc. If Boulder Bikes, Inc. acquires 75 percent of the shares of Backpacks and Bikes, Inc., and both continue to operate, Backpacks and Bikes would be classified as which of the following?
   A. A merger.
   B. A subsidiary.
   C. A split-off.
   D. A reorganization.

8. Grumpy, Dumpy, and Sneezy form the HIHO Corporation. A few months later, Sneezy's ailment gets the best of him and he dies of pneumonia. Which of the following statements is correct?
   A. Ownership of Sneezy's shares of the corporation will vest in Snow White upon his death.
   B. Sneezy's death does not affect the existence of the corporation.
   C. Grumpy and Dumpy will become the sole shareholders of the corporation.
   D. The secretary of state of the state of incorporation is required by law to withdraw the charter of the corporation within six months of Sneezy's death.

9. The shareholders of Cautious Corporation have been approached by Aggressive Corporation who made a tender offer for their shares. With respect to a tender offer, which of the following statements is incorrect?
   A. The tender offer must remain open for fifteen business days following the making of the offer.
   B. If an increased price is offered after certain shares have been purchased, shareholders who previously tendered their shares are entitled to additional consideration.
   C. If the tender offer is not for all of the shares outstanding, the offeror must purchase shares on a first in basis.
   D. The offer must be extended for ten days if the number of shares or the price offered is changed following the initial offer.

10. Which of the following is not a method used to fight a tender offer?
   A. Adopting a poison pill.
   B. Selling a crown jewel.
   C. Issuing additional stock.
   D. None of the above.

V. Answers to Objective Questions

Terms:

1. *Registered agent.* This person is usually an individual attorney or law firm who has an established relationship with the company. When businesses seek to establish themselves out of the jurisdiction in which they are incorporated, most states will request the business to designate a registered agent under its jurisdiction.
2. *De jure corporation.* The term *de jure* denotes a condition in which there has been a total compliance with all the requirements of the law. In corporate law, this means the corporation has been created as a result of compliance with all of the constitutional and statutory requirements of the state of incorporation. The RMBCA has eliminated many of the former distinctions between a *de jure* and a *de facto* corporation.
3. *Ratification.* Ratification is but one of several methods by which contracts originally entered into by promoters with third parties are shifted over to the corporation once it comes into existence legally. Other methods include novation and adoption.
4. *Business judgment.* There must still be a showing that the transaction was made with due care and good faith.
5. *Resign; register his/her dissent.* They must take themselves out of the decision-making loop.
6. *Self-dealing.* This may exist any time a person in a fiduciary relationship uses the property of another for his or her own personal benefit.
7. *Target.* This entity is the object of the tender offer.
8. *Greenmail.* Because of the widespread use of this payment, it is now subject to a special tax as specified in the Internal Revenue Code, Sec. 5881.

9. *Articles of dissolution.* Because the corporate form is created by law, the law allows its existence to come to a voluntary end.
10. *Poison pills.* This is another example of the increasing colorful, yet hostile, language that has come to be identified with this area of law.

True/False:

1. False. A corporation can only be incorporated in one state even though it can do business in all other states in which it qualifies.
2. False. Although the purpose must be stated in the articles of incorporation, a corporation can be formed for any lawful purpose, and no specific purpose need be stated.
3. False. A corporation becomes liable on a promoter's contract only if it agrees to become bound to the contract by ratification, adoption, or novation.
4. C. Directors and officers cannot engage in activities that compete with the corporation unless full disclosure is made and a majority of the disinterested shareholders approve the activity. Ms. Beacon's part ownership of the cruise ship should be disclosed to Court, Inc.
5. D. State corporation statutes generally impose the requirement in order to make sure that these sorts of lawsuits are not abused.
6. A. Only a controlling shareholder is saddled with fiduciary duties and the corresponding liabilities for a breach thereof.
7. False. The shares of a deceased, incompetent, or otherwise legally disabled person may be voted by the legal representative of such an individual.
8. True. Usually, directors serve a one-year term. However, if no election is held, the directors serve until a successor is appointed.
9. False. Unless otherwise provided, only a majority of shares entitled to vote need be represented to have a quorum.
10. False. The board of directors is responsible for formulating the policy decisions affecting the management, supervision, and control of the operation of the corporation.

Multiple Choice:

1. C. A *de facto* corporation occurs where there has been a good faith attempt to organize the corporation pursuant to statute. A *de jure* corporation occurs where there has been complete or substantial compliance with the statute. These common law rules have been changed by those states that have adopted the RMBCA.
2. C. A promoter must contract in her own name because the corporation is a nonexistent entity. Therefore, unless the corporation adopts, ratifies, or enters a novation with respect to such contract, Margaux remains liable.
3. B. A promoter owes a fiduciary duty to disclose all relevant information. Promoters are liable for any secret profits made by or damages caused by a breach of this duty. Lord Byron is liable to Tort Mining Corporation for the amount of the profit.
4. D. The articles of incorporation should contain the corporation's name, purpose, powers, and other important information in addition to the amount and types of stock authorized. These rules vary in specificity from state to state.
5. D. Promoters are jointly liable on their contracts in those situations where the corporation never comes into existence.
6. D. A third party who deals with the enterprise as if it were a corporation is estopped from challenging its validity.

7. B. When one corporation owns a substantial number of the shares of another corporation, the acquiring corporation is known as the parent corporation and the acquired corporation is known as the subsidiary corporation.
8. B. Unlike the death of a partner in a partnership setting, the death of a shareholder does not affect the existence of the corporation.
9. C. If the tender offer is not for all of the shares outstanding, the offeror must purchase shares on a *pro rata* basis.
10. D. All of the methods listed are commonly used to avoid tender offers.

VI. Answers to Essay Question

It appears that Mr. Fossa did at least initially prepare the proper documentation for the incorporation of Relievo Brand Incorporated. The state of Kannabras has an incorporation statute that can be used to incorporate a new business in that state. As with all such statues, certain substantive and procedural measures must be taken in order to properly comply with the statute before articles or a certificate of incorporation will be issued. These requirements include a number of items of information that must be supplied to the state such as names and addresses of the key parties as well as agents for service of process. In addition, the purpose for which the business is incorporated must be a lawful one. In this case, the underlying purpose of the business, to engage in organic farming, is not unlawful. But, because it is an activity that can affect the health and welfare of its citizens, the state of Kannabras can regulate the activity under its general police power.

Had Harvey and Harry complied with the state's request to submit their products to health inspection, they most likely would have had *de jure* corporation. With the creation of a *de jure* corporation, they would have been entitled to one of the biggest benefits of using the corporate form--limited personal liability. A *de jure* corporation was not formed here.

Because they did not form a *de jure* corporation under the common law, Harvey, Harry, and the directors might be personally liable for the debts of the corporation unless they can show some other public policy-based excuse from that liability. Several possibilities exist under the common law.

They can first argue that they sought to get incorporated in good faith and should now be treated as a *de facto* corporation. Or in the alternative, they can argue that it would be an unjust enrichment to third party creditors to allow such creditors to go after them personally when they thought they were dealing with a corporation. On both accounts, the ultimate answer will rest on how the court defines the good faith efforts of Harvey, Harry, and the board to incorporate. Given the fact that they intentionally refused to submit their products to health inspectors in direct violation of a regulatory requirement of the state, it appears unlikely that they will be able to avoid personal liability.

The RMBCA provides that corporate existence begins when the articles of incorporation are filed. The Secretary of State's filing of the articles of incorporation is conclusive proof that the incorporators satisfied all the conditions of incorporation. After that, only the state can bring a proceeding to cancel or revoke the incorporation or involuntarily dissolve the corporation. Third parties cannot thereafter challenge the existence of the corporation or raise it as a defense against the corporation.

The corollary of this rule is: failure to file articles of incorporation is conclusive proof of the nonexistence of the corporation. If a failure to file has taken place, the purported shareholders of the corporation are exposed to personal liability for the debts and obligations of the purported corporation. In this case, if the Kannabras Secretary of State preconditioned the filing to compliance with the health regulations, the filing may be deemed not to have ever taken place. Under the RMBCA, this would then result in personal liability for Harvey and Harry. If, in turn, the filing was allowed to go through by the Secretary of Sate in spite of the problems with the health code, then the RMBCA would deem the filing to be conclusive proof of incorporation. With that proof, Harvey, Harry, and the members of the board of directors could claim limited personal liability from the debts and obligations of the corporation.

The corporation would be classified as a domestic, for profit, closely held corporation.

Assuming the corporation is formal, the contracts with the various suppliers would be assumed by the corporation. Whether or not the promoters would continue to remain personally liable would depend on the format used. If the contracts are ratified or renegotiated by way of a novation, then the promoters would be relieved of any further liability. Under an adoption, they would continue to be liable unless released by the suppliers.

Assuming that Harvey and Harry will be the only owners of common stock in this closely held corporation, they will have to provide legally acceptable consideration for the shares. As owners of common shares, they will be entitled to vote for board members and receive dividends declared by the board. Under this case scenario, no debt financing was used by Relievo Brand Incorporated.

The basic rights of the shareholders, directors, and managers upon formation of the corporation will be based on the articles of incorporation and the applicable laws of Kannabras as illustrated with regard to the common law versus RMBCA rules on filing.

Chapter 30
DIRECTORS, OFFICERS, AND SHAREHOLDERS

One of the secrets of life is that all that is really worth the doing is what we do for others.
Lewis Carroll

I. Overview

Erosion is defined as a gradual wearing away or disintegration. The wall of limited liability for corporations and their main protagonists has suffered from steady erosion over the past several decades. The floodwall is still holding, but the holes in the dike seem to get bigger every year. What is interesting about these reductions in limited liability is that they are often initiated at the behest of other members of the same corporate family rather than outside parties. Disadvantaged shareholders, disgruntled corporate officers, and quixotic directors have all done their share of diminishing the liability immunities once held so sacred in the corporate scheme of things.

There are two basic theories under which most of the limitations to corporate entity and/or corporate participants' liability are restricted or eliminated. First, public policy never has allowed the corporate form to act as a total shield from liability to third parties. From the very inception of corporate law, courts and legislators have clung to the back door option of holding someone personally liable where circumstances deemed it appropriate. A creature of the law should not be allowed to become a tool of circumvention of the law. Thus in the areas of crimes, torts, and generally undesirable behavior, a number of corporate shield-bursting mechanisms have always been in place. These have ranged from piercing the corporate veil to outright revocation of the continued existence of the corporation. In all these cases, the underlying premise is that the corporate form should not allow a person to do something that would otherwise be prohibited by public policy.

The second theory is found in the intracorporate workings of the organization itself. Just as the corporate form should not be allowed to defraud creditors and the like, so too should it not be allowed to be used by various members of corporate organizations to harm each other. The cornerstone of relationships between shareholders, directors, and officers is founded in the law of fiduciaries, loyalty, and mutual support. Even though not all aspects of these interdependent relationships could ever realistically be expected to be harmonious, some common good is expected to be interwoven into their behavior towards each other. As illustrated by the cases in this chapter, some of the most vociferous battles of all are fought in the confines of the corporate boardrooms around the country.

In both external relationships and intracorporate disagreements, the willingness of the contesting parties to go public with their grievances with more frequency than ever has led to more exceptions to the rule of limited corporate liability. Time will tell whether or not this erosion bodes well for the economy. If it means a crippling of the economic engine of corporate vitality relative to comparable overseas entities, it does provide cause for alarm. If, however, it holds real wrongdoers responsible for harm, who would otherwise be shielded by the false front of a legal stage set, then it is time for the house of cards to come down.

II. Hypothetical Multi-issue Essay Question

Mr. Clyde Catorski worked as a business manager for the famous rock group Cha Cha and the Potato Heads. The group successfully combined the best elements of Latin Disco sounds with the laid back qualities of Couch Potatoism. Clyde organized the business as a corporation under the laws of California in order to assure the group the benefits of limited liability against lawsuits. Several parents have sued the group alleging that their music turned their children into "potato heads," and that the music fostered potato cults. Under the articles of incorporation, Clyde was a minority shareholder (10 percent), an officer of the corporation, and a member of the board of directors.

Last week, the band played in Erfurt and the warm-up band, named Oral Hygiene, really stole the show. Their unique blend of flashy gold teeth and ear-piercing rock sent young fans into an eczematous itching frenzy. Clyde saw the writing on the wall and immediately solicited a business management deal with the lead singer, Harry Cycloid. Their conversation resulted in a secret deal between Clyde and Harry wherein Clyde would book Oral Hygiene into concerts rather than Cha Cha and the Potato Heads. Over the next several months Cha Cha and the Potato Heads lost concert dates and tee-shirt sales worth over $1 million. Clyde, however, is doing better than ever with Oral Hygiene. Last week, they played in the Root Canaldome. Cha Cha wants to sue. What results?

III. Outline

- Management of a Corporation
  - Shareholders
    - Owners of the corporation who vote on the directors and other major actions to be taken by the corporation
  - Board of directors
    - Responsible for making policy decisions and employing the major officers for the corporation
  - Officers
    - Responsible for the day-to-day operation of the corporation

- Methods of Electing Directors
  - Straight voting
  - Cumulative voting
  - Supramajority voting

- Shareholder Lawsuits
  - Direct lawsuits can be brought against a corporation to
    - Enforce the right to vote
    - Enforce preemptive rights
    - Compel payment of declared but unpaid dividends
    - Inspect books and records of the corporation
    - Enjoin the corporation from committing an ultra vires act

Compel dissolution of the corporation
Derivative lawsuits can be brought by a shareholder against an offending party on behalf of the corporation when the corporation fails to bring the lawsuit

Rights of Directors
The board of directors of a corporation is responsible for formulating the policy decisions affecting the management, supervision, and control of the operation of the corporation

Fiduciary Duties of Corporate Directors and Officers
Duty of obedience
Duty to act within the authority conferred by law and the corporation
Duty of care
Duty to use care and diligence when acting on behalf of the corporation
Duty of loyalty
Duty to subordinate personal interests to those of the corporation and its shareholders

Piercing the Corporate Veil
If a shareholder dominates a corporation and uses it for improper purposes, a court of equity can disregard the corporate entity and hold the shareholder personally liable for the corporation's debts and obligations

Insurance and Indemnification

IV. Objective Questions

Terms:

1. A corporation has ________________ ________________ to conduct those activities necessary to operate as a business whether or not expressly provided elsewhere.

2. When a corporation performs an act that is outside its limited purpose clause or beyond the applicable statutory limitations, the corporation is said to have committed a(n) ________________ ________________ act.

3. The officers and directors of a corporation owe fiduciary duties to the corporation and its shareholders. The ________________ ________________ ________________ imposes upon the officers and directors an obligation to exercise care and diligence when acting on behalf of the corporation.

4. The rule, which relieves corporate officers and directors from personal liability to the corporation and the shareholders for honest mistakes of judgment, is known as the ________________ ________________ rule.

5. A written document that a shareholder signs that authorizes another person to vote his or her shares at a shareholders' meeting is called a ________________ ________________.

6. The required number of shares that must be represented in person or by proxy to hold a shareholders' meeting is called a ________________.

7. A corporation is liable for the crimes of its directors, officers, and employees. This is a concept under the law of _______________.

8. The resulting suit when a shareholder brings a lawsuit that belongs to the corporation is known as a(n) _______________ lawsuit.

9. The right of a director or officer to be repaid for personal litigation expenses incurred in successfully defending himself is known as _______________ _______________.

10. The lawsuits of many shareholders against the corporation for the same wrong can be joined into one action known as a_______________ _______________ lawsuit.

True/False:

1. ____ A corporation typically has express power to borrow money, incur liabilities, and issue notes and other obligations, but not to lend money to assist its employees.

2. ____ If a corporation attempts to engage in an *ultra vires* act, a shareholder can sue for an injunction to prevent the corporation from engaging in such act.

3. ____ The standard used when determining whether a director or officer has acted within the duty of care owed the corporation is whether that person used that level of care that a reasonable director would have used in the conduct of the business of another.

4. ____ A director who relied in good faith upon false or misleading information presented by a reliable employee cannot be held liable for her actions taken in reliance upon such information.

5. ____ A dissenting director may register her dissent by entering it in the minutes of the meeting at which the action in question was taken.

6. ____ An officer may enter into any transaction with the corporation.

7. ____ A director who usurps a corporate opportunity has breached her duty of loyalty owed to the corporation.

8. ____ If a corporation chooses not to sue a director for the return of secret profits made by the director, a shareholder may bring the suit and retain the secret profit recovered.

9. ____ A corporation can be held liable for the torts, but not the crimes, of its agents and employees.

10. ____ Before a shareholder is entitled to bring a derivative suit, he is generally required to show that he owned shares of the corporation at the time the wrong occurred.

Multiple Choice:

1. Ms. Nancie Jenner, a shareholder of The Rusty Corp., has properly commenced a derivative action against The Rusty Corp board of directors. Ms. Jenner alleges that the board breached its fiduciary duty and was negligent by failing to independently verify the financial statements prepared by management upon which Michael & Co., CPAs issued an unqualified opinion. The financial statements contained inaccurate information that the board relied upon in committing large sums of money to capital expansion. This resulted in The Rusty Corp. having to borrow money at extremely high interest rates to meet current cash needs. Within a short period of time, the price of The Rusty Corp. stock declined drastically. Which of the following statements is correct?
   A. The board is strictly liable, regardless of fault, because it owes a fiduciary duty to both the corporation and the shareholders.
   B. The board is liable because any negligence of Michael & Co., CPAs is automatically imputed to the board.
   C. The board may avoid liability if it acted in good faith and in a reasonable manner.
   D. The board may avoid liability in all cases where it can show that it lacked *scienter*.

2. Assume the same facts as in question 1. If the court determines that the board was negligent and the board seeks indemnification from the corporation, which of the following statements is correct?
   A. The board may not be indemnified since a presumption that the board failed to act in good faith arises from the judgment.
   B. The board may not be indemnified unless The Rusty Corp.'s shareholders approve such indemnification.
   C. The board may be indemnified only if The Rusty Corp. provides liability insurance for its officers and directors.
   D. The board may be indemnified by The Rusty Corp. if the court deems it proper.

3. The Elizabeth Corporation decided to acquire certain assets belonging to The Ernie Corporation. As consideration for the assets acquired, the Elizabeth Corporation issued 20,000 shares of its no-par common stock with a stated value of $10 per share. The value of the assets acquired subsequently turned out to be much less than the $200,000 in stock issued. Under the circumstances, which of the following is correct?
   A. It is improper for the board of directors to acquire assets other than cash with no-par stock.
   B Only the shareholders can have the right to fix the value of the shares of no-par stock exchanged for assets.
   C. In the absence of fraud in the transaction, the judgment of the board of directors as to the value of the consideration received for the shares shall be conclusive.
   D. Unless the board obtained an independent appraisal of the acquired assets' value, it is liable to the extent of the overvaluation.

4. Ms. Kathleen Beacon is a director of both Court, Inc., and Pearl Street Corporation. Which fact requires Ms. Kathleen Beacon to make a disclosure to Court's other directors?
   A. Ms. Beacon is on Pearl Street Corporation's board of directors even if Court and Pearl Street Corporation are not in competition and do no business together.
   B. Pearl Street Corporation is contemplating a major purchase from Court at Court's preset prices.
   C. Ms. Beacon is a limited partner in a partnership that owns a cruise ship that Court is contemplating purchasing.
   D. A majority of disinterested shareholders disapprove of the activity.

5. Johnny Mephisto travels the country looking for shareholder derivative actions that can and should be brought for the betterment of mankind. Which of the following statements is correct with respect to Johnny's right to bring a derivative action?
   A. Johnny must persuade at least one other shareholder to participate before he can bring a derivative action.
   B. Johnny can acquire stock of a corporation in order to file a derivative action.
   C. Johnny can bring a derivative action without trying to persuade the officers and directors to take action to correct the matter.
   D. Johnny did not have to be the owner of stock of the corporation at the time of the alleged offensive action.

6. Assume the same facts as in question 5. Johnny acquires one share of stock in Dawgdoo Corporation, a large multinational corporation, and comes to you for advice concerning his rights and duties as a shareholder. Which of the following is not a right or duty of Johnny as a shareholder of Dawgdoo?
   A. Johnny owes fiduciary duties to the corporation and the other shareholders when exercising his rights as a shareholder.
   B. Johnny may be liable to creditors of the corporation if a court disregards the corporate entity.
   C. Johnny has the right to vote in elections of directors and other elections involving essential corporate changes.
   D. Johnny does not have the right to bring a derivative suit.

7. Mike Moore, an employee of Billy's Brewing Corporation, was involved in an automobile accident while on his way to an alcoholic's cure meeting one evening after work. Vicki Victim has brought suit against Mike and Billy's Brewing Corporation. Which of the following statements is correct?
   A. Mike is liable for his own negligence whether or not the corporation is also liable.
   B. The corporation is liable for Mike's negligence because it was a foreseeable consequence of Mike's employment.
   C. The corporation is not liable for Mike's negligence because the accident occurred after business hours.
   D. Both A and C.

8. Phil Playboy was elected to the board of directors of HiTech Corporation several years ago. Phil is much more concerned about the prestige of being a corporate director than he was about the duties involved. He has not attended a board meeting in two years. During that time, the president made some very bad loans resulting in serious financial loss to the corporation. Which of the following statements is correct?
   A. Phil has no liability because he merely exercised poor business judgment by missing the meetings.
   B. Phil has no liability because missing the meetings was an honest mistake.
   C. Phil can be held liable if it can be proven that he was negligent in performing his duties.
   D. Phil has no liability.

9. In which of the following situations will a shareholder be deemed to owe a fiduciary duty to minority shareholders?
   A. When the shareholder owns at least two-thirds of the shares of the corporation.
   B. When the shareholder owns more than 50 percent of the shares of the corporation.
   C. When the shareholder owns a sufficient number of shares to effectively control the corporation.
   D. None of the above.

10. In which of the following situations would a court be least likely to pierce the corporate veil?
    A. A shareholder employed by the corporation maintains the same checking account for his personal use and for the corporate checking account.
    B. A corporation involved in asbestos removal is capitalized with 3,000 shares of stock that it sold for $.50 per share.
    C. A corporation sells corporate assets to shareholders for fair market value.
    D. None of the above.

VI. Answers to Objective Questions

Terms:

1. *Implied powers*. These are sometimes referred to as "common sense" powers.
2. *Ultra vires*. Under case law, this definition has been expanded to cover situations where the corporation had the power to act, but the exercise of the power was improper or irregular.
3. *Duty of care*. As this is a rather open ended term, this duty is interpreted on a case-by-case basis.
4. *Business judgment*. There must still be a showing that the transaction was made with due care and good faith.
5. *Proxy card.*. This is the way a shareholder appoints another person as their agent to vote their shares at the shareholders' meeting.
6. *Quorum*. This is the required number of shares that must be represented in person or by proxy to hold a shareholders' meeting.
7. *Agency*. The law of agency governs this.
8. *Derivative*. This action is called derivative because it is based on the primary rights of the corporation but is being asserted by the shareholder on its behalf.
9. *Mandatory indemnification*. These statutes require that the director be made whole or held harmless for these costs.
10. *Class action*. Where a class is properly recognized as having a common legal action vis-à-vis an opposing party, efficiency is served by combining the multiple proceedings into one action.

True/False:

1. False. A corporation generally has express authority to borrow and to lend.
2. True. In addition, the corporation can sue for damages or the state attorney general can seek a dissolution of the corporation.
3. False. The appropriate standard of care is that level of care that an ordinary person in a comparable position would use under similar circumstances.
4. True. Only if the director had knowledge that would cause reliance to be unwarranted can she be held liable.
5. False. A dissenting director may register his or her dissent either by entering it in the minutes, by filing a written dissent prior to adjournment, or by forwarding a written dissent immediately following adjournment of the meeting.
6. False. Such transactions may be entered into if fair and reasonable to the corporation. In addition, such transactions may be entered if fully disclosed and validly approved by the directors or the shareholders.
7. True. If usurping is proven, the director must return the opportunity to the corporation and return any profits made as a result of the usurpation.
8. False. A shareholder is entitled to bring a derivative suit in this instance, but any profit recovered must be returned to the corporation.

9. False. A corporation can be held liable for both the torts and the crimes of its agents and employees. The normal criminal sanction imposed on a corporation is the assessment of monetary fines or loss of legal privileges.
10. True. The shareholder must have either owned the shares or acquired them by operation of law.

Multiple Choice:

1. C. The board of directors owes a duty of care to the corporation and the shareholders. In exercising this duty, the directors may rely in good faith upon the reports of reliable and competent officers and employees and upon professionals such as public accountants as to matters within their expertise.
2. D. A court may order indemnification if it finds that the director or officer is reasonably and fairly entitled to such indemnification. B is the second best answer, but to be the correct answer, it must also have shown that the director or officer acted in good faith and believed his or her actions to be in the best interest of the corporation.
3. C. In the exercise of its discretion, the business judgment rule provides that the board is protected from liability to the corporation and the shareholders for honest mistakes in judgment. As a practical matter, the rule is necessary to avoid second-guessing the fluctuations of the marketplace.
4. D. A majority of the disinterested shareholders approve the activity. Ms. Beacon's part ownership of the cruise ship should be disclosed to Court, Inc.
5. D. Without owning stock of the corporation at the time of the alleged offensive action, Johnny has no right to bring a derivative suit.
6. D. Only a controlling shareholder is saddled with fiduciary duties and the corresponding liabilities for a breach thereof.
7. D. A corporate employee is personally liable for his own torts. In addition, a corporation is liable for the torts of its agents and employees, but only if such torts occurred while such person was acting within the scope of employment for the corporation.
8. C. A director breaches the duty of care if he is negligent in the performance of his duties.
9. C. Majority ownership is not required to effectively control a corporation. A shareholder who effectively controls the corporation owes a fiduciary duty to minority shareholders.
10. C. However, if the corporation sold the assets for less than fair market value, the court would likely pierce the corporate veil.

VI. Answers to Essay Question

It appears that the Cha Cha and the Potato Heads' Corporation was duly organized and granted a corporate charter under the laws of California. Under its articles of incorporation, it could pursue most any sort of legal entertainment activity. Assuming its music was not obscene, did not cause persons to riot, and did not create any other clear and present danger to society, the band could play on. If any of the parents had been able to prove that music was creating a potato head cult, the state many have considered revoking their corporate charter. That outcome does not appear likely here.

Here we have a situation where a key person in the corporation has acted in breach of his fiduciary duty. Clyde is an officer of the company as well as a member of the board of directors. In both capacities Clyde is held to a fiduciary duty of loyalty to the corporation and the other shareholders to act in the best interests of the corporation. In undertaking a secret contractual relationship with a direct competitor, he has committed several corporate sins. He has engaged in self-dealing, usurping corporate opportunities for himself, directly competing with the corporation, and generally acting against the corporation's best interests. Based on all these violations of his basic duty of loyalty, he can be liable to the corporation for damages based on the above listed actions.

The other shareholders can look to the Cha Cha and the Potato Heads corporate board of directors to seek these remedies in the name of the corporation. If the board of directors fails to act, the shareholders may seek to bring a derivative lawsuit on behalf of the corporation. In any event, Clyde should have left Cha Cha first before getting involved with Oral Hygiene. A key element of corporation law has always been loyalty to one master at a time. Any good cocker spaniel could have taught Clyde that lesson.

Chapter 31
MERGERS AND TAKEOVERS OF CORPORATIONS

In economics, the majority is always wrong.
John Kenneth Galbraith

I. Overview:

The world of corporate combinations through mergers and the like has been turned upside down in recent history. Several factors have contributed to the frenzy of activity on the street, not all of it well motivated. As with so many of these sorts of fundamental changes to how society orders its affairs, there is no one real underlying motive or rhyme or reason to it. As with any major shift in economic alignment, arguments pro and con can be heard through the halls of academia, government agencies and Wall Street.

The proponents for these changes first argue mergers, takeovers, and consolidation are simply marketplace adjustments that reflect attempts to correct inefficiencies in the marketplace. Where a company is poorly managed, running on "cruise control," and is not sufficiently lean and mean, it should be swallowed up by more efficient competitors. This view literally adopts the law of the jungle where only the strong survive. A really open marketplace rewards the most efficient in economic terms, and drags on that efficiency are doomed to failure and absorption in the end. The law and economics, or "Chicago School" movement, proponents argue that mechanisms of law should not only allow this natural business evolution to take place, but that to stifle it would do harm to society in the end.

On the other side of the coin, a number of commentators have argued that runaway activity in the areas of mergers, acquisitions, buyouts, and the like breed a whole new generation of unethical, short term thinking, financial charlatans whose only real interest is their personal bank accounts. Critics of the merger phenomenon point to the endless game-playing going on in this arena. They argue that what has been created is a big time casino where the chips are bigger than ever. Out of that gaming mentality, the larger society has had to pay for numerous financial scandals because the original players were not really motivated by the long-term good of society. In addition, long term growth, research, and development of essential business fundamentals has been cast aside in the name of greenmail, corporate spin-offs, tax manipulation, and golden parachutes. The net result of this sort of short term thinking is that while the financial market manipulators are playing high stakes poker, our industrial, financial, and transportation infrastructure has sunk into a quagmire of noncompetitiveness in the world marketplace.

The truth probably lies somewhere in between these poles when it comes to these business combinations. Time will ultimately tell if this recent history has reflected a transition cost to stay competitive as a nation, or if we are frittering away our industrial base at the gaming table. However it all turns out, each and every one of us will ultimately share in the processes' ultimate benefit or burden.

II. Hypothetical Multi-issue Essay Question

Truly Useless Corporation has been the envy of the widget manufacturing trade for years. Their crystal widgets have been both the price and quality leader in widget markets all over the world ever since their invention by the corporate founder, Mr. Lyaeus Lycisca. Lyaeus married Lotis Lovely, and they had three children, Libitina, Lycaon, and Lelaps. Lyaeus and Lotis wanted to see the children go into the family business, so they went to their attorney, Linus Lethe, and he drew up a corporate business plan wherein the parents retained 25 percent of the shares in the company and the children each got 25 percent.

Things were cooperative between all the family members for a number of years until Libitina could no longer stand Lelap's bad habits. Libitina convinced her parents and her siblings to sell the company to No Redeeming Value Incorporated for $10 a share (book value). Lelaps objects and wants to stop the pending sale. He contends that crystal widgets are doing better than ever and that the acquiring corporation is really just a shell entity owned by the other family members who want to freeze him out because of his questionable personal habits. What results?

III. Outline

What Is a Merger?
   A merger occurs when one corporation is absorbed into another corporation and ceases to exist

Example of a Merger

What Is a Consolidation?
   A consolidation occurs when two or more corporations combine to form an entirely new corporation

Example of a Consolidation

What Is a Share Exchange?
   A share exchange occurs when one corporation acquires all the shares of another corporation and both corporations retain their separate legal existence

Example of a Share Exchange

What Is a Tender Offer?
   A tender offer is an offer that an acquirer makes directly to a target corporation's shareholders in an effort to acquire the target corporation

Illustration of a Tender Offer

Defensive Strategies and Tactics for Fighting a Tender Offer
   Persuasion of shareholders
   Delaying lawsuits
   Selling a crown jewel
   Adopting a poison pill
   White knight merger
   Pac-man or reverse tender offer
   Issuing additional stock
   Creating an employee stock ownership plan (ESOP)
   Flip-over and flip-in rights plans
   Greenmail and standstill agreements

State Antitakeover Statutes

State antitakeover statutes are statutes enacted by state legislatures that protect corporations incorporated in or doing business in the state from hostile takeovers

IV. Objective Questions:

Terms:

1. Where shareholders challenge the incumbent management and both sides solicit proxies from shareholders, this process is generally referred to as a _______________ _______________.

2. Business combinations can occur in several different forms. A _______________ occurs when one corporation absorbs another and the existence of the absorbed corporation ceases.

3. A part owner of a corporation who objects to a proposed merger even though it has been properly approved is called a _______________ _______________.

4. A shareholder who properly objects to a merger is entitled to a cash payment equivalent to the _______________ _______________ of his shares.

5. The procedure that may be used to combine a parent corporation with a subsidiary corporation is called a(n) _______________ _______________ _______________.

6. The procedure involving the combination of two or more corporations to form an entirely new corporation is known as _______________.

7. The organization whose shares a purchaser seeks to buy in a tender offer is known as the _______________ corporation.

8. The payment by a target corporation of a premium over the fair market value of its shares to buy back a block of its stock previously acquired by the tendering corporation is known as _______________.

9. These documents must be filed with the secretary of state of the state of incorporation to effectuate a voluntary dissolution; they are known as the _______________ _______________ _______________.

10. In an attempt to fight off hostile takeover attempts, many corporations build provisions known as _______________ _______________ into their articles of incorporation and other relevant documents that provide for adverse consequences if corporate ownership changes hands.

True/False:

1. ____ One who violates the antifraud provisions of the federal securities laws with respect to the solicitation of proxies is subject to civil and criminal actions by the federal government, and to civil actions brought by private individuals wronged by such violation.

2. ____ In all situations in which the incumbent management becomes involved in a proxy contest, the corporation must reimburse such persons for expenses incurred in that contest whether they win or lose.

3. ____ Following a merger, the surviving corporation gains all the rights, privileges, powers, duties, obligations, and liabilities of the merged corporation.

4. ____ An ordinary merger requires approval of the board of directors of each corporation and of a supramajority of the shareholders of the merged corporation.

5. ____ Approval of a short-form merger requires only the consent of the board of directors of the parent corporation.

6. ____ In a short form merger, shareholders of both the parent corporation and the subsidiary corporation have dissenting shareholder's appraisal rights.

7. ____ A consolidation requires the approval of the board of directors of both corporations and the affirmative approval of a majority of the shareholders of each corporation.

8. ____ In a tender offer, the target corporation is absorbed into the tendering corporation.

9. ____ Directors are shielded from liability for breach of their fiduciary duties if the breach occurs in an effort to defend against a tender offer.

10. ____ Under the Williams Act, a tender offer must remain open for fifteen days after commencement of the offer, and must be extended for ten days if the number of shares or the price offered changes.

Multiple Choice:

1. In a short-form merger:
   A. The approval of the shareholders of both corporations is needed.
   B. The approval of the board of directors of both corporations is needed.
   C. A short form merger is not allowed under the RMBCA.
   D. All of the above are false.

2. Under which of the following circumstances would a corporation's existence terminate?
   A. The death of its sole owner shareholder.
   B. Its becoming insolvent.
   C. Its legal consolidation with another corporation.
   D. Its reorganization under the federal bankruptcy laws.

3. Which of the following would be grounds for the judicial dissolution of a corporation on the petition of a shareholder?
   A. Refusal of the board of directors to declare a dividend.
   B. Waste of corporate assets by the board of directors.
   C. Loss operations of the corporation for three years.
   D. Failure by the corporation to file its federal income tax returns.

4. Little Dog, Inc., and Big Dog, Inc., are considering a merger. Little Dog is a small, closely held corporation, and Big Dog is publicly traded on a national exchange. A merger would require:
   A. Approval of the SEC.
   B. Creditors of the corporation which does not survive to file claims within thirty days or lose their rights.
   C. That dissenting shareholders have their stock purchased at its fair value.
   D. Approval by at least a two-thirds majority of the board of directors of each corporation, without shareholder approval.

5. Boulder Bikes, Inc. united with Backpacks and Bikes, Inc. to form the Outdoor Shop, Inc. This particular transaction would be classified as which of the following?
   A. A merger.
   B. A subsidiary.
   C. A consolidation.
   D. A reorganization.

6. In question 5, if Boulder Bikes, Inc. united with Backpacks and Bikes, Inc. retaining the name Boulder Bikes, Inc., this particular transaction would be classified as which of the following?
   A. A merger.
   B. A subsidiary.
   C. A consolidation.
   D. A reorganization.

7. In question 5, if Boulder Bikes, Inc. acquires 75 percent of the shares of Backpacks and Bikes, Inc., and both continue to operate, Backpacks and Bikes would be classified as which of the following?
   A. A merger.
   B. A subsidiary.
   C. A split-off.
   D. A reorganization.

8. Grumpy, Dumpy, and Sneezy form the HIHO Corporation. A few months later, Sneezy's ailment gets the best of him and he dies of pneumonia. Which of the following statements is correct?
   A. Ownership of Sneezy's shares of the corporation will vest in Snow White upon his death.
   B. Sneezy's death does not affect the existence of the corporation.
   C. Grumpy and Dumpy will become the sole shareholders of the corporation.
   D. The secretary of state of the state of incorporation is required by law to withdraw the charter of the corporation within six months of Sneezy's death.

9. The shareholders of Cautious Corporation have been approached by Aggressive Corporation who made a tender offer for their shares. With respect to a tender offer, which of the following statements is incorrect?
   A. The tender offer must remain open for fifteen business days following the making of the offer.
   B. If an increased price is offered after certain shares have been purchased, shareholders who previously tendered their shares are entitled to additional consideration.
   C. If the tender offer is not for all of the shares outstanding, the offeror must purchase shares on a first in basis.
   D. The offer must be extended for ten days if the number of shares or the price offered is changed following the initial offer.

10. Which of the following is not a method used to fight a tender offer?
    A. Adopting a poison pill.
    B. Selling a crown jewel.
    C. Issuing additional stock.
    D. None of the above.

V. Answers to Objective Questions

Terms:

1. *Proxy contest.* The term proxy is derived from the Latin word "procuracy" meaning the writing that authorizes an agent to act for another.
2. *Merger.* This is the amalgamation of two corporations pursuant to statutory requirements.
3. *Dissenting shareholder.* Many corporation statutes provide for appraisal remedies for dissenting shareholders.
4. *Fair value.* This is the term used to define the appraisal remedy.
5. *Short-form merger.* It is called short form because of the relatively brief amount of paper work involved.
6. *Consolidation.* Both of the original entities no longer legally exist, and they are replaced by a new entity.
7. *Target.* This entity is the object of the tender offer.
8. *Greenmail.* Because of the widespread use of this payment, it is now subject to a special tax as specified in the Internal Revenue Code, Sec. 5881.
9. *Articles of dissolution.* Because the corporate form is created by law, the law allows its existence to come to a voluntary end.
10. *Poison pills.* This is another example of the increasing colorful, yet hostile, language that has come to be identified with this area of law.

True/False:

1. True. A shareholder who is injured by a material misrepresentation or omission in proxy solicitation materials can recover damages.
2. False. The corporation is only required to reimburse the incumbent management for expenses incurred in a proxy contest if the issue is one of corporate policy.
3. True. It literally steps into the legal shoes of the nonsurviving corporation.
4. False. Unless stated otherwise in the articles or bylaws, a simple majority of the shareholders of each corporation is all that is required. The approval of the shareholders of the surviving corporation is only required if the merger increases the number of voting shares by more than 20 percent.
5. True. This is specified by statute.
6. False. Shareholders of the parent corporation do not have dissenting shareholder's appraisal rights because their rights are not significantly affected.
7. True. These approvals are required because this is considered an organic change.
8. False. Both the tendering corporation and the target corporation retain their legal status.
9. False. In attempting to defend against takeovers, the directors and officers must adhere to their fiduciary duties.
10. True. The Williams Act specifies these terms.

Multiple Choice:

1. D. Under Section 11.04 of the RMBCA, all that is required is the approval of the parent board of directors if the ownership percentage of 90 percent is met.
2. B. A consolidation occurs when two or more corporations combine to form an entirely new entity, and the original corporation's existence will terminate.
3. B. A shareholder can institute an action for judicial dissolution where the corporate assets are being misapplied or wasted.
4. C. Shareholders are provided a statutory right to dissent. A shareholder who properly registers her dissent is entitled to have her shares purchased at their fair value, as provided in the RMBCA.
5. C. A consolidation occurs when two or more corporations combine to form an entirely new corporation.
6. A. A merger occurs when one corporation is absorbed into another corporation and ceases to exist.
7. B. When one corporation owns a substantial number of the shares of another corporation, the acquiring corporation is known as the parent corporation and the acquired corporation is known as the subsidiary corporation.
8. B. Unlike the death of a partner in a partnership setting, the death of a shareholder does not affect the existence of the corporation.
9. C. If the tender offer is not for all of the shares outstanding, the offeror must purchase shares on a *pro rata* basis.
10. D. All of the methods listed are commonly used to avoid tender offers.

VI. Answers to the Essay Question

From the facts as given, there appears to be no question that the shareholders in this case each have a *pro rata* right to vote their shares with regard to normal functions such as election of the board of directors. What is unclear is if a supramajority vote would be required for an organic change to the corporate structure like a merger or acquisition. The likelihood is that such a supramajority would be required by most articles of incorporation. If not, then a simple majority of shareholders entitled to vote is all that would be needed to approve a board of directors resolution to merge or be acquired.

If the merger were approved by the requisite majority, or supramajority if applicable, then Lelaps can have his dissent registered and seek to get an appraisal of value for his shares. This appraisal would be conducted according to court procedures pursuant to the state corporation statute and on the articles of incorporation on the bylaws of the corporation. Most corporate valuations are higher than "book" value in the marketplace, and it appears likely that the $10 per share value may be raised by the court.

In the alternative, Lelaps may want to argue for an involuntary dissolution and liquidation of the corporation. To succeed he would have to prove to the court that there has been some sort of improper behavior on the part of the board of directors and/or the majority shareholders which would be deemed oppressive as to him. Dissenting shareholders ordinarily do not have the right to recover the fair market value of their shares upon dissolution of a corporation unless such evidence of wrongful acts is proven. In any event, it looks like Lelaps is out of the new No Redeeming Value corporate scheme of things.

Chapter 32
INVESTOR PROTECTION AND ON-LINE SECURITIES TRANSACTIONS

SEC. Securities and Exchange Commission, a federal agency devoted to providing young lawyers with the experience necessary to secure jobs advising corporations on how to circumvent the regulations of the SEC
D. Robert White

I. Overview

Assume a sporting event were to be contested under the following conditions:

1. All the players were well trained.
2. The rules of the game were fully explained to the players.
3. Those rules are fairly and evenhandedly applied to all participants.
4. An even playing field is used as a site for the contest.

With all these suppositions in place, can you rest assured your team will win? Or can you, at best, hope that, win or lose, your team was engaged in a fair contest?

In the broadest sense, the buying and selling of securities is indeed similar to an athletic event. Each participant goes into the game with his or her own self-interest in mind. And all the fair rules in world will not change one essential truth of these or any other contests—there will be winners and there will be losers. That reality must always be kept in mind from the outset by anyone seeking to make his or her fortune through the sale or purchase of securities. Risk is inherent to the nature of this activity, and anyone who fails to appreciate that simple fact should not be there in the first place.

It is most difficult for professionals to master the ins and outs of the financial markets, let alone the casual investor. Yet the lure of playing this game is so strong that every year millions of people invest hard-earned money with nothing more than high hopes and a prayer. Securities law was designed to at least give some substance to those hopes and prayers. That substance is public information upon which investment choices can be rationally made. These laws are not designed to assure a win in this high-risk game, but rather to provide a more even playing field.

The great financial stock market crash of 1929 and the ensuing Depression brought on by that calamity brought to the fore the need to create a greater governmental role in securities markets. Prior to that period, the sale of stocks in corporations remained essentially unregulated except for the common law doctrines of fraud and the like. Manipulative and unscrupulous trading practices coupled with a lot of hopes and unrequited prayers all pointed to a need for a better set of ground rules by which this game could be played.

The basic rules of the game go back almost seventy years to the Securities Act of 1933 and the Securities Exchange Act of 1934 that created the Securities and Exchange Commission. Over the years, the Commission's role has greatly increased with the advent of new technologies like programmed trading and with the need to expand its regulatory framework into the financial services arena. Because of recent scandals in this sector of the economy, a number of new white-collar crimes have been added to the government's arsenal for dealing with abuses in this area. All in all, it has made the specialized practice of securities law or SEC accounting more difficult, yet more challenging, than ever. As future business leaders, you have more than just a self-survival need to stay abreast of these changing rules of the game. As a nation we cannot afford the losses created by the manipulative abuses recently seen in these critical markets. If we do not watch out, 1929 can be revisited upon us, this time only worse. The key question checklist will itemize the applicable statutory provisions.

II. Hypothetical Multi-issue Essay Question

Mr. Bouteilles Gerbeuses was suffering from a common Twenty-First century malaise, working at a good paying job that provided him only with material but not spiritual satisfaction. He wanted to break away from his golden ball and chain and indulge in his real fantasy of spirits—wine making. He has not had any formal wine training, but he knew what he liked when he tasted it. He decided to start his own winery called *Cuvees de Prestige* specializing in sparkling wines produced in the Noir Valley of Massazona.

He could not afford to capitalize the entire operation himself, so he came up with a novel idea. Why not sell the individual rows of vines to "partners"? They would each get his expertise in producing the wine, and individually marked barrels with their own labels. These barrels could be consumed by the investors or sold for a tidy profit.

Mr. Gerbeuses convinced several hundred partners to buy one row each for $10,000 and proceeded to start up production of *Cuvees de Prestige* wines. Unfortunately what he didn't tell his partners was that he didn't know how to tell the difference between a *blanc de blancs*, a *blanc de noir*, and a *pinot noir*. Having failed to master the art of malolactic fermentation, the wine tasted more like vinegar than the nectar of the gods Mr. Gerbeuses had hoped for. His partners are also fermenting and look to see if they can get out of this tannic sedimentation of a deal. What results?

III. Outline

Definition of a Security

- An interest or instrument that is common stock, preferred stock, a bond, a debenture, or a warrant
- An interest or instrument that is expressly mentioned in securities acts
- An investment contract

Securities Act of 1933

- A federal statute that primarily regulates the issuance of securities by corporations, partnerships, associations, and individuals

Registration Statement

- Document that an issuer of securities files with the SEC that contains required information about the issuer, the securities to be issued, and other relevant information

Securities Exempt from Registration
- Securities issued by any government in the U.S.
- Short-term notes and drafts that have a maturity date that does not exceed nine months
- Securities issued by nonprofit issuers
- Securities of financial institutions
- Securities issued by common carriers
- Insurance and annuity contracts issued by insurance companies
- Stock dividends and stock splits
- Securities issued in a corporate reorganization where one security is exchanged for another security

Preventing Transfer of Restricted Securities
- To protect the nontransferability of restricted securities, the issuer must
  - Require the investors to sign an affidavit stating that they are buying the securities for investment, acknowledging that they are purchasing restricted securities, and promising not to transfer the shares in violation of the restriction
  - Place a legend on the stock certificate describing the restriction
  - Notify the transfer agent not to record a transfer of the securities that would violate the restriction

Securities Exchange Act of 1934
- A federal statute that primarily regulates the trading in securities

Tipper-Tippee Liability
- Tipper
  - A person who discloses material nonpublic information to another person
- Tippee
  - The person who receives material nonpublic information from a tipper

Section 10(B) and Section 16(B) Compared

Other Federal Securities Laws
- Securities Enforcement Remedies and Penny Stock Reform Act of 1990
- Market Reform Act of 1990
- Racketeer Influenced and Corrupt Organizations Act
- Private Securities Litigation Reform Act

IV. Objective Questions

Terms:

1. Two federal statutes designed to require disclosure of information to investors and prevent fraud are known as the __________ __________ __________ __________ and the __________ __________ __________ __________ __________.

2. Generally, a __________ is present when an investor invests money in a common enterprise with the expectation of making a profit from the significant efforts of others.

3. The federal administrative agency empowered to administer federal securities laws is called the __________ __________ __________ __________.

4. A written document used as a selling tool by the issuer, which enables prospective purchasers of securities to evaluate certain risks of a particular investment, is known as the ________________.

5. When an issuer engages in public relations campaigns that tout the prospects of the company and an upcoming securities offering, this is called ________________ ________________ ________________.

6. The time between when a registration statement is filed with the SEC and when the registration statement becomes effective is called the ________________ period.

7. The document which must be delivered to investors at the time of confirming a sale or when sending a security to a purchaser is known as the ________________ prospectus.

8. Each state is entitled to enact its own laws, known as ________________ ________________ laws, regulating the offer and sale of securities within its borders.

9. In certain situations, employees and advisors of an issuer of securities use material, nonpublic information to make a profit by buying or selling securities of the issuer. This practice is known as ________________ ________________.

10. When certain insiders buy and sell securities within a six-month period of time, any gains realized are called ________________ ________________ profits.

True/False:

1. ____ During the prefiling period, the issuer is permitted to offer, but not to sell, securities of the proposed offering.

2. ____ During the waiting period, the issuer may distribute a preliminary prospectus and publish tombstone ads that, in effect, condition the market for the upcoming issuance of securities.

3. ____ Certain transactions in securities are exempt from the registration requirements of the federal securities laws as are certain types of securities.

4. ____ During the posteffective period, the issuer may close the transactions solicited previously, but may not solicit any new transactions.

5. ____ Securities issued by colleges and universities are typically exempt from the registration requirements of the federal securities laws.

6. ____ Securities issued using a private placement exemption may only be sold to 35 investors.

7. ____ Violation of the federal securities laws can result in civil and criminal actions by the SEC as well as private civil actions by individual investors.

8. ____ Certain individuals associated with a securities offering can avoid liability by proving that, after reasonable investigation, they had reasonable grounds to believe, and did believe that at the time the registration statement became effective, the statements therein were true and that there was no omission of a material fact.

9. ____ An insider who trades on public information is subject to prosecution for violating the federal securities laws.

10. ____ If an insider provides nonpublic information to another who knows the information is not public information and who trades the related securities in reliance on that information, both the insider and the trading party have committed violations of the federal securities laws.

Multiple Choice:

1. In general, which of the following is least likely to be considered a security under the Securities Act of 1933?
   A. General partnership interests.
   B. Warrants.
   C. Limited partnership interests.
   D. Treasury stock.

2. Under the Securities Act of 1933, the registration of securities that are offered to the public in interstate commerce is:
   A. Directed toward preventing the marketing of securities which pose serious financial risks to the prospective investor.
   B. Not required unless the issuer is a corporation.
   C. Mandatory unless the cost to the issuer is prohibitive as defined in the SEC regulation.
   D. Required unless there is an applicable exemption.

3. Which of the following is subject to the registration requirements of the Securities Act of 1933?
   A. Public sale of its bonds by a municipality.
   B. Public sale by a corporation of its negotiable five year notes.
   C. Public sale of stock issued by a common carrier regulated by the Interstate Commerce Commission.
   D. Issuance of stock by a corporation to its existing stockholders pursuant to a stock split.

4. The principle purpose of the registration requirements of the Securities Act of 1933 is to:
   A. Prevent public offerings of securities in which management fraud or unethical conduct is suspected.
   B. Provide the SEC with the information necessary to determine the accuracy of facts presented in the financial statements.
   C. Assure that investors have adequate information upon which to base investment decisions.
   D. Provide the SEC with the information necessary to evaluate the financial merits of the securities being offered.

5. Shadee Corporation wishes to sell securities to raise $850,000 for the purpose of financing several corporate expansion projects. The securities will be offered under Rule 504. Which of the following statements is correct with respect to the transaction?
   A. The transaction cannot be completed without registration because Rule 504 only allows the issuance of $750,000 of securities.
   B. The securities can be sold to accredited investors only.
   C. The securities can be sold to an unlimited number of accredited investors and to 35 unaccredited investors.
   D. The securities can be sold to an unlimited number of accredited unaccredited investors.

6. Ricky Risktaker was of the theory that to make it big, he had to take big risks in his investments. He took a huge risk in buying stock in Worthless Corporation. He is now suing the CPA firm that audited the financial statements including in the registration statement in an attempt to recover his investment. Which of the following statements is correct?
   A. The CPA firm is not liable if it can prove that it acted with due diligence in conducting its audit.
   B. The CPA firm is strictly liable for any material defects in the registration statement.
   C. Ricky must prove that he relied on the registration statement before he will be entitled to recover.
   D. None of the above.

7. Drink-Rite Corporation, a corporation whose shares are publicly traded, was having a phenomenal year. Profits were rolling in because rolls were coming off of its customers. However, the fun is about to end. The Food and Drug Administration has notified the president of Drink-Rite that the corporation will be forced to stop selling its product because of a recent discovery of significant health risks created by the product. The president immediately sells his shares and calls his father who also sells his shares. In addition, the president's father calls a friend who also sells his shares. Who has made an illegal sale in this situation?
   A. The president.
   B. The president's father.
   C. The friend of the president's father.
   D. A and B.
   E. All of the above.

8. Hometown Corporation, a Colorado corporation, hopes to raise capital by issuing new securities. Which of the following facts would prevent the issuance of securities pursuant to an intrastate offering exemption?
   A. The principal office of Hometown is located in Omaha, Nebraska.
   D. The securities are offered to several residents of New Mexico, but ultimately purchased by only Colorado residents.
   C. Hometown earns 30 percent of its gross revenues in Wyoming.
   D. A and C.
   E. All of the above.

9. Assume the same facts as in question 8, except that Hometown hopes to issue the securities using a Rule 505 Small Offering Exemption. Which of the following facts would prevent the issuance of securities pursuant to a Rule 505 Small Offering Exemption?
   A. Securities totaling $4,500,000 are sold in the transaction.
   B. The securities are sold to 37 accredited investors.
   C. The securities are offered in newspapers and magazines throughout the region.
   D. The securities are sold to investors whom the issuer knows will transfer the securities within three years of purchase.
   E. C and D.

10. Which of the following statements about a private placement not involving a public offering of securities is incorrect?
   A. Resale of such securities is restricted.
   B. General solicitations of purchasers is prohibited.
   C. Securities may only be offered to corporations and partnerships.
   D. None of the above.

V. Answers to Objective Questions

Terms:

1. *Securities Act of 1933; Securities Exchange Act of 1934.* These laws are the two keystones of today's securities law system.
2. *Security.* The definition of securities has expanded to cover more and more sorts of ventures, including some limited partnership offerings.
3. *Securities and Exchange Commission or SEC.* This is the securities equivalent of the IRS.
4. *Prospectus.* Note that this is not a contract *per se.*
5. *Conditioning the market.* There are limitations on this practice.
6. *Waiting.* This period is designed to give time to act if someone may want to object.
7. *Final prospectus.* There is time to act based on this information.
8. *Blue-sky laws.* This is nicknamed after fraudulent sales of securities that had no underlying value beneath the "Blue Sky."
9. *Insider trading.* Much needed reforms have come based on recent scandals.
10. *Short-swing.* These transactions are now targeted by reforms to securities laws.

True/False:

1. False. During the prefiling period, the issuer is not permitted to offer or sell securities.
2. True. Although actual sales are prohibited during the waiting period, the issuer may engage in certain offering activities.
3. True. Congress took cost considerations into account when these exemptions were allowed.
4. False. During the posteffective period, the issuer may close the transactions arranged previously and solicit new offers and sales.
5. True. Securities issued by nonprofit institutions, such a colleges and universities, are exempt from registration.
6. False. Securities sold using a private placement exemption may be sold to an unlimited number of accredited investors, but may only be sold to 35 nonaccredited investors.
7. True. Depending on the violation, a particular action could result in all of the mentioned actions.
8. True. Such a due diligence defense is available to a nonexpert with respect to that portion of a registration statement prepared by nonexperts.
9. False. If the information is truly "public" in nature, then an insider is free to trade without committing any violations.
10. True. Unfortunately, much of this still goes on.

Multiple Choice:

1. A. Under the *Howey* test, a security exists where the investor invests money in a common enterprise with the expectation of making a profit from the significant efforts of *others*. A general partnership interest would result in profits from the efforts of each general partner including the partner's own efforts as coowner of the enterprise..
2. D. Unless a security or transaction qualifies for an exemption, Section 5 of the Securities Act of 1933 requires securities offered to the public through the use of the mails or any facility of interstate commerce to be registered with the SEC by means of a registration statement and an accompanying prospectus. This rule is designed to afford maximum coverage of the 1933 Act.
3. B. Unless a security or the transaction is exempt, the issuance and sale must be registered. Answer B is not an exempt security because commercial paper of a corporation must have a maturity of nine

months or less to qualify as an exempt security. A and C are examples of exempt securities. In answer D, there is no offer or sale of a security because the securities are received in a stock split and not for value given in exchange.

4. C. The primary purpose of the Securities Act of 1933 is to assure that securities issuers provide potential investors with full and fair disclosure of all material information needed to make a prudent investment decision. The Act does not assure that the investor will make a profit on his or her investment.
5. D. Rule 504 exempts the sale of up to $1 million of securities during a 12-month period from registration. The securities may be sold to an unlimited number of accredited and unaccredited investors, but general selling efforts to the public are not permitted.
6. A. To be shielded from liability, the CPA firm only needs to prove that it acted with due diligence with respect to the portions of the registration statement it prepared, i.e., only the financial statements.
7. E. All three parties have made illegal sales in this situation because all knew that they were trading on material nonpublic information.
8. E. To qualify for the intrastate offering exemption, the issuer must be a resident of the state in which the exemption is sought. In addition, the issuer must be doing business in that state, i.e., earn 80 percent of its gross revenues in that state. Finally, the shares may not be offered or sold to a resident of another state.
9. D. To qualify for a Rule 505 Small Offering Exemption, general selling efforts, such as advertising in a newspaper or magazine, are forbidden.
10. C. Securities may be sold to an unlimited number of accredited investors and to 35 unaccredited investors.

VI. Answers to Essay Question:

Mr. Gerbeuses may have a barrel full of real problems on his hands. The production of alcoholic products has always been a highly regulated industry even before the enactment of the securities laws of the 1930s. Assuming he got through the regulatory maze of the Treasury Department's Alcohol, Tobacco, and Firearms Division, he must now worry about possibly having run afoul of the Securities and Exchange Commission. Under the 1933 Act, newly issued securities must comply with the registration requirements of the Act unless they fall into specific exemptions outlined in the 1937 Act. The first issue revolves around the definition of security for purposes of the 1933 Act. Under the court decision of *S.E.C. v. Howey,* any contract that calls for an investment with the expectation of income from the efforts of a promoter or third person may be deemed to be a security. Here the most likely interpretation is that the sales contracts for rows of vines would be declared a security. Calling it a partnership would not make any difference.

Next, how was this security sold? If there is use of any facility of interstate commerce or the mail, it may be covered under the 1933 Act. Here with sales to several hundred people, lawyers would indicate a very likely use of the mail for purposes of selling the securities.

Then decide if any of the exemptions from registration apply. The securities themselves do not fall into any of the exempt categories. It is not a nonprofit organization, nor is this a sale by a governmental entity. It is not a reorganization or stock split. In addition, it does not appear that any of the transaction exemptions apply. It is not stated as only an intrastate offering. Nor does it appear that Mr. Gerbeuses tried to qualify the sales as a private placement or a small offering under either Section 504 or 505. Because he failed to comply with the 1933 Act, he may now be subject to both civil and criminal sanctions by the government. In addition, the private parties hurt by these violations may sue him.

In addition to the problems raised by the 1933 Act, there are problems that may arise under the 1934 Act if Mr. Gerbeuses engaged in subsequent trading of his partnership interests after they had first been issued. This is particularly a problem for him under Rules 10(b) and 10(b)-5 that are designed to go after fraudulent transfers of securities. These violations could lead to both civil and criminal sanctions against him by both the government and private parties.

Mr. Gerbeuses may be liable under the RICO statute or one or more of the newer federal securities laws recently passed as a result of scandals in the securities markets. He also may have violated California's Blue Sky Securities Law. All in all, Mr. Gerbeuses probably should have stayed in his office and fantasized about his winemaking abilities over a glass of *Dom Perignon*. It would have cost him far less.

Chapter 33
ADMINISTRATIVE LAW AND CONSUMER PROTECTION

An expert is one who knows more and more about less and less.
Nicholas Murray Butler

I. Overview

A labyrinth is identified as a structure or garden characterized by an extremely complex maze with tortuous dead ends and blind alleys. Anyone who has dealt with a large governmental bureaucracy can readily appreciate the frustrations of trying to get through that maze with sanity intact. Government's burgeoning growth of administrative agencies at every level is indeed cause for concern for its constituents. According to statistics published by the U.S. Congress, the federal government alone has over three million civilian employees. In spite of constant calls to reduce the size of government's role in the average person's affairs, that role has grown tremendously. The media headlines may be focused on the goings on in the capitol, but the real functions of government are carried out in the trenches by this "fourth branch of government" (as distinguished from the role of the press popularly known as the "fourth estate") every day. This chapter seeks outlines the basic ground rules about how the agencies are created, how they are authorized to act, and what controls have been put in place so as to protect the rights of both the citizenry and the government.

The basic function undertaken by these administrative agencies is to carry out the ministerial functions necessary to the operation of the government. These functions are first authorized by what are called organic statutes, which create the agency, and enabling statutes, which delegate certain powers to the agency to act for the executive, legislative, or judicial branches of government. It is interesting to note at the outset that the "clean functional lines" of executive, legislative, and judicial can and often do become blurred when examining the breadth and scope of administrative agency activities.

Once the existence of the agency is settled upon and its scope of authority is established, it is then decided whether it is acting within that scope vis-à-vis the particular issue at hand. Remember the basic assumption here is that the executive branch, legislative branch, or judicial branch has chosen to designate and delegate a certain portion of its authority to act. This delegation is based on the presumption that the agency can be expected to have certain levels of expertise, scales of economy, and attention to detail that could not be readily expected of the policy makers. The next step is to see if the power in question was in fact truly delegated, and if so, is it being properly exercised by the agency?

The mechanisms for control of agency powers are relatively sparse given the scope of agency activity. The key provisions for control of agency powers are found in the executive branch chain of command and in the overview powers vested in the judiciary. In addition, there have been a number of specific information access statutes, such as the Freedom of Information Act, Government in the Sunshine Act, and the Administrative Procedure Act to help persons dealing with these agencies to get through the labyrinth. Like it or not, what it all points to is that we live in an age of specialization. The days of dealing directly with your elected representative are simply long gone, if in fact, they ever were there in the first place. We need this specialization in many ways in order for the functions of government to be truly effective.

This chapter also covers the fourth major set of venues within a quadripartite of remedies available to a wronged or injured consumer. First, there is criminal law. Victims of consumer fraud and similar offenses have always been able to seek state-supported sanctions against wrongdoers. This venue may provide some ephemeral satisfaction for the victim and may even, at least temporarily, protect society from further harm. But criminal law does not truly make the victim whole. As a matter of fact, most of the miscreants convicted of consumer fraud are also judgment proof, i.e., they have no assets from which civil judgments can be satisfied.

The second area of consumer protection is found in tort law and its permutations of intentional tort, negligence tort, and strict liability. These remedies can and do provide meaningful substance to civil correction of wrongdoing where the defendant is found to have some financial means. As seen in the prior discussions of these areas, tort law generally and products liability specifically are ripe with controversy and a great deal of uncertainty in today's legal environment. The major drawback to both the criminal law and tort law methods of consumer protection is that they represent *after-the-fact* remedies to harm already done. They are reactive remedies as opposed to proactive forms of prevention of harm. It has been argued that large civil judgments act as societal signals that are designed to discourage repetition of undesirable behavior.

The third side to our quadrilateral picture is found in contracts. Contract law has the advantage of providing the consumer with the opportunity to anticipate any problems before they befall him or her. This notion is traditionally found in the doctrine of *caveat emptor,* which courts of another age used with cavalier abandon. Both the common law of contracts and its progeny, the Uniform Commercial Code, have come a long way from the days of "Let the buyer beware."

In spite of all this progress in the areas of crime, torts, and contract, the gap between consumer harm and consumer protection continues to remain unfilled. Legislators at all levels of government have sought to help fill this void with a number of consumer protection measures. These measures often incorporate elements from both civil and criminal areas of enforcement. In addition, these measures can and often do have a prophylactic effect on many potential harms to the consumer. Unfortunately, another hallmark of many of these measures is that they are the end product of a trail of harm that had reached a crisis or disastrous level. Consider how long it took to take certain dangerous prescription drugs or unsafe toys off the market. Where these laws do provide a measure of safety, some consumer comfort may be found in "at least better late than never." Protection from harm has come a long way, but there is still no light at the end of the tunnel.

II. Hypothetical Multi-Issue Essay Question

Mr. Joe Logrie was the proud owner of a new litter of cocker spaniel puppies. The parents, T. Bear, and Margaux, are AKC Champion show dogs. They have also garnered additional accolades by being temperament tested and certified as tracking companion dogs for field trials. The puppies names are: Enzo, Bordeaux, Harry, Felicity, Patience, Toulouse, Byron, Sonia and Tort. Margaux is one tired mama!

Joe decides that the expenses related to the breeding and raising of these dogs are much the same as caring for any dependent and goes down to the Backwater District Office of the Internal Revenue Service to get some advice from the IRS. Taxpayer Service Representative, Mr. Oliver Officious. Oliver told Joe he also raised cocker spaniels and felt that they made wonderful dependents. As dependents, Oliver saw no reason why they shouldn't be listed as such on Joe's 1040. Joe proceeded to file his tax return on April 15th and listed eleven dependents, all of them having long ears and names like Tort and Enzo. He even got social security numbers for them, hoping that they would eventually work for their keep.

Alas, poor Joe was audited by the IRS on April 16th and was told he could not claim his dogs as dependents in spite of their long-eared needs to be fed by him. Joe points to Oliver's advice as his defense. What results?

III. Outline

Federal Administrative Agencies
- Administrative agencies that are part of the executive or legislative branch of government

Legislative Powers of Administrative Agencies
- Substantive rule making
  - Adopting rules that advance the purpose of the statutes that the agency is empowered to enforce
- Interpretive rule making
  - Adopting rules the interpret statutory language
- Statements of policy
  - Announcing proposed courses of action the agency plans to take in the future
- Licensing
  - Granting licenses to applicants and suspending or revoking licenses

Executive and Judicial Powers of Administrative Agencies
- Executive
  - Administrative subpoena
  - Administrative searches
- Judicial
  - Adjudicating cases through an administrative proceeding
    - Procedural due process
    - Substantive due process

Judicial Review of Administrative Agency Actions
- Conditions for review
  - Case must be ripe for review
  - Petitioner must have exhausted all administrative remedies
- Standards for review
  - Questions of law
  - Questions of fact
    - Arbitrary, capricious abuse of process test
    - Substantial evidence test
    - Unwarranted by the facts test (or de novo standard of review)

Public Disclosure of Agency Actions and Protection of Individual Rights

Caveat Emptor
- "Let the buyer beware"
- The traditional guideline of sales transactions

Federal Food, Drug, and Cosmetic Act
- Administered by the Food and Drug Administration
- Provides a basis for the regulation of the testing, manufacture, distribution, and sale of foods, drugs, cosmetics and medicinal products in the US

Acts That Regulate Product Safety
- Consumer Product Safety Act
- Fair Packaging and Labeling Act
- Poison Prevention Packaging Act
- Magnuson-Moss Warranty Act
- Flammable Fabrics Act
- Child Protection and Toy Safety Act
- Refrigerator Safety Act

Unfair and Deceptive Practices
- Prohibited by Section 5 of the Federal Trade Commission Act
- Unfair and deceptive practices include:
  - False and deceptive advertising
  - Bait and switch
  - Door-to-door sales
  - Unsolicited merchandise

Federal Consumer-Debtor Protection Acts
- Truth in Lending Act
- Consumer Leasing Act
- Fair Credit and Charge Card Disclosure Act
- Equal Credit Opportunity Act
- Fair Credit Reporting Act
- Fair Credit Billing Act
- Fair Debt Collection Practices Act
- Uniform Consumer Credit Code

IV. Objective Questions:

Terms:

1. Administrative agencies use ________________ ________________ ________________ to interpret the statutes that the agency is authorized to enforce.

2. An act that establishes certain administrative procedures that federal administrative agencies must follow in conducting their affairs is the ________________ ________________ ________________.

3. A doctrine that says when an administrative agency is created, it is delegated certain powers and the agency can only use those legislative, judicial, and executive powers that are delegated to it is the ________________ ________________.

4. A rule or regulation issued by an administrative agency that has much the same power as a statute—it has the force of law and must be adhered to by covered persons and businesses—is a ________________ ________________.

5. The adoption of a substantive rule that must be carried out by a trial-like hearing is called _______________ rule making.

6. A type of deceptive advertising that occurs when a seller advertises the availability of a low-cost discounted item but then pressures the buyer into purchasing more expensive merchandise is known as _______________ _______________ _______________.

7. The act that requires creditors to make certain disclosures to debtors in consumer transactions that do not exceed $25,000 is called the _______________ _______________ _______________ Act.

8. An amendment to the TILA that requires disclosure of certain credit terms on credit and charge card solicitations and applications is known as the _______________ _______________ _______________ _______________ _______________ Act of 1988.

9. A credit card sent to a person without the person having requested it is called a _______________ credit card.

10. Information about a subject's character, reputation, personal traits, and mode of living compiled from personal or telephone interviews with neighbors, friends, and business associates can be found in an _______________ _______________ _______________.

True/False:

1. ____ Administrative agencies can only be created by the federal government.

2. ____ Most federal administrative agencies are part of the legislative branch of government.

3. ____ The statutes enacted by administrative agencies are the substantive law enforced by the agencies.

4. ____ Violators of administrative agency statutes may be subject to civil, but not criminal liability.

5. ____ Neither public notice nor participation is required when an agency decides to issue an interpretive rule.

6. ____ Under Section 5 of the Federal Trade Commission Act, actual deception must be proven to show deceptive or unfair advertising.

7. ____ Under the Postal Reorganization Act, a person who receives unsolicited merchandise through the mail may use the merchandise without incurring an obligation to pay for it or return it.

8. ____ The Truth In Lending Act covers all creditors, including those who do not impose a finance charge.

9. ____ Creditors must notify their customer/debtors of the debtor's right of recission on credit transactions where the customer's house was used as security for an extension of credit.

10. ____ A cardholder whose credit card is stolen is liable for all unauthorized charges made on the card before the issuer is notified that the card has been stolen.

Multiple Choice:

1. Very Important Agency (VIA), an agency recently set up by the federal government, has some questions about how it would conduct its affairs. Where should VIA look for guidance?
   A. The Government in the Sunshine Act.
   B. The Administrative Procedure Act.
   C. The Equal Access to Justice Act.
   D. None of the above. VIA can conduct itself in any reasonable manner.

2. Assume VIA is set up and running properly. Which of the following powers may it have if properly delegated?
   A. Executive powers.
   B. Legislative powers.
   C. Judicial powers.
   D. All of the above.

3. VIA decides there are several important issues that must be examined and developed into rules. Assume VIA can establish these rules by notice and comment rule making. Which of the following must VIA do?
   A. Publish a notice and invite and consider interested comment.
   B. Have the interested parties present evidence and conduct cross-examination.
   C. Both of the above.
   D. None of the above.

4. VIA decides it is necessary to search the business premises of one of its licensees, a company called On The Edge (OTE). Under what circumstances will a search be considered reasonable?
   A. A warrantless search is conducted in an emergency situation.
   B. OTE is in a business where warrantless searches are automatically considered valid.
   C. OTE voluntarily agrees to the search.
   D. All of the above.

5. VIA has been delegated power to adjudicate cases through an administrative proceeding. Nonconforming Niel has allegedly violated several agency rules. Which of the following is correct?
   A. Niel must receive notice of the charges against him.
   B. Niel must be allowed to present evidence on his own behalf.
   C. Both of the above.
   D. None of the above.

6. Collector Carrie buys a bronze sculpture from Exquisite Exclusives Gallery (EEG). The sculpture costs $5,000, and Carrie cannot purchase it outright. Therefore, EEG allows Carries to buy the sculpture by paying for the sculpture in two equal payments without a finance charge. Under the Truth-in-Lending Act, which of the following is correct?
   A. EEG must disclose its annual percentage rate.
   B. EEG must disclose any prepayment penalties.
   C. Both of the above.
   D. None of the above. EEG is not subject to TILA.

7. Driver Daisy decides to lease a car from Linda's Leasing. The lease is for five years and a total cost of $10,000. Under the Consumer Leasing Act, what must Linda's Leasing include in the lease agreement?
   A. Any express warranties on the car.
   B. Daisy's liability at the end of the five-year term.
   C. An itemization of all lease and finance charges.
   D. All of the above.

8. Careless Chris leaves his wallet sitting on a restaurant counter when he leaves to use the telephone. While he is gone, Slimy Sylvester steals Chris' credit card. Chris is unaware of the theft when he returns to his seat. In fact, Chris does not discover his card has been stolen until five days later. By that time, Sylvester has charged $2,000 on the card. Chris reports the stolen card as soon as he discovers it. Which of the following is correct?
   A. Chris is liable for the $2,000 in unauthorized charges.
   B. Chris is not liable for any of the unauthorized charges.
   C. Chris is liable for $50 of the unauthorized charges.
   D. Chris is liable for one half, or $1,000, of the unauthorized charges.

9. Assume the same facts as in question 8 except that Chris discovers the card missing and reports it immediately. However, Sylvester manages to charge $2,000 on the card anyway. Which of the following is correct?
   A. Chris is liable for the $2,000 in unauthorized charges.
   B. Chris is not liable for any of the unauthorized charges.
   C. Chris is liable for $50 of the unauthorized charges.
   D. Chris is liable for one half, or $1,000, of the unauthorized charges.

10. Ripped-Off Rhonda purchased a new microwave from Ample Appliances with her credit card. However, the microwave is obviously defective, and Rhonda does not want to pay for the microwave until it is repaired or replaced. Which of the following is correct?
    A. Rhonda can withhold payment to her credit card company until the dispute is resolved.
    B. Rhonda must pay the credit card company and seek a refund from Ample Appliances.
    C. Either of the above.
    D. None of the above. A consumer has no recourse if a faulty product is purchased on a credit card.

V. Answers to Objective Questions

Terms:

1. *Rules and regulations.* Like it or not, ours is a highly regulated society and government agencies have been delegated as the agents of law enforcement in the broadest sense of the term.
2. *Administrative Procedure Act.* This Act was passed in 1946 and is intended to provide a basic set of guidelines by which federal administrative agencies must conduct their affairs.
3. *Delegation doctrine.* An agency cannot be imbued with more power than the original delatator had to give. If an agency exceeds the limits of its delegated powers, it is an unconstitutional act.
4. *Substantive rule.* Before a federal administrative agency proposes to adopt a substantive rule, it must follow the procedures set forth in the Administrative Procedures Act.
5. *Formal.* This procedure may sometimes be required under 5 U.S.C. §556 and usually involves contentious issues in highly regulated activities.
6. *Bait and switch.* Remedies such as rain checks are designed to offset some of the harmful effects of these sorts of practices.

7. *Truth-in-Lending.* As the average price of autos continues to go up, Congress is expected to raise the threshold coverage of this act.
8. *Fair Credit and Charge Card Disclosure.* This act reflects a trend in the law to make legal documents easier to comprehend by lay people.
9. *Unsolicited.* Few financial institutions do this anymore after having been burned by wrongful users of unsolicited cards.
10. *Investigative consumer report.* The debtor is entitled to request and examine these reports.

True/False:

1. False. Administrative agencies are created by federal, state, or local governments.
2. False. Most federal administrative agencies are part of the executive branch of government.
3. True. The statutes enacted by administrative agencies are the substantive law, as opposed to the procedural law, enforced by the agencies.
4. False. Violators of administrative agency statutes may be subject to either civil or criminal liability depending on the violations.
5. True. Generally, this is true, although many agencies will hold hearings in order to elicit the opinions of interested parties.
6. False. Under Section 5 of the FTCA, actual deception need not be proven to show deceptive or unfair advertising.
7. True. The sender puts him or herself at risk of not being paid if he or she engages in this practice.
8. False. TILA covers many creditors. However, certain guidelines set which types of creditors are and are not covered.
9. True. There are time-line limitations on the use of this right.
10. False. A cardholder whose credit card is stolen is liable only for $50 on all unauthorized charges made on the card before the issuer is notified that the card has been stolen. If the cardholder notifies the issuer before the stolen card is used, the cardholder is not liable for any unauthorized charges.

Multiple Choice:

1. B. The Administrative Procedure Act sets out certain administrative procedures that federal administrative agencies must follow in conducting their affairs. A is incorrect because the Government in the Sunshine Act was enacted to open certain federal administrative agency meetings to the public. C is incorrect because the Equal Access to Justice Act was enacted to protect persons from harassment by federal administrative agencies.
2. D. If properly delegated, an administrative agency can have executive, legislative, and judicial powers.
3. A. Notice and comment rule-making, or informal rule-making, involves publishing a general notice of the proposed rule making, giving interested people the opportunity to comment, and reviewing input. B is incorrect because a trial-like procedure is used in formal rule making. C is incorrect because B is incorrect. D is incorrect because A is correct.
4. D. Administrative agency searches are proper under all the circumstances of A, B and C.
5. C. VIA must both give Niel notice of the charges against him and allow him to present evidence on his own behalf. These requirements must be satisfied in order to afford Niel procedural due process.
6. D. Since EEG neither charges Carrie a finance charge nor requires payment in more than four installments, it is not subject to TILA. A and B are incorrect because they only apply to creditors covered by TILA.
7. D. The Consumer Leasing Act requires all three.

8. C. Chris is liable for $50 of the $2,000 in charges since he did not notify the credit card issuer before Sylvester made unauthorized charges on the card. However, Chris is not liable for more than $50. A is incorrect because the credit card holder is liable for a maximum of $50 in this situation. B is incorrect because Chris did not contact the credit card issuer before unauthorized charges were made on the card. D is incorrect for the same reason A is incorrect.
9. B. Since Chris notified the credit card issuer immediately, he is not liable for any of the unauthorized charges.
10. A. B is incorrect because Rhonda does not have to pay the credit card company. She may withhold payment until the dispute is settled. D is incorrect because a consumer does have recourse if a faulty product is purchased on a credit card.

VI. Answers to Essay Question

This issue is being examined by an administrative agency within the executive branch of the federal government. The general power to tax is in the control of Congress, but the actual administration of tax laws is vested in the powers of the executive branch. Within that branch, the Treasury Department is the cabinet level agency delegated to oversee tax administration for the federal government. That agency has within it the Internal Revenue Service that is the designated administrative arm of the executive branch empowered with the duty to collect taxes.

The basic legislation that allows the IRS to act as our country's tax collector is the Internal Revenue Code. It is obviously a very technical act, subject to constant revision and debate. Under that code, the IRS is allowed to issue numerous rules and regulations relating to both substantive and procedural aspects of the tax laws. The IRS must comply with the Administrative Procedures Act and all other statutes relating to providing adequate notice to the public of its proposed rules and the like. In addition, unless it is information specifically exempted, such as law enforcement information, it must also comply with the Freedom of Information Act and similar administrative statutes.

Assuming these rules have been complied with, it is most likely that the IRS would take the position that its revenue rules do not allow dogs to be taken as personal dependents on the return, no matter how cute they are. This may seem cruel and cold hearted, but that is the way the old dog biscuit crumbles.

What can poor Joe do? He was advised by an IRS agent to do what he did. Shouldn't he be able to rely on that advice? Unfortunately, he cannot use that reliance as a defense in this case. As the Supreme Court said in a case dealing with administrative agencies: "Whatever the form in which the government functions, anyone entering into an arrangement with the government taking the risk of having accurately ascertained that he purports to act for the government stays within the bounds of his authority." (*Federal Crop Inc. Corp. v. Merrill*, 332 U.S. 380 (1947)). Here Oliver gave Joe bad advice, but that fact does not excuse Joe's duty to find out for himself what the Internal Revenue Code really said on the issue of who qualifies as a dependent.

To make matters worse, Joe probably cannot hold Oliver liable for negligence because it would not be deemed to be an intentional tort by Oliver. The advice was given in the course of his employment as a government agent. Agency law and governmental immunity would protect Oliver from personal liability. Is it any wonder that so many people have an overriding sense of frustration when dealing with their Uncle Sam?

Chapter 34
ENVIRONMENTAL PROTECTION

We all live under the same sky, but we don't all have the same horizon.
Konrad Adenauer

I. Overview

A large chemical company recently aired a TV commercial to announce that it plans to use double-lined tanker ships to transport its products. In the ad, this news is shown being greeted with jubilation by all sorts of wildlife, while Beethoven's stirring Ninth Symphony swells in the background. The Twenty-First century ushered in a greater concern for environmental issues than ever before. One would hope this is not just the latest politically correct marketing fashion but rather a realization that our natural resources can no longer be taken for granted. Like it or not, our advances in technology have finally given us the power to permanently alter nature. This power has been used to feed millions and save lives. And it has been used to destroy endangered species, burn holes in the ozone, and uproot entire ecosystems.

The law of environmental protection to balance the needs and desires for technological advancement with a realistic perspective of what natural limitations our environment places on these capacities. The age of growth *per se* as a good thing is changing to: "Is the growth worth it?" The environmental movement is not new. Conservationists have a long and proud history of protecting our natural resources. What is new are the threats raised by oil spills, hazardous wastes, nuclear accidents, and unmitigated depletion of irreplaceable resources. The emergence of environmental legislation which has a sense of urgency and response to these issues is only now in its third decade—a mere second on the ecological clock.

Because this legislation is so relatively new, the growing pains of new social engineering continue to be felt by business. These laws have especially impacted industries that have traditionally relied heavily on utilization of natural resources such as mining, oil, gas exploration, and timber. Consider also the auto industry. For years automakers outside the U.S. have complained about our pollution controls being too costly and cumbersome. It is no small point of pride that the world is catching up with us on this issue by finally adopting our standards for safety and emissions.

II. Hypothetical Multi-issue Essay Question

Bailout Bank of Metropolis is seeking to extend its operations to every corner of the great state of Louisiark. The state legislature just passed a law allowing branch banking Bailout can now open branches outside of Metropolis. Bailout's Vice-President in charge of real estate is Mr. Mortimer M. Misque. Mortimer's cousin, Ms. Sallie S. Spillaway, is Vice President in charge of real estate for Dryhole Oil Company. Dryhole's business has been terrible ever since it converted all its self-service operations into combination gas stations and mini-refineries. Dryhole thought it could save a lot of money by not having to ship gasoline from its out-of-state refineries to retail outlets.

Mortimer and Sallie hit upon a great idea. Why not sell the Dryhole gasoline refinery outlets to Bailout and convert them to branch banks? The deal went through, and Bailout proceeded to renovate the gas stations. The old refinery and pump facilities were simply capped into mothballs in case the bank ever decided to go into the business.

Unfortunately, several years after these renovations, Bailout's money began to take on a funny odor. In addition, the residents of the neighborhoods adjoining Bailout branches began to have trouble with their drinking water. What can be done?

III. Outline

Environmental Protection
- Environmental Protection Agency
  - An administrative agency created by Congress in 1970 to coordinate the implementation and enforcement of the federal environmental protection laws
- National Environmental Policy Act
  - A federal statute enacted in 1969 that mandates that the federal government consider the adverse impact a federal government action would have on the environment before the action is implemented

Clean Air
- Clean Air Act
  - A federal statute enacted in 1963 to assist states in dealing with air problems
- National ambient air quality standards
  - Standards for certain pollutants set by the EPA that protect human beings and vegetation, matter, climate visibility, and economic values
- Toxic air pollutants
  - Air pollutants that cause serious illness or death to humans

Clean Water
- Water pollution
  - Pollution of lakes, rivers, oceans, and other bodies of water
- Point sources
  - Sources of water pollution such as paper mills, manufacturing plants, electric utility plants, and sewage plants
- Thermal pollution
  - Heated water or material discharged into waterways that upsets the ecological balance and decreases the oxygen content
- Wetlands
  - Areas that are inundated or saturated by surface water or ground water that support vegetation typically adapted for life in such conditions

Toxic Substances
Chemicals used for agricultural, industrial, and mining uses that cause injury to humans, birds, animals, fish, and vegetation

Hazardous Waste
Hazardous waste
Solid waste that may cause or significantly contribute to an increase in mortality or serious illness or pose a hazard to human health or the environment if improperly managed
Land pollution
Pollution of the land that is generally caused by hazardous waste being disposed of in an improper manner

Nuclear Waste
Radiation pollution
Emissions from radioactive wastes that can cause injury and death to humans and other life and can cause severe damage to the environment

Noise Pollution
Unwanted sound from planes, manufacturing plants, motor vehicles, construction equipment, stereos, and the like

Environmental Protection Laws
Air pollution
Clean Air Act
Water pollution
Clean Water Act
Toxic substances
Federal Insecticide, Fungicide, and Rodenticide Act
Toxic Substances Control Act
Hazardous waste
Resource Conservation and Recovery Act
Comprehensive Environmental Response, Compensation, and Liability Act
Nuclear waste
Nuclear Regulatory Commission
Preservation of wildlife
Endangered Species Act
Noise pollution
Noise Control Act

IV. Objective Questions

Terms:

1. In 1970, Congress created the ________________ ________________ ________________ that coordinates the implementation and enforcement of federal environmental protection laws.

2. The document that must be prepared for all proposed federal legislation that would significantly affect the quality of the human environment is called an ________________ ________________ ________________.

3. The major source of mobile air pollution is the _______________.

4. The agency responsible for enforcing pollution standards for airplanes is known as the _______________ _______________ _______________.

5. Asbestos, mercury, vinyl chloride, and benzene are examples of _______________ _______________.

6. The EPA can establish water pollution control standards under the _______________ _______________ _______________.

7. Areas that are inundated or saturated by surface or groundwater that support vegetation typically adapted for life in saturated soil conditions are called _______________.

8. The act that extended environmental protection to the oceans is known as the _______________ _______________ _______________ _______________ _______________ Act.

9. The act that provides for the creation of a fund to finance the cleanup of hazardous wastes is popularly known as _______________.

10. Licenses for the construction and opening of commercial nuclear power plants are granted by the _______________ _______________ _______________.

True/False:

1. ____ Under common law, individuals could bring a private civil suit based on private nuisance to recover damages from a polluting party.

2. ____ The Environmental Protection Agency has adjudicative powers to hold hearings, make decisions, and order remedies for violations of federal environmental laws.

3. ____ It is not necessary to prepare an environmental impact statement for all proposed legislation that may affect the environment.

4. ____ Once an environmental impact statement is prepared, the public has sixty days in which to submit comments to the EPA.

5. ____ States establish national ambient air quality standards for pollutants.

6. ____ The EPA can require automobile manufacturers to recall and repair or replace pollution control equipment that does not meet air quality control standards.

7. ____ The first regulation to prevent and control water pollution was enacted in 1886.

8. ____ The Clean Water Act authorizes the U.S. government to clean up oil spills and spills of other hazardous substances in ocean waters within six miles of the shore.

9. ____ The Resource Conservation and Recovery Act defines a hazardous waste as one that may cause or significantly contribute to an increase in mortality or serious illness or pose a hazard to human health or the environment if improperly managed.

10. ____ The EPA can order a responsible party to clean up a hazardous waste.

Multiple Choice:

1. Which of the following is contained in an environmental impact statement?
   A. Description of the affected environment.
   B. No description of alternatives to proposed action.
   C. List of resources that will be committed to the action.
   D. A and C, but not B.

2. Regions that do not meet national ambient air quality standards for certain pollutants are described as nonattainment areas. Which is the worst category?
   A. Moderate.
   B. Severe.
   C. Extreme.
   D. Serious.

3. The Clean Air Act requires new automobiles to meet which of the following air quality control standards?
   A. Leaded fuel is permitted for highway use after 1995.
   B. Cars must meet air quality standards prior to 1995, five years, or 50,000 miles, whichever comes first.
   C. Production of engines that use leaded fuel is prohibited after model year 1990.
   D. The EPA cannot require automobile manufacturers to recall automobiles that do not meet air quality standards.

4. Which of the following is not a point source of water pollution according to the Clean Water Act?
   A. Toilets in homes.
   B. Mines.
   C. Paper mills.
   D. Municipal sewage plants.

5. The Superfund requires the EPA to:
   A. Identify sites in the U.S. where hazardous wastes have been disposed.
   B. Rank sites based of severity of risk.
   C. Put the highest-ranking sites on a National Priority List.
   D. All of the above.

6. The EPA cannot recover the cost of a cleanup from:
   A. The generator who deposited the waste.
   B. Workers who deposited the waste under the direction of their employer.
   C. The owner of the site at the time of the disposal.
   D. The current owner and operator of the site.

7. The Superfund imposes liability without fault; the name for this doctrine is:
   A. *Res ipsa loquitur*.
   B. Contributory negligence.
   C. Strict liability.
   D. *Respondeat superior*.

8. Under the Superfund's right-to-know provision, businesses are required to:
   A. Disclose monthly the presence of certain listed chemicals to the community.
   B. Disclose annually the emissions of chemical substances released into the environment.
   C. Immediately notify the government of spills, accidents, and other emergencies involving hazardous substances.
   D. None of the above.

9. Which of the following is not a federal statute that protects wildlife?
   A. Resource Conservation and Recovery Act.
   B. Bald Eagle Protection Act.
   C. Fishery Conservation and Management Act.
   D. Wild Free-Roaming Horses and Burros Act.

10. Which of the following noises is not regulated by the Noise Control Act?
    A. Aircraft.
    B. Railroads.
    C. Barking dogs.
    D. Trucks.

V. Answers to Objective Questions

Terms:

1. *Environmental Protection Agency.* The EPA is growing in both economic and political importance. President Clinton has recommended that the office of EPA Administrator be raised to cabinet rank.
2. *Environmental impact statement.* Many states have followed suit and now call for local versions of these statements.
3. *Automobile.* An interesting footnote on this issue is that most auto-related pollution is being generated by the minority of autos built prior to the issuance of pollution guidelines.
4. *Federal Aviation Administration.* The FAA has sought to reduce pollution from airplanes by coordinating takeoffs on a national scale in order to cut down on wasteful holding patterns over major airports.
5. *Toxic pollutants.* As thresholds of scientific discovery expand, this list gets longer every year.
6. *Clean Water Act.* This effort is coordinated with state and local water quality control authorities.
7. *Wetlands.* Arriving at a precise definition of land to qualify for this designation continues to be a political powder keg.
8. *Marine Protection, Research and Sanctuaries.* Much of the work done under this act is coordinated with state and local agencies.
9. *Superfund.* The actual title of this legislation is the Comprehensive Response Compensation and Liability Act.
10. *Nuclear Regulatory Commission.* As compared to other environmentally sensitive activities, the federal government exclusively controls this activity.

True/False:

1. True. These actions are traceable to old common law property rights.
2. True. This pattern of powers is typical under the rights of the executive branch.
3. False. An environmental impact statement must be prepared.
4. False. The time period is thirty days.

5. False. The EPA establishes NAAQS for pollutants.
6. True. Recalls can be very expensive for manufacturers who often fight over the scope and breadth of the recall.
7. True. This process can take years if there is a dispute.
8. False. The distance is twelve miles off shore.
9. True. This definition has been fought over in the courts because of all the possible interpretations that may arise from it.
10. True. If the party refuses, the EPA may clean it up and sue for the cost of the cleanup.

Multiple Choice:

1. D. Alternatives to the proposed actions are also listed in the environmental impact statement.
2. C. Extreme is considered to be at the near emergency state and has a high priority for corrective action.
3. B. These protections supersede any protections given by the manufacturer's warranty.
4. A. These are assumed to be normal activities that are handled by local authorities.
5. D. All of these activities are required under the Act.
6. B. This is an example of the application of agency law to an environmental statute.
7. C. This is an example of the application of a tort law doctrine to an environmental statute.
8. C. Where the wrongdoer is the business itself, expecting total self-regulation appears unrealistic, which is why informants are often the source of this sort of information.
9. A. This Act is focused on natural resources other than wildlife, although it collaterally benefits wildlife living in protected areas.
10. C. It is expected that local authorities will have control over this issue.

VI. Answers to Essay Question

The jurisdiction over environmental issues is generally given to the Environmental Protection Agency (EPA). On a number of issues, the EPA shares this jurisdiction with other federal or state and local agencies. One of the primary regulatory aspects of the EPA function is to gather information on key environmental concerns. State and local agencies often have similar information gathering functions.

One common method of gathering such information is through the use of environmental impact statements. Normally the EPA will not require such a statement on proposed projects not covered by the National Environmental Policy Act, but most state and local governments do require such disclosures. Here there appears to be no evidence of Bailout or Dryhole having provided such a statement to any government agency. They may be held liable for such a failure.

It appears that there may be some damage to the water table in the areas underneath the former gasoline outlets *cum* refineries. This sort of problem has become relatively common around the country as older gasoline outlet facilities are closed or converted to other uses. The old tanks are often left in the ground and allowed to rust. As they rust out, toxic chemicals are released from them into the ground water table. These emissions are often in violation of the Clean Water Act. Various state and local agencies may seek both equitable injunctive relief and money damages for the cost of cleanup and possibly even criminal sanctions, where appropriate, for the violation of these statutes.

In addition to the violation of the Clean Water Act, this situation may involve imposition of the provisions of the Comprehensive Environmental Response Compensation and Liability Act (Superfund). Under this Act, the costs of cleaning up a designated site may be passed on to any owners of the property who had an active part in the management of the property. There is an exception from liability for those parties whose only interest in the property arose out of a security interest. Here the bank may try to argue that defense but would lose on the merits. Bailout Bank became an active owner of the property when it decided to convert the stations to branch banks. As such, it is subject to joint and several liability with Dryhole Oil to pay for the cost of the cleanup. The EPA can order them to clean it up directly. If they fail to do the cleanup, the EPA can clean it up and seek recovery of the cost from Bailout, Dryhole, or both. Interestingly enough, both Mortimer and Sallie are now in other lines of work. They took jobs at nuclear power plants.

Chapter 35
ANTITRUST LAW

Things do not change; we do.
Henry David Thoreau

I. Overview

Most people do not think of the trust device as a business tool. As seen in other chapters, today's use of trusts centers on the need to hold property for the benefit of others. In another era, however, the business trust was notoriously used as a device to eliminate competition. In the late 1800s, it was common to have key commodities and the industries related to those products controlled by large corporate enterprises. These entities would band together into a form of common trust ownership. The trustee, in turn, was able to control the prices, territories of distribution, and the like of the product. For example, prior to the enactment of antitrust laws, industries like oil, cotton, sugar, and whiskey were all dominated by such trusts. Probably the best known of these trusts was Standard Oil. In 1890 the Standard Oil Trust controlled over ninety percent of the market for oil products in the U.S. By the time the trust was "busted" in 1911, over thirty companies were ordered separated from the parent firm. This sort of monopolization of the marketplace led to the landmark antitrust legislation in 1890, the Sherman Antitrust Act. The act has two main objectives:

1. To prevent combinations in trust or otherwise, which act in restraint of trade, i.e., illegal joining together to restrain trade.
2. To control markets thought to have a monopoly, i.e., illegal domination so strong as to *ipso facto* restrain trade.

These objectives are set out in Sections 1 and 2 of the Act and will be described in more detail in the key question checklist. What is interesting about this act is that Congress used very broad language to give the Justice Department maximum latitude in seeking enforcement of its provisions. This latitude has, in turn, not been consistently used. There appears to have been a constant shift in the enforcement strategies used by various administrations over the years. Those who favor strict enforcement essentially adhere to the notions originally proffered by Senator John Sherman and his cohorts. Under this traditional philosophy, competition is best served by having as many players in the arena of commerce as possible. Conversely, large concentrations of power in commerce are thought to be inherently bad.

On the other side of the enforcement spectrum, the more modern view says that large concentrations of economic power are not evil *per se* as long as they are efficient and are still fighting for a competitive position on a world-based economic playing field. Recent administrative decisions have clearly favored this more lenient view as evidenced by the large number of mergers, acquisitions, leveraged buyouts, and consolidations that have taken place on Wall Street without objection.

The federal courts, in turn, have taken the middle road. Under their rules of interpretation, two main classifications of offenses have evolved. The *Per se Rule* is used to strike down restraints that courts deem to be so inherently anticompetitive that they cannot be allowed as a matter of law, regardless of any claimed justifications. On the other hand, the *Rule of Reason* has given courts latitude to accept restraints of trade on a case-by-case basis where legitimate concerns are overriding.

As strong and powerful a tool in the fight against monopolization and restraints of trade as the Sherman Act is, it has proven to be only a partial remedy to the problem. The Sherman Act sets the basic goals and objectives of keeping marketplaces open to competition. The Clayton Act and the Federal Trade Commission Act are designed to provide tools of implementation to those basic public policy objectives. As compared to the almost philosophical tenor of Sections 1 and 2 of the Sherman Act, the Clayton Act and more particularly, the Robinson-Patman Amendment to it, speak to much more specific objectives. These objectives arose out of discriminatory practices aimed at getting the little guy.

The biggest problem with the Clayton Act, and to a lesser extent with the FTC, is the government's commitment to enforcement combined with some very problematic aspects of the statutes themselves. On the issue of governmental level of commitment to enforcement, there is no question that things have changed in the geo-global scheme of economic competition. The market factors that were protected in the early part of the Twentieth Century are different as we enter the Twenty-First Century. A free and open market is not measured now on regional or even national scales, but rather on worldwide competitive position. These changes have provided the philosophical underpinnings for the much more tolerant view taken by the government towards mergers, acquisitions, combinations, and the like. Yet the basic economic principles of monopolization, restraint of trade, and unfair trade practices have not changed. So government finds itself in a dilemma. It is trying to recognize the need to allow our economy to stay competitive on a worldwide playing field, but it must continue to keep the rules fair in that game.

The second factor involves questions that have been raised about the economic sense of the Clayton Act itself. Many critics of the act have argued that while provisions seeking to prevent price discrimination look good on the sheet music, they flop in the concert hall. The reason these particular measures have failed to live up to their billing is that some price volume costs, incentives are all part of the competitive edge that all players are constantly looking for. To deny the reality of those competitive needs not only frustrates real competition, it may give noncompetitive parties an unwarranted edge against more efficient competitors by way of officious intermeddling on the part of government.

Where antitrust laws do not work, they actually become part of the problem. No one really wants to go back to the bad old days of robber barons and unmitigated jungle warfare in our economic system. Yet the time has probably come for a wholesale reexamination of our antitrust laws. This examination should focus on two main goals: (1) rewrite the laws so as to really provide a level playing field for all competitors and (2) get the government to act when needed instead of being an immobilized Hamlet who cannot decide what to do.

All in all, antitrust law is in a greater state of flux than ever. Out of this debate, many are now arguing that the time has come for an overhaul of the entire antitrust law structure. Given the sporadic history of enforcement, who will enforce the law as intended no matter how it may be rewritten?

II. Hypothetical Multi-issue Essay Question

Spookey's Scareys specializes in Halloween costumes in Salem, Mazazona. Spookey's has 11 percent of a declining market. Its competitors are Ghosts and Goblins with 29 percent, Witches and Wanderers with 31 percent, Midnight Costumes with 20 percent, and several other Mom and Pop operations with the remaining 9 percent. On an average costume selling for $100, Spookey's has an overall cost of $90, including $75 paid to its main supplier, Bad Taste Haberdashery.

Because the market has been bad, Spookey's is looking for a buyer. It entered into a deal with Ghost and Goblins to sell a 51 percent interest in the store. Ghosts and Goblins would continue to operate both stores under their separate names. In addition, because of its combined buying power, the newly merged company is now able to get costumes from Bad Taste at $50 apiece and pass on the savings with lowered retail prices. Are there any antitrust problems with the above transactions?

III. Outline

The Sherman Act
- Section 1
  - Prohibits contracts, combinations, and conspiracies in restraint of trade. To violate Section 1, the restraint must be found to be unreasonable under either of two tests:
    - Rule of reason
    - Per se rule
  - Requires the concerted action of two or more parties
- Section 2
  - Prohibits the act of monopolizing, and attempts and conspiracies to monopolize. Can be violated by the conduct of one firm

Restraints of Trade
- Section 1 of the Sherman Act outlaws certain anticompetitive activities
  - It outlaws contracts, combinations, and conspiracies in restraint of trade
  - It applies to two or more parties

Rules to Determine Lawfulness of a Restraint
- Rule of reason
  - A rule that holds that only unreasonable restraints of trade violate Section 1 of the Sherman Act
- Per se rule
  - A rule that is applicable to those restraints of trade considered inherently anticompetitive

Horizontal Restraint of Trade

Group Boycott by Sellers

Group Boycott by Purchasers

Vertical Restraint of Trade

Defenses to Section 1 of the Sherman Act
- Unilateral refusal to deal
- Conscious parallelism
- Noerr doctrine

Monopolization
Section 2 of the Sherman Act prohibits the act of monopolization and attempts or conspiracies to monopolize trade
Can be violated by the conduct of one firm

Mergers
Section 7 of the Clayton Act provides that it is unlawful for a person or business to acquire the stock or assets of another "where in any line of commerce or in any activity affecting commerce in any section of the country, the effect of such acquisition may be substantially to lessen competition, or to tend to create a monopoly"

Types of Mergers
Horizontal
Vertical
Backward vertical
Forward vertical

Potential Reciprocity Theory

Tying Arrangements
A tying arrangement is a restraint of trade where a seller refuses to sell one product to a customer unless the customer agrees to purchase a second product from the seller
Section 3 of the Sherman Act prohibits tying arrangements involving goods, services, intangible property, and real property

Price Discrimination
Price discrimination is charging different prices to different customers for the same product without any justification
Price discrimination is governed under Section 2(a) of the Robinson-Patman Act

Unfair Methods of Competition
Section 5 of the Federal Trade Commission Act prohibits unfair methods of competition and unfair and deceptive acts or practices

IV. Objective Questions

Terms:

1. Three times the actual damages ordinarily awarded is called ________________ ________________.

2. The rule that holds that only unreasonable restraints of trade violate Section 1 of the Sherman Act is called the ________________ ________________ ________________.

3. The standard that is applicable to those restraints of trade that are considered inherently anticompetitive is called the ________________ ________________ rule.

4. A restraint of trade that occurs when two or more parties on different levels of distribution enter into a contract, combination, or conspiracy to restrain trade is called a ________________ ________________.

5. The doctrine that states that if two or more firms act the same but no concerted action is shown, there is no violation of Section 1 of the Sherman Act is called _______________ _______________.

6. A merger between two or more companies that compete in the same business and geographical market is called a _______________ merger.

7. The test for determining the lawfulness of horizontal mergers which finds horizontal mergers presumptively illegal under Section 7 if (1) the merged firm would have thirty percent or more market share in the relevant market and (2) the merger would cause an increase in concentration of thirty-three percent or more in the relevant market is called the _______________ _______________ test.

8. A merger that does not fit into any other category or a merger between firms in totally unrelated businesses is a _______________ merger.

9. A defense to a Section 7 action that says a competitor may merge with a failing company if (1) there is no other reasonable alternative for the failing company and (2) no other purchaser is available is called the _______________ _______________ _______________.

10. A form of price discrimination that is less readily apparent than direct forms of price discrimination is _______________ _______________ _______________.

True/False:

1. ____ All of the major federal antitrust acts have both civil and criminal sanctions.

2. ____ Under the *per se* standard of determining whether the Sherman Act prohibits conduct, there is no balancing of pro- and anticompetitive effects.

3. ____ Whether price fixing is a violation of Section 1 of the Sherman Act is determined by the Rule of Reason.

4. ____ To show a violation of the Sherman Act, two or more firms acting the same way must be acting in concert. That is, if each firm independently reaches the same decision, there is no violation.

5. ____ Under Section 2 of the Sherman Act, a plaintiff must show a defendant's specific intent to monopolize a market; the showing of mere deliberate or purposeful conduct is not sufficient to support a Section 2 action.

6. ____ A natural monopoly is a defense to the charge of monopolizing even if the natural monopoly operates in a predatory way.

7. ____ Under Section 7 of the Clayton Act, there must be proof of an actual decrease in competition for the court to forbid a merger.

8. ____ The merger guidelines of the Herfindahl-Hirschman Index are law, and any increase in the index resulting from a merger will raise a presumption of illegality.

9. ____ Conglomerate mergers are always legal since they involve mergers between firms that conduct totally unrelated business.

10. ____ Under the meeting the competition defense to a Section 2(b) of the Robinsion-Patman Act charge, a seller can lower his selling price below that of his competitor to be competitive in a market.

Multiple Choice:

1. Elvira, Inc., sells various interrelated automotive products that it manufactures. One of the items, an antilock braking system, was manufactured almost exclusively by Elvira, Inc. Elvira, Inc. realized the importance of the product to its purchasers and decided to capitalize on the situation by requiring all purchasers to take at least two of its other automotive brake products if they wished to obtain the antilock braking system over which it has almost complete market control. Which of the following best describes the situation?
   A. The plan is both ingenious and legal and should have been resorted to long ago.
   B. The arrangement is an illegal tying agreement and is *per se* illegal.
   C. Since Elvira, Inc., did not have complete market control over the unique product in question, the arrangement is legal.
   D. As long as the other products which must be taken are sold at a fair price to the buyers, the arrangement is legal.

2. The Hilmer Duck Decoy Company entered into an agreement with retail merchants whereby they agreed not to sell beneath Hilmer Duck Decoy's minimum suggested retail price in exchange for Hilmer's agreeing not to sell its duck decoys at retail in their respective territories. The agreement does not preclude the retail merchants from selling competing duck decoys. What is the legal status of the agreement?
   A. It is legal if the product is sold under a trade name or is trademarked.
   B. It is legal if the power to fix maximum prices is not relinquished.
   C. It is illegal unless it can be shown that the parties to the agreement were preventing cutthroat competition.
   D. It is illegal even though the price fixed is reasonable.

3. Sharlene's Firm and Bob's Firm, who have been competitors for years, contract to restrain trade so as to increase their own profits. Which of the following is correct?
   A. The behavior of these two firms is governed by the National Labor Relations Act.
   B. To be liable under the Sherman Act, the contract must be in writing.
   C. The two firms have engaged in a horizontal restraint of trade.
   D. All of the above.

4. High Tech Sound (HTS) and Super Sonics (SS) are both producers of stereo equipment. HTS conducts a market research study and decides to set the prices of its stereos at $200, $300, and $400 according to the various levels of quality. SS is located in a different state and does its own market research study. SS also decides, without knowledge of HTS' prices, to set its stereo prices at $200, $300, and $400 depending on the quality of the stereo. Which of the following is correct?
   A. This is price fixing and therefore, a *per se* violation of Section 1 of the Sherman Act.
   B. This is price fixing and will be examined under the Rule of Reason.
   C. This is price fixing and therefore, a *per se* violation of Section 2 of the Sherman Act.
   D. Under the doctrine of conscious parallelism, this conduct is permitted and there is no violation of the Sherman Act.

5. Superior Systems (SS) has devoted an enormous amount of resources to research and development of its product, water purification systems. In doing this, SS has clearly developed a superior product and, consequently, a monopoly in the purification market. Which of the following is correct?
   A. SS has violated Section 2 of the Sherman Act.
   B. SS has violated Section 1 of the Sherman Act.
   C. SS has not violated the Sherman Act since its monopoly is a result of a superior product.
   D. SS has not violated the Sherman Act because a single firm cannot be considered a monopoly.

6. Fantastic Feet and Showtime Shoes, both shoe retailers, have proposed a merger. Both companies sell nationwide. What type of merger is this?
   A. A vertical merger.
   B. A conglomerate merger.
   C. A horizontal merger.
   D. A market extension merger.

7. Assume the same facts as in question 6. How is this proposed merger likely to be treated under Section 7 of the Clayton Act?
   A. Nothing will happen until the merger has taken place since there must be actual proof of a lessening of competition under Section 7.
   B. Nothing will happen under Section 7 since mergers between the same types of companies are always permitted if they both conduct business on a national level.
   C. The merger will be subjected to strict review under Section 7 since this will certainly increase the concentration in the relevant market.
   D. None of the above.

8. Assume the same facts as in question 6 except that now instead of merging with Showtime Shoes, Fantastic Feet wants to merge with National Shoe Manufacturers. This merger would allow the newly formed company to both produce and sell the shoes. What type of merger is this for Fantastic Feet?
   A. A forward vertical merger.
   B. A backward vertical merger.
   C. A conglomerate merger.
   D. None of the above.

9. Assume the same facts as in question 8. How is this proposed merger likely to be treated under Section 7?
   A. Nothing will happen until the merger has taken place since there must be actual proof of a lessening of competition under Section 7.
   B. Nothing will happen under Section 7 since mergers between two companies which are not in the same market are always permitted if they both conduct business on a national level.
   C. The merger will be subjected to strict review under Section 7 since this will certainly increase the concentration in the relevant market.
   D. None of the above.

10. Technophonics, a large manufacturer of car phones, has proposed a merger with Traveling Talkers, another car phone manufacturer. Though Technophonics is successful in this competitive market, Traveling Talkers is on the verge of collapse. In an effort to salvage what it can, Traveling Talkers has sought other companies with which to merge, but Technophonics is the only interested firm. A merger between these two companies would violate Section 7 of the Clayton Act. What is Technophonic's best defense?
    A. The Small Company Doctrine.
    B. The Failing Company Doctrine.
    C. The Potential Reciprocity Theory.
    D. The *De Novo* Theory.

V. Answers to Objective Questions

Terms:

1. *Treble damages.* Under Section 4 of the Clayton Act, single damages found by a jury are tripled in amount for antitrust violations.
2. *Rule of Reason.* If Section 1 of the Sherman Act were literally applied, it would prohibit many business contracts. In order to mitigate against that literal interpretation, the Rule of Reason is designed to allow the courts to weigh a number of factors when deciding if a contract question is in restraint of trade.
3. *Per se.* No balancing is required if the restraint is considered to be inherently anticompetitive. If so determined, a court will not permit any defenses or justifications to save it.
4. *Vertical restraint.* This restraint can take the form of a contract, combination, or conspiracy between the parties and is subject to the scrutiny of both the Rule of Reason and the *per se* rule interpretations of Section 1 of the Sherman Act.
5. *Conscious parallelism.* It must be shown that the parties acted independently of each other and arrived at their decisions separately.
6. *Horizontal.* This can involve a merger of one company with another company producing a similar product and selling it in the same geographic market.
7. *Presumptive illegality.* This test was adopted by the U.S. Supreme Court in the landmark antitrust case of *United States v. Philadelphia National Bank*, 374 U.S. 321 (1963).
8. *Conglomerate.* Where the merger may not fit into any other category, it may still be a violation of Section 7 of the Clayton Act if it creates an unfair advantage, hurts potential competition, or creates potential reciprocity between competitors.
9. *Failing company doctrine.* Unfortunately, there are far too many examples of this defense on the economic landscape in a period of prolonged recession.
10. *Indirect price discrimination.* Any number of methods have been devised to favor certain customers. They can include favorable credit terms, freight charges, and the like.

True/False:

1. False. The Sherman Act is the only federal antitrust act that has criminal sanctions. The other federal antitrust acts allow only civil sanctions.
2. True. The balancing of pro- and anticompetitive effects is used in the Rule of Reason test, not the *per se* standard of determining whether conduct is prohibited by the Sherman Act.
3. False. Price fixing is a *per se* violation of the Sherman Act.
4. True. Under the doctrine of conscious parallelism, two or more firms must be acting in concert to be in violation of the Sherman Act.

5. False. The showing of deliberate or purposeful conduct is sufficient to support a Section 2 action. Showing a defendant's actual specific intent to monopolize a market is not required.
6. False. Innocent acquisition and natural monopoly are defenses to the charge of monopolizing. These defenses are lost if the firm acts in a predatory or exclusionary way.
7. False. Under Section 7 of the Clayton Act, if a merger is likely to substantially lessen competition, the court may prevent the merger. Actual showing of the lessening of competition is not required.
8. False. The merger guidelines of the HHI are only guidelines, not law. In addition any increase in the index does not raise a presumption of illegality. Depending on the degree of the change in the index, a merger proposition may be challenged by the Justice Department.
9. False. Conglomerate mergers are not always permitted under the Clayton Act. The lawfulness of such mergers are examined under several different factors.
10. False. Under the meeting the competition defense, a seller can lower his selling price to match that of his competitor. However, the seller cannot set his price below that of his competitor.

Multiple Choice:

1. B. The situation described here is a tying arrangement and is *per se* illegal. A is incorrect because the plan is illegal and definitely not ingenious. C is incorrect because tying arrangements are illegal even if the seller does not have complete market control over the unique product. D is incorrect because tying arrangements are *per se* illegal even if the other products are sold at a fair price. Elvira, Inc., cannot use market dominance of antilock braking systems in this illegal manner.
2. D. The situation described here is a price fixing arrangement and is illegal even though the price is reasonable. A is incorrect because price fixing is illegal whether the product is trademarked or sold under a trade name. B is incorrect because price fixing is illegal whether it is for a minimum or maximum price. C is incorrect because there is no defense for price fixing. Thus, the Hilmer Duck Decoy Company has to shoot down this bad idea.
3. C. When two or more firms at the same level of distribution contract for such an arrangement, this is horizontal restraint of trade. A is incorrect because the behavior of these two firms is governed by Section 1 of the Sherman Act. B is incorrect because such a contract is illegal whether it is oral, written, or implied from the circumstances.
4. D. Since there was no intentional effort by these firms, nor did they even know of the other's pricing system, there was no intent to fix prices. Therefore, under the doctrine of conscious parallelism, this conduct is not prohibited under the Sherman Act. A, B, and C are incorrect because since neither firm knew of the other, there is no price fixing and this activity is not a violation of the Sherman Act.
5. C. A company that acquires a monopoly because of superior business acumen, skill, and foresight has an innocent acquisition of a monopoly. This is a defense to the charge of monopolizing. Therefore, SS has not violated the Sherman Act. A and B are incorrect for the reason discussed above. D is incorrect because a single firm can be considered a monopoly.
6. C. A merger where both companies are in the same level of distribution is a horizontal merger. Here, both companies are shoe retailers. A is incorrect because a vertical merger involves an integration of the operations of a supplier and a customer. B is incorrect because a conglomerate merger is between two companies in unrelated businesses. D is incorrect because a market extension merger is a merger between two companies in similar fields whose sales do not overlap. Here, both companies sell nationwide. Therefore, their fields do overlap.
7. C. Where two companies that compete in the same geographical market merge in a horizontal merger, the merger is subjected to strict review under Section 7 because it will clearly result in an increase in concentration in the relevant market. A is incorrect because Section 7 actions may be brought based on the probability of a lessening of competition. Actual proof is not necessary. B is incorrect because horizontal mergers of this type are almost always strictly scrutinized under Section 7.

8. B. Where a retailer merges with a producer, this is a backward vertical merger. A is incorrect because a forward merger would involve a company lower on the chain of distribution merging with a company higher on the chain of distribution. C is incorrect because a conglomerate merger is a merger between two companies in unrelated fields.
9. D. A is incorrect because actual proof of a lessening of competition is not required since Section 7 actions are based on the probability of a lessening in competition. B is incorrect because such mergers are not always automatically permitted. C is incorrect because such mergers do not increase the concentration in the relevant market since there are different markets involved. Vertical mergers will be examined considering such factors as the past history of the firms, the trend toward concentration in the industries involved, the barriers to entry, the economic efficiencies of the merger, and the elimination of potential competition caused by the merger.
10. B. The Failing Company Doctrine is a defense to Section 7 actions if (1) there is no other reasonable alternative for the failing company and (2) no other purchaser is available. Here, both these requirements are met. A is incorrect because the Small Company Doctrine is a defense to Section 7 actions for two or more small companies which wish to merge to compete more effectively with a large company. C and D are incorrect because they are theories for enjoining mergers, not defenses to Section 7 actions.

VI. Answers to Essay Question

The basic policy presumes that competition is best protected in an open and fair marketplace. Monopolization and concentrations of economic controls are contrary to an open and fair marketplace and may be anticompetitive.

The main statutes that may be used are the Sherman Act of 1890, the Clayton Act of 1914, and the Robinson-Patman Act of 1930. In this case, the second two acts provide the best possibility for putting the actions of Spookeys and Ghost and Goblins to the antitrust law litmus test.

The partial acquisition of a competitor in the same geographical area may injure competition. In examining this horizontal merger, the Justice Department will use its Herfindal-Hirshman guidelines to see if the new arrangement will create an illegal concentration of costume sales in Salem in one controlled company. As a defense to this possible challenge, Ghost and Goblins can raise the argument that Spookey's was a failing company. That defense appears weak because there was no evidence of it losing money or that it was about to go bankrupt. Thus the merger could most likely be disallowed under Section 7 of the Clayton Act.

With regard to the new pricing arrangement with Bad Taste, the problems with Robinson-Patman become more apparent. Here the ultimate price charged to the consumer is lower and it would appear on its face (mask?) that competition is actually fostered by the new arrangement. The other costume stores in Salem might argue, however, that they have been discriminated against by predatory pricing. The best defense here would be to show that the new prices reflect efficiencies realized from higher volume purchases, i.e., lower costs. In addition, if Bad Taste can show that any buyer who purchased the higher volume would get that same lower price, price discrimination would be much more difficult to prove.

From the problem as presented, it appears that no unfair trade practices were used. If for example, faulty products were fraudulently substituted for originals or some sort of false advertising were used by Ghosts and Goblins, then the FTC would have to assume the role of "Ghostbusters!"

Chapter 36
PERSONAL PROPERTY AND BAILMENTS

The most common and durable source of faction has been the various and unequal distribution of property.

James Madison

I. Overview

Courses in business law and the legal environment have consistently been at the top of the charts in popularity among business students. There are numerous explanations for this, not the least of which is the dedication of law professors who have foregone more lucrative options to work in the academy. One of the other big reasons for the high level of interest in law classes is the relevance the subject matter brings to the students' personal lives. It is all well and good to teach how major corporations strategically plan their affairs. But students find it equally important to learn how to plan their own personal affairs through the law. The laws of contracts, property, and torts all have not only significant roles in business but also vital roles in how one arranges his or her personal affairs. Personal property is one such key area.

The study of personal property revolves around being able to answer three key questions:

1. How is this property classified?
2. How is this property acquired or transferred?
3. What are the legal consequences of the answers to questions 1 and 2?

All personal property falls into one classification or another. The type of classification used will have important consequences on how property will be treated in the eyes of the law. For example, consider the basic distinction between real and personal property. Sales of real property transactions generally come under the purview of the common law of contracts. Sales of goods, however, are generally covered by the UCC. If you are selling trees on the land, that property is classified as real, and the common law of contract controls. If the trees have been cut and are being sold to a mill, the UCC will now call the shots. If that lumber becomes part of a house, the common law of contracts again controls because the house is treated as real property.

In addition to classification, the acquisition and transfer of rights and duties to property are of key personal and business importance. Most property is transferred by way of contract with some sort of reciprocal exchange of consideration. You work, get paid, and that money is exchanged for property. You may be lucky, however, and find it, or inherit, or just have it given to you. In all these events, the acquisition or transfer must be made in compliance with the elements required by law. Once you have acquired the property, what are the rights and duties that arise out of that ownership? What if you found it? What if others have claims against it? These issues are of key importance in both business and private lives.

The second teaching objective this chapter focuses on the law of bailments. The term bailment is derived from the French word *baillier* meaning to deliver. In the legal sense, a bailment is a delivery of personal property with a special purpose. That purpose involves some particular use or deposit of that property with the expectation that it be redelivered to the person who originally delivered it or otherwise dealt with at his or her direction. What is not expected to occur in a bailment is any sale, gift, or transfer of title.

One of the aspects of the study of bailments that surprises first-time students of the law is just how common this sort of transaction is. Drop your clothes off at the dry cleaners, it is a bailment. Have your car parked, it is a bailment. Check your bags with an airline, it too is a bailment. Bailments are indeed an everyday occurrence. What is not as common is that things can go wrong in bailments. Just like insurance claims, many people find out about the rules of the bailment game after there has been a loss.

Bailments are essentially a special kind of contract. Normal contract rules would apply except that a number of special circumstances have arisen often enough to warrant legislative responses to the problems raised by those situations. Consider the example of common carriers. Here the public wants and expects a high level of reliance in being assured that goods entrusted to the common carrier, for a fee, will be protected. The public's expectation has been reflected in the insurance-styled strict liability of carriers for goods in their possession.

In addition, another element that often shows up in bailment cases is the interaction between the law of torts and contracts. Almost invariably what has happened is that there is an attempt to limit, by contract, the liability that may otherwise have inured to one of the parties by reason of tort. Courts are then confronted with the age-old problem of choosing between freedom of contract and public policy. On the contract side, good, clear limitations of extent of liability serve everyone well if they are fairly arrived at between the parties. Conversely, contract limitations should not be allowed to become a false front behind which all sorts of negligent behavior may be somehow hidden or excused.

The other materials covered in this chapter are part and parcel of the bailment story. In the commercial setting, warehouses, common carriers, and innkeepers are involved in special bailments. The various rights, duties, and liabilities of these special situations are also reviewed.

II. Hypothetical Multi-Issue Essay Question

Bloke and Crissy Carrington have been married for the past 50 years (the first 20 in California) and live on the Worthmore Ranch in Aspen, Colorado. Bloke met Ms. Misty Blue (age 21) at the annual "How's your Aspen?" cookout on the ranch. He decides it's now or never for his mid-life crisis and runs off with Misty to Miami, telling Crissy: "My lawyer will call your lawyer about the divorce." Three days later Bloke expires on the beach from overstimulation. He did not leave a will but told Misty just before he died, "I leave you everything! It was worth more."

Crissy comes to you for advice as to the following properties acquired during the marriage (she was always a faithful housewife):

1. The Worthmore Ranch, held as tenancy in the entirety.
2. Bank accounts in California and Colorado. They had been held in joint tenancy, but Bloke tried to wire the money to Miami the day before he died. The money had not yet been wired to Miami.
3. Jewelry worth over $1 million that Bloke had handed to Misty on the plane to Miami.

What results?

III. Outline

Two Kinds of Property
- Personal
  - Property that consists of tangible property, such as automobiles, furniture, and jewelry, and intangible property, such as securities, patents, and copyrights
- Real
  - The land itself as well as buildings, trees, soil, minerals, timber, plants, and other things permanently affixed to the land

Acquisition of Personal Property
- Possession
- Purchase or production
- Gift
- Will or inheritance
- Accession
- Confusion
- Divorce

Ownership Rights in Mislaid, Lost, and Abandoned Property
- Mislaid property
  - Owner of premises where property is mislaid is entitled to possession
  - He or she holds it as an involuntary bailee until the owner reclaims it
- Lost property
  - Finder acquires title to the property against the whole world except the true owner
  - Owner may reclaim his or her property from the finder
- Abandoned property
  - Finder acquires title to the property, even against its original owner

Elements Necessary to Create a Bailment

Types of Bailments
- Ordinary bailments
  - For the sole benefit of the bailor
  - For the sole benefit of the bailee
  - For the mutual benefit of the bailor and bailee
- Special bailments
  - Common carriers
  - Warehouse companies
  - Innkeepers

Bailees and Bailors
- Bailee's rights
  - Exclusive possession of the bailed property
  - Use of the bailed property, or
  - Compensation for work done or services provided
- Bailor's duties
  - Pay agreed-upon compensation to bailee
  - Not to interfere with bailee's possessory interest
  - Notify bailee or any defects in the bailed property

Special Bailees
  Common carriers
  Warehouse companies
  Innkeepers

IV. Objective Questions

Terms:

1. Property that consists of tangibles such as automobiles, furniture, and jewelry, and intangibles such as securities, patents, and copyrights are all classified as ________________ ________________.

2. When property is owned by two or more persons at the same time, there are four types: (1) joint tenancy, (2) tenancy in common, (3) tenancy by the entirety, and (4) community property. All these are forms of ________________ ________________.

3. All real property and physically defined personal property such as buildings, goods, animals, minerals, and such are deemed to be ________________ ________________.

4. The effectiveness of the gift can be inferred from the circumstances or language used by the donor. These factors are used by the courts to help determine ________________ ________________.

5. A gift that is made in contemplation of death is called a ________________ ________________ ________________.

6. When an owner discards property with the intent to relinquish her rights to it or when an owner of mislaid or lost property gives up any further attempts to locate it, it is called ________________ ________________.

7. A bailment for the sole benefit of the bailor or the bailee is called a ________________ ________________.

8. The duty of care that goes beyond ordinary care that says common carriers and innkeepers have a responsibility to provide security to their passengers or guests is called the duty of ________________ ________________.

9. A contractual provision that relieves one or both parties to the contract from tort liability is called an ________________ ________________.

10. Common carriers can limit their liability for goods that are lost, damaged, destroyed, or stolen. These limitations are set out by ________________ ________________.

True/False:

1. ____ The method of acquiring ownership in property by capture is very important and widely used in today's society.

2. ____ A gift is only valid if the donee gives some consideration for the property that the donor gives.

3. _____ Under the Uniform Gifts to Minors Act, a gift that complies with the act is revocable until the donor's death.

4. _____ A gift *causa mortis* is only effective if consistent with the donor's written will.

5. _____ The owner of the premises when mislaid property is left has rights to the property that are superior to the rights of the person who finds the property.

6. _____ The finder of lost property need not return the property to the rightful owner if the owner's identity is known, but he has not demanded the return of his property.

7. _____ In bailment, like a sale, title to the goods transfers to the bailee.

8. _____ In a bailment for the sole benefit of the bailor, the bailee owes a duty of ordinary care.

9. _____ A bailor who does not warn the bailee of any defects in the bailed property is liable for damages that result from such defects.

10. _____ If one party terminates a bailment at will, the party is liable for any damages resulting from the termination.

Multiple Choice:

1. Farmer Fran owns several acres of farmland in Nebraska. At present, much of the land is planted with corn that is growing. How would Farmer Fran's property be characterized?
   A. Both the land and corn are personal property.
   B. The land is personal property, and the corn is real property.
   C. The corn is personal property, and the land is real property.
   D. Both the land and corn are real property.

2. Aunt Adrian wishes to give her niece Nancy a gift. Adrian is in good health. Adrian gives Nancy a diamond ring that has been in the family for generations. Adrian tells Nancy that since she had no daughters, she wants Nancy to have the ring. Nancy graciously accepts the ring. Which of the following is correct?
   A. Nancy has received a gift *causa mortis*.
   B. Nancy has received an *inter vivos* gift.
   C. Nancy has received an inheritance.
   D. None of the above.

3. Homeowner Helen contracts with Builder Bob to have Bob build a deck on to her house while she is on vacation for two weeks. Helen gives Bob her address and instructions about the deck. Helen then leaves town. However, Bob mistakenly builds the deck on Hank's house. Hank is Helen's neighbor and is also out of town for two weeks. The deck is a permanent structure, firmly secured to one side of the house. Which of the following is correct?
   A. Hank is an owner by production.
   B. Hank is an owner by accession.
   C. Hank is an owner by purchase.
   D. Hank is an owner by gift.

4. Loser Larry loses his gold watch in City Park. Stroller Stella finds the watch while walking in the park one day. Stella discovers Larry's name and address on the back of the watch. Which of the following is correct?
   A. Stella may keep the watch since it is lost property.
   B. Stella may keep the watch since Larry has not demanded its return.
   C. Stella must return the watch to Larry since she knows his identity.
   D. Stella must leave the watch where she finds it.

5. Guest Gary takes off his diamond studded ring to help wash dishes at his friend Sarah's house after a dinner party. Gary places the ring on top of a cabinet so it will not be lost. However, Gary leaves Sarah's house and forgets the ring. Mary, Sarah's housekeeper, discovers the ring one day while cleaning. Which of the following is correct?
   A. Mary has superior rights against everyone but Gary since she found the ring.
   B. Sarah has superior rights against everyone but Gary since the ring was misplaced in her house.
   C. Mary is a bailee of the ring.
   D. None of the above.

6. Diner David patronizes his favorite restaurant, Creative Caribbean Cuisine. Tonight, David plans to propose to his girlfriend. When David enters the restaurant, he gives his coat to the coat clerk. The coat clerk gladly accepts the coat. However, the clerk is unaware that David has a diamond ring in the pocket of the coat. Which of the following is correct?
   A. There has been a bailment of David's coat.
   B. There has been a bailment of the diamond ring.
   C. Both of the above.
   D. None of the above.

7. Allen asks his friend Brenda if he can borrow her car. Brenda agrees to allow Allen to use her car for the day. Which of the following is correct?
   A. This is a gratuitous bailment.
   B. Brenda is a bailor here.
   C. Allen owes a duty of utmost care to protect the car.
   D. All of the above.

8. Distinguished Daniel takes his pocket watch to Wilfred's Watch Repair (WWR) to have the watch repaired. When Daniel gets his watch back, the crystal has been cracked. Which of the following is correct?
   A. WWR owed a duty of utmost care.
   B. WWR is presumed to have breached its duty of care since the crystal is cracked.
   C. WWR is an involuntary bailee.
   D. WWR is strictly liable.

9. Mel asks Ned if he can borrow Ned's motorcycle for the weekend. Ned agrees but tells Met it will cost $50 for weekend use of the motorcycle. Mel picks up the motorcycle on Friday afternoon. However, Ned forgets to tell Mel that the motorcycle cannot be driven over 40 mph or it will overheat. In addition, although Ned does not know this, the back wheel is loose. This problem could have been easily discovered, but Ned has not really had time to inspect the bike lately. On Saturday, Mel is riding the bike at 55 mph when it overheats and severely burns Mel's leg. Mel is trying to get to a doctor when the back wheel flies off, and Mel falls from the bike and breaks his arm. Which of the following is correct?
   A. Ned is liable for Mel's burns caused by the overheating.
   B. Ned is liable for Mel's broken arm caused by the loose back wheel.
   C. Both of the above.
   D. None of the above.

10. Designer Duds, a clothes manufacturer, delivers an order of clothes to Trans Country Transport (TCT), a common carrier, for delivery to retail outlets around the country. TCT agrees to ship the clothes for $1,000. Which of the following is correct?
   A. This is a mutual benefit bailment.
   B. TCT is strictly liable for damage to the clothes.
   C. TCT can limit its liability to a stated dollar amount.
   D. All of the above.

V. Answers to Objective Questions

Terms:

1. *Personal property.* Technically, personal property is any property that cannot be classified as real property. Remember, however, that property that may have started as personal may be converted to real (such as a fixture) or vice versa (such as timber which has been severed from the land).
2. *Concurrent ownership.* Coownership can take place over both real and personal property. It can be between only two persons, such as a marital community property, or between thousands of persons, as seen in large publicly traded companies.
3. *Tangible property.* This is property having a physical form, capable of being touched. Sometimes the physical evidence of an intangible property is called a semitangible, such as a stock certificate. This is a tangible paper, but its real value is the share of ownership that it represents in the corporation.
4. *Donative intent.* Intent is one of the three key elements when examining possible gift transfers. The other two are delivery and acceptance on the part of the donee.
5. *Gift causa mortis.* A gift *causa mortis* is established when the donor makes a gift in anticipation of approaching death from some existing condition and then dies without having revoked the gift.
6. *Abandoned property.* When these two elements are in place, the property is considered discarded in the eyes of the law, and it will belong to the first person who claims it after the abandonment.
7. *Gratuitous bailment.* These bailments are for the sole benefit of the bailor and most often arise when the bailee is requested to care for the bailor's property as a favor. The bailee owes only a slight duty of care in these scenarios.
8. *Strict liability.* This means that the law imposes a very high duty of care because of the special nature of the bailment. With common carriers, only a limited set of circumstances will act as defenses. Innkeepers liabilities have been more restricted by state statutes.
9. *Exculpatory clause.* These are contract clauses that seek to release one of the parties from his or her liability for a wrongful act. The scope of the clauses is limited by overriding public policy. For example, such clauses cannot be used to excuse a crime.

10. *Federal law.* These laws require that where common carriers are allowed to limit their liability, they must offer shippers the opportunity to pay a premium and declare a higher value for the goods.

True/False:

1. False. In today's society, there is very little property that is not already owned. Acquiring ownership in property by capture was much more common during the early years of our country's history.
2. False. Lack of consideration is what distinguishes a gift from a purchase. If consideration is required, it is not a gift.
3. False. If a gift complies with the Uniform Gifts to Minors Act, the gift is irrevocable.
4. False. A gift *causa mortis* need not be consistent with a donor's written will, and it takes precedent over a conflicting will.
5. True. The rationale for this rule is that the true owner may return to the premises to find the mislaid property. If this happens, the owner of the premises must return the property.
6. False. If the finder knows the true owner's identity, the finder must return the lost property. That is, the finder cannot keep the property until the true owner demands its return if the finder knows the true owner's identity.
7. False. A bailment is different from a sale or gift because title to the goods does not transfer to the bailee, but remains with the bailor.
8. False. A bailment for the sole benefit of the bailor is a gratuitous bailment. In such a bailment, the bailee only owes a duty of slight care.
9. True. The specific defects a bailor must warn a bailee about vary depending on the type of bailment. However, the bailor must at least warn the bailee of any known defects in the bailed property.
10. False. A bailment at will is not for any fixed term and may usually be terminated by either party at any time. If a bailment is for a fixed term, a party who terminates the bailment prior to the end of the term will be liable for resulting damages.

Multiple Choice:

1. D. Land and property permanently attached to it are real property. In addition, crops that have not yet been harvested and removed from the land are also real property. Therefore, both the land and corn are real property. A is incorrect because personal property is everything which is not real property. Since both the land and corn are real property, they are not personal property. B is incorrect because the land is real property. C is incorrect because the corn is real property here since it has not been removed from the land.
2. B. Nancy has received an *inter vivos* gift from Adrian. Adrian was alive and well at the time the gift was made. In addition, Adrian had donative intent to give the ring to Nancy, Adrian actually delivered the ring to Nancy, and Nancy accepted the ring. A is incorrect because a gift *causa mortis* is made by a dying donor and is conditional upon the donor's death. C is incorrect because an inheritance is received under a will when the donor has already died.
3. B. Since Bob mistakenly made improvements on Hank's house and such improvements cannot be easily removed, Hank is an owner by accession. A is incorrect because ownership by production is where a person uses raw materials to produce a finished product and therefore owns the product. C is incorrect because ownership by purchase is where a person buys property. D is incorrect because ownership by gift is where a person voluntarily transfers property to another. Here, Bob did not intend to give Hank a new deck.

4. C. If a finder knows the identity of the true owner, she must return the property. A is incorrect because since Stella knows Larry's identity, she must return the watch. B is incorrect because Stella must return the watch since she knows Larry's identity even if Larry does not demand it be returned. This is because Larry may not know Stella has found the watch. D is incorrect because Stella may take the watch, but must return it under the circumstances discussed above.
5. B. Sarah has superior rights to the ring as to everyone except Gary since it was lost at her house. A is incorrect because the owner of the premises has rights to the mislaid property, not the finder. C is incorrect because Sarah, not Mary, is the involuntary bailee of the ring.
6. A. There has been a bailment of David's coat because the clerk knowingly accepted the coat. B is incorrect because the clerk was unaware of the diamond ring and did not knowingly accept the ring. C is incorrect because B is incorrect.
7. D. A is correct beçause Brenda is allowing Allen to use her car as a favor and is not charging him anything. B is correct because Brenda is the owner of the car and is loaning the car to Allen. Therefore, Brenda is the bailor and Allen is the bailee. C is correct because in a bailment that is for the sole benefit of the bailee, the bailee owes a duty of utmost care to protect the bailed property.
8. B. The law presumes that if bailed property is damaged while in the possession of the bailee, it is because of lack of proper care by the bailee. However, this presumption is rebuttable. A is incorrect because WWR owes a duty of ordinary care, not a duty of utmost care. C is incorrect because WWR voluntarily took the pocket watch to repair it and is therefore not an involuntary bailee. D is incorrect because strict liability is imposed on common carriers, but not ordinary bailees.
9. C. Since Mel paid $50 for weekend use of the car, this is a mutual benefit bailment. In such bailments, the bailor must warn the bailee of defects that the bailor actually knows of, as well as those which could have been discovered by reasonable inspection. Therefore, Ned is liable for damages resulting from both the overheating problem as well as the loose back wheel.
10. D. A is correct because both parties benefit from this arrangement. B is correct because common carriers are strictly liable for bailed goods. C is correct because although TCT cannot avoid strict liability, it can limit its liability to a stated dollar amount if this is brought to the bailor's attention and it is not against public policy.

VI. Answers to Essay Question

First, the properties in question must be generally classified. The ranch is clearly classified as real property. This classification would include the surface, mineral, and air rights associated with land. It would also include any permanent improvements on the land, including buildings, fixtures, and the like.

The bank accounts would be classified as personal property. Because the accounts may be represented by passbooks or some other indicia of ownership such as a certificate of deposit, these assets would be classified as intangibles in that the real value of the property is found in what these documents represent, i.e., monies deposited with the financial institution. The jewelry is classified as tangible personal property. All the property involved is privately held.

Because all the properties were presumed to be acquired during a fifty-year marriage, it must be assumed that in the hands of Bloke and Crissy, the property was either purchased or somehow otherwise legally acquired by them. The more important immediate issue is: was any of the property in question transferred to Misty in the problem as presented? Of overriding concern is the attempt by Bloke to deprive Crissy of her legitimate marital claims to the properties. Here the laws of Colorado, Florida, and California will all come into play to one extent or another.

With regard to the ranch, under the laws of tenancy by the entirety, the property will pass automatically at death to the surviving spouse. This right of survivorship is by operation of law, and the property will not pass by way of probate, with or without a will. Transfer of the property must have been made with the consent of both tenants and must be evidenced by some sort symbolic written document, i.e., a deed. None of those occurred here. Some states sever the tenancy by entirety in case of divorce. Here the divorce had not yet taken place, so Crissy takes the ranch as surviving tenant by the entirety.

As for the bank accounts, many of the same issues apply. Joint tenancy rules are such that a transfer of the property must be made with the consent of both parties. Some states allow a unilateral severance of the joint tenancy but limit the ability to pull out the money to the one half claimed by the severing joint tenant. Because there was no actual or constructive delivery of these monies to Misty under the purported gift *causa mortis*, this attempted transfer also appears to have failed. If the accounts are still deemed to be held in joint tenancy at the time of Bloke's death, Crissy will take these assets as surviving joint tenant. The property will pass outside of probate.

If the accounts are deemed to have been severed out of joint tenancy, Crissy can still claim her half. The other half would become part of Bloke's estate. As part of his estate, it would be subject to the probate laws of the states involved. Because he did not leave a will, the laws of intestacy would apply. Under most of these laws, the surviving spouse would be the first taker of the estate.

As for the jewelry, it appears that a completed gift may have taken place. But even here, Misty's claim to the jewelry may be subject to Crissy's marital property claims. This may have been a fraudulent attempt to deprive Crissy of marital property.

It appears that there are no intended bailments in the ordinary sense of the word. Most bailments involve some sort of commercial purpose where there is an expected mutual benefit to both parties, such as goods delivered for repairs or a rental of some personal property. The only real possible bailment here is the possession of the jewelry. If Crissy is able to prove that Bloke committed a fraud on her marital property vis-à-vis the transfer of possession to Misty Blue, then she can argue that Misty is a wrongful bailee of her property. As such, Misty would have a duty to return the goods, if not directly to Crissy, at least to the control of the court for its final disposition of the matter.

All in all, let's hope that Misty Blue has other assets to rely on. Crissy looks like a clear winner here, as well it should be.

Chapter 37
REAL PROPERTY AND LANLORD-TENANT RELATIONSHIP

The mind covers more ground than the heart but goes less far.
Chinese Proverb

I. Overview

This chapter is designed to introduce students to the law of real property from two key perspectives: first, ownership and the rights and duties that arise out of the ownership of real property, and second, use of real property and the respective rights and duties can arise out of that use vis-à-vis others.

Real property represents the largest single outlay most people make in the course of their earning years. Even if they choose to rent, the price of keeping a roof over one's head will still probably be their biggest expense. Real estate is not only necessary as a matter of physical survival, it is critically important to our economic system because of this large dollar outlay. One of the most basic terms used in the law of real estate is "fee simple absolute." It connotes the highest form of recognized ownership in real property. The term is originally derived from the words *feud* or *fief* and *fief d'haubert* meaning a fee held by tenure of a knight's service to the lord of the manor. It is infinite, with no limitation on inheritability, and does not end upon the happening of any event. Think of fee simple as the whole pie. That pie, in turn, may be sliced and diced into all sorts of smaller morsels.

Another way to look at real estate as a circular object is in the physical shape of the earth. It is round, and each ownership of land has a unique wedge-shaped slice of that round body. The basic parameters of that ownership start with the surface rights as defined by the surveyed metes and bounds in the legal description. In addition to those rights, real estate extends theoretically to the center of the earth in minerals below the surface and in development of air rights. Both these rights are subject to use limitations and the rights of other owners of adjoining properties.

The other interesting aspect of this chapter goes into more detail on forms of coownership of property. Most of us, sooner or later, will get involved with coownership of property in one way or another. Anyone who is married is a likely coowner. Anyone who shares property interests by gift, inheritance, or earnings is likely to be a coowner. Even if one's property is entirely his or her own, he or she will need to know the rules of the coownership game for purposes of credit, finance, business planning, and the like. How, when, and where coownership rights and duties are created is as important as the basic terms of real property law itself.

An interesting comparative measure of the U.S. way of life is the percentage of home ownership versus rental. Our economy has traditionally boasted of a high percentage of home ownership relative to most any other part of the world. In most economically backward countries, the percentage of home ownership is very low. Even in the more highly developed parts of the world like Western Europe, the percentage has been much lower than in the U.S. This achievement has been one of the foundations of the great middle class dream of Americans and certainly a stabilizing influence in our society.

The disturbing aspect of this particular economic and sociological measure is that recent history has witnessed a decline in the number of Americans who can realistically aspire to own their own homes. As the percentage goes down, the laws involving the rights, duties, and obligations of persons involved in nonfreehold estates of all sorts become increasingly important. With over half of our population now living under someone else's roof, the law of leasing needs constant and fair reexamination and updating. As in any such process, the legitimate claims of both sides must be listened to.

The landlord is entitled to a fair return on his or her investment. This includes not only a financial return but also a return on the investment. Income property should be protected from waste and harm just like any other property. Too many people seem to have adopted an attitude that says: ""f it's not mine, then the hell with it."

Conversely, the bargaining power between landlord and tenant has never been entirely equal. Between market limitations of supply and the tenant's lesser economic bargaining position, the common law can hardly have been accused of being the tenant's friend in the past. More modern attitudes and statutory enactments have given the tenant a more even playing field, but not real parity. Whether we like it or not, nonownership is still looked upon as second-class citizenship.

This chapter also examines some of the main land use statutes used in this country. Here, the law moves away from its first and second tier treatment based on ownership and looks instead more to the common good. The days of wide-open spaces are long gone, and along with them the notion that an owner can do whatever he or she wants with his property. Land use is simply too important and too interwoven with the rights of others to adapt a *laissez-faire* attitude. Various branches of government must perform the constant balancing act. On the owners and user's side of the equation, public policy wants free, quiet, environmentally sensitive use of land. Good zoning controlled growth and other governmental regulations can and do provide enlightened measures towards those goals. Properly executed, good land use law literally translates into good quality of life for users.

Conversely, where government sets up measures that are unreasonable and even confiscatory, the larger society is badly served. The results are not only higher costs of doing business, but also depreciation of the American Dream. Zoning laws and the like have been pointed to by many commentators on the American scene as an invidious subterfuge to extend governmental control too far into our personal lives. The government does have the duty to assure the health, safety, and welfare of its citizenry through the use of its police power on land use. It should be used to help us live well but not to tell us how to live.

II. Hypothetical Multi-Issue Essay Question

This question is designed to help students identify some of the key issues arising out of a break up of a substantial amount of property due to divorce. Unfortunately, with a national divorce rate of over 50 percent, such scenarios are an everyday occurrence in our society. As with so many other aspects of the legal environment in the law of real property, it is fine to hope for the best, but best to plan for the worst.

Dufus Duffee and his wife, Loquacia, were married in Boulder, Colorado on April 1, 1970 by the Rev. Tammy Bigbucks. They still live and own a home in Boulder even though they have become very wealthy. Dufus realized that even though most people were not to the manor born, they still wanted to have a family crest other than toothpaste. Dufus devised a great marketing gimmick. Why not emboss luxury goods with a Royal Dog named Lord Throckmorton T. Thurston the Third? The Big T. Brand Holding Company now nets over $15 million annually by selling everything from fashion goods to hardware, all identified by the Big T. status symbol.

Dufus and Loquacia also own, in fee, homes in California, New Mexico, and New York City. All of them are patterned after the Newport Cottage and are valued at $10 million each. In addition, the Big T. Brand Holding Company owns and operates plants in Montana, Louisiana, and Illinois. All the family assets, including 100 percent of the shares in Big T. Brand Holding Company, are held in Dufus' name alone. Loquacia stayed at home to raise their six children, Otter, Bubba, Rollo, Misty, Dufus II, and Dufus III.

Dufus decides to seek a divorce because he falls in love with the young and mysterious Margaux de Mollin while visiting the islands of Myorca and Yourorca. What can he claim as his besides Margaux's affection?

III. Outline

What is Real Property?
- The land itself as well as buildings, trees, soil, minerals, timber, plants, and other things permanently affixed to the land

Freehold Estates
- Fee simple absolute
  - Highest form of ownership of real property
  - Ownership is infinite in duration, has no limitation on inheritability, and does not end upon the occurrence or nonoccurrence of any event
- Fee simple defeasible
  - Grants owner all of the incidents of a fee simple absolute except that it may be taken away if a specified condition occurs or does not occur
- Life estate
  - Interest in property for the life of a specified person
  - A life estate terminates upon the death of the named person and reverts back to the grantor or his or her estate or other designated person

Future Interests
- Future interest
  - The interest that the grantor retains for himself or herself or a third party
  - Reversion
    - Right to possession of real property returns to the grantor after the expiration of a limited or contingent estate
  - Remainder
    - Right to possession of real property goes to a third person upon the expiration of a limited or contingent estate

Concurrent Ownership
- Joint tenancy
  - Deceased tenant's interest automatically passes to co-tenants
  - Tenant may transfer his or her interest without consent of co-tenants; transfer severs joint liability
- Tenancy in common
  - Deceased tenant's interest passes to his or her estate
  - Tenant may transfer his or her interest without the consent of co-tenants; transfer does not sever tenancy in common
- Tenancy by the entirety
  - Deceased tenant's interest automatically passes to his or her spouse
  - Neither spouse may transfer his or her interest without the other spouse's consent
  - Community property
    - Surviving spouse automatically receives 1/2 of community property; other one half passes to heirs

Neither spouse may transfer his or her interest without the other spouse's consent

How Ownership of Real Property Is Transferred
- Sale
- Tax sale
- Gift, will, or inheritance
- Adverse possession

Deeds and Recording Statutes
- Deed
  - A writing that describes a person's ownership interest in a piece of real property
- Recording statute
  - A state statute that requires the mortgage or deed of trust to be recorded in the county recorder's office of the county in which the real property is located

Nonpossessory Interests in Real Property
- Easement appurtenant
  - Easement over a servient estate that benefits a dominant estate
  - Easement runs with the land
- Easement in gross
  - Easement that grants a person a right to use another's land
  - Does not run with the land
- License
  - Grants a person the right to enter upon another's real property for specified event or time
- Profit
  - Grants holder the right to remove something from another's real property

What Is a Landlord-Tenant Relationship?
- A relationship created when the owner of a freehold estate (landlord) transfers a right to exclusively and temporarily possess the owner's property to another (tenant)

Types of Tenancies
- Tenancy for years
  - Continues for the duration of the lease and then terminates automatically without notice
  - Does not terminate on the death of either party
- Periodic tenancy
  - Continues from payment interval to payment interval
  - May be terminated by either party with adequate notice
  - Does not terminate on the death of either party
- Tenancy at will
  - Continues at the will of the parties and may be terminated by either party at any time with adequate notice
  - Terminates on the death of either party
- Tenancy at sufferance
  - Arises when a tenant wrongfully occupies real property after the expiration of another tenancy or life estate
  - Continues until the owner either evicts the tenant or holds him or her over for another term
  - Terminates on the death of the tenant

Landlord's Duties
- To deliver possession of the leased premises
  - Duty to deliver possession
  - Duty not to interfere with tenant's right to quiet enjoyment
- To maintain the leased premises
  - Building codes
  - Implied warranty of habitability
- Duty of reasonable care to tenants and third parties not to cause them injury

Tenant's Duties
- Duty to pay rent
- Duty not to use leased premises for illegal or nonstipulated purposes
- Duty not to commit waste
- Duty not to disturb other tenants
- Duty of reasonable care to persons who enter upon the leased premises

Transferring Rights to Leased Property
- Assignment of the lease
  - A transfer by a tenant of his or her rights under a lease to another
  - Assignor
    - The party who transfers the rights
  - Assignee
    - The party to whom rights have been transferred
- Sublease
  - When a tenant transfers only some of his or her rights under the lease
  - Sublessor
    - The original tenant in a sublease situation
  - Sublessee
    - The new tenant in a sublease situation

Regulation of Land Use
- Private
  - Private nuisance
  - Restrictive covenants
- Public
  - Public nuisance
  - Zoning
  - Eminent domain

IV. Objective Questions

Terms:

1. Real property includes some items of personal property that are affixed to real property which are called ________________.

2. Common law provided that a surface owner had possessory rights above the surface to the ________________.

3. Plant life and vegetation growing on the surface of land is considered to be _______________ property.

4. The highest form of ownership in real property that grants the owner the fullest bundle of legal rights is called a _______________ _______________ _______________.

5. An interest in real property that lasts for the life of a specified person is called a _______________ _______________.

6. An eviction that occurs when the landlord causes the leased premises to become unfit for their intended use is called _______________ _______________.

7. Discrimination in the sale and rental of real property based on race, color, national origin, sex, or religion is illegal under the _______________ _______________ Act and the _______________ _______________ Act.

8. A warranty that provides that the leased premises must be fit, safe, and suitable for ordinary residential use is a _______________ _______________ _______________ _______________.

9. A private agreement between landowners that restricts the use of their land is a _______________ _______________.

10. The taking of private property by the government for public use, provided just compensation has taken place is called _______________ _______________.

True/False:

1. ____ Owners of land do not possess mineral rights to the earth located beneath the surface of the land.

2. ____ When determining whether personal property has become a fixture, courts can look at the intent of the parties.

3. ____ Leases are classified as freehold estates.

4. ____ A reversion is a right of possession that returns to the grantor after the expiration of a limited or contingent estate.

5. ____ Land is the most common form of real property.

6. ____ Under a lease assignment, once the assignor assigns his rights to an assignee, the assignor is released from all obligations he owes to the landlord.

7. ____ Under a sublease, the landlord can sue either the sublessor (tenant) or the sublessee to recover rent payments.

8. ____ A restrictive covenant, although it is a private agreement between landowners, binds all subsequent landowners as well.

9. ____ Generally, the landlord, not the tenant, must maintain and repair common areas.

10. ____ Generally, a tenant is generally liable for injuries to a third person if the injuries were caused by the tenant's negligence.

Multiple Choice:

1. Which of the following is not one of the three types of freehold estates?
   A. Fee simple absolute.
   B. Reversion.
   C. Fee simple defeasible.
   D. Life estate.

2. A owns a life estate to real property so long as B is alive. A owns a:
   A. Fee simple absolute.
   B. Fee simple defeasible.
   C. Simple life estate.
   D. Estate *pour autre vie*.

3. Which of the following is not an example of a fixture?
   A. Furniture.
   B. Wall to wall carpeting.
   C. Dishwasher.
   D. Door knobs.

4. Which of the following is true of a fee simple absolute?
   A. It has no limitation on inheritability.
   B. It is infinite in nature.
   C. It does not end upon the happening of any event.
   D. All of the above.

5. Which of the following is *not* one of the four unities necessary for a joint tenancy?
   A. Interest.
   B. Possession.
   C. Unity of marriage.
   D. Title.

6. Mr. and Mrs. Yuppie are looking for the perfect lot on which to build their new house. One major consideration is the zoning restrictions that they must comply with. The Yuppies find a great lot with a beautiful view. The area requires that only houses, but no businesses, may be built. In addition, all the houses must be no less than 3,000 square feet. Finally, there must be some brick in the exterior architecture of the building. The Yuppies have fallen in love with a lot that has:
   A. Use of property zoning restrictions.
   B. Height, bulk, and area zoning ordinances.
   C. Architectural limit zoning ordinances.
   D. All of the above.

7. Assume the same facts as in question 6. Also assume the Yuppies can only afford to build a house that is 2,200 square feet. What should the Yuppies do if they still want to build on the lot?
   A. Apply for a variance.
   B. Apply for spot zoning.
   C. Either of the above.
   D. None of the above. The Yuppies have no choice but to comply if they want to build on the lot.

8. Under the terms of a triple net lease, the tenant is responsible for:
   A. Taxes.
   B. Insurance.
   C. Utilities.
   D. All of the above.
   E. A and B only.

9. Neighbor Nancy and Neighbor Ned agree that neither one shall ever build a house higher than three stories on his or her land. Nancy and Ned intend to have this covenant run with the land, and they put the covenant in writing. However, Nancy and Ned are not in privity of estate. If Nancy or Ned sells their land to a purchaser who is aware of this agreement, what is the remaining neighbor's best argument for binding the purchaser to the covenant?
   A. This is a restrictive covenant.
   B. This is an equitable servitude.
   C. A house over three stories is a private nuisance.
   D. None of the above.

10. Now assume Nancy is Ned's wealthy aunt. Nancy gives Ned the property adjacent to hers as a gift. When Nancy gives the property to Ned, the two agree never to build a house higher than three stories on their land. Nancy and Ned intent to have this covenant run with the land and they put the covenant in writing. If Nancy or Ned sells their land to a purchaser, what is the remaining neighbor's best argument for binding the purchaser to the covenant?
    A. This is a restrictive covenant.
    B. This is an equitable servitude.
    C. A house over three stories is a private nuisance.
    D. None of the above.

V. Answers to Objective Questions

Terms:

1. *Fixtures*. Certain personal property is so closely associated with real property that it can become part of the realty. Under the common law, several methods can be used by the courts to decide if personal property has become a fixture. These include the intent of the parties, attachment or annexation, or the use of item in question.
2. *Heavens*. Air rights include the airspace above the property theoretically to the heavens. In reality this right is limited not only in practical terms but even more by zoning laws and airspace navigation rules. Air rights are, however, still very important in congested areas and often are used as bargaining chips in the zoning process such as set-backs for office buildings.
3. *Real*. Plant life and vegetation growing on the surface is considered real property. This includes both natural and cultivated plant life. When the land is sold, these assets are considered part of the land unless agreed otherwise. Once severed, they are, however, considered personal property.

4. *Fee simple absolute.* This is the highest form of ownership in real property because it grants the owner the fullest bundle of rights that a person can hold in real property. The term is traceable to the fee paid by the lord of the manor to his knights for services to the realm.
5. *Life estate.* This interest in real property is measured by the life of a specified person. This method of ownership is very often used in conjunction with a will or trust and is designed to take advantage of certain favorable provisions of wealth transfer tax laws.
6. *Constructive eviction.* While not depriving the tenant of actual possession, the premises have been rendered uninhabitable by the landlord's acts or omissions.
7. *Civil Rights; Fair Housing.* These statutes prohibit discrimination in the rental or sale of real property.
8. *Implied warranty of habitability.* This warranty has been imposed by most jurisdictions even if the parties to the lease have failed to expressly mention terms like habitability or fit for ordinary use.
9. *Restrictive covenant.* These covenants must be in compliance with both public policy and not in violation of civil rights statutes.
10. *Eminent domain.* Under the Fifth Amendment to the U.S. Constitution, just compensation must be paid to the owner of the property that has been taken for public use.

True/False:

1. False. Mineral rights are part of the bundle of sticks presumed owned unless otherwise agreed upon by the parties.
2. True. This is one of the key considerations taken into account by the courts on this issue. Oftentimes commercial fixtures appear very permanent, yet can be removed based on the contractual agreements between the parties.
3. False. Leases are not classified as freehold estates because they are contract-based use not ownership estates.
4. True. This is one of the categories of interests in property that is classified as a future interest.
5. True. Most people think of real property in the surface sense. In fact, real property rights include mineral rights below the surface and air rights above the surface.
6. False. The assignor is only released from his obligations to the landlord if both parties come to an agreement releasing the assignor. If there is no such agreement, the assignor is still obligated to the landlord.
7. False. Under a sublease, the landlord can only sue the sublessor (who is the original tenant), but not the sublessee.
8. True. It can run with the land, which means that the covenant will follow or accompany the title.
9. True. This is the general rule, which can be modified by the terms of the lease. If the lease is silent on the issue, the general rule will apply.
10. True. The tenant owes a duty of reasonable care to persons who enter upon the leased premises.

Multiple Choice:

1. B. The reversion is a future interest and is thus not considered a freehold estate until such time as it vests. The other estates are all classified as freehold estates.
2. D. An estate *pour autre vie* is a life estate measured in the life of a third party. A's life estate is measured in terms of B's life, and it will end at B's death.
3. A. Furniture is usually, but not always, movable and therefore not classified as part of the real property. Courts can look at intent, use, or annexation to see if an item of personal property has been converted to a fixture.
4. D. A, B, and C are all elements found in the definition of a fee simple absolute.

5. C. Marriage is a required element for the creation of a tenancy by the entirety. A joint tenancy does not require marriage as a prerequisite.
6. D. A is correct because restricting the land to only residential use is a use of property zoning ordinance. B is correct because the square foot requirements are a height, bulk, and area zoning restriction. C is correct because the brick requirement is an architectural limit zoning restriction.
7. C. The Yuppies may try to make an exception to the zoning requirements by either a variance or spot zoning. However, spot zoning is not likely to be permitted. D is incorrect because the Yuppies have some, although few, options here.
8. D. All of the above are generally included in the terminology used in triple net leases.
9. B. Because Nancy and Ned are not in privity of estate here, there is not a restrictive covenant. However, if the purchaser was aware of this agreement, the remaining neighbor could try to argue this is an equitable servitude. A is incorrect because there is no privity of estate here. C is incorrect because this is an unreasonable assertion.
10. A. Here, there is privity of estate since Nancy is giving the land to Ned. B is incorrect because with privity of estate, restrictive covenant is a stronger argument. C is incorrect because this is an unreasonable assertion.

VI. Answers to Essay Question

It appears from the description of the real properties that they are held in fee simple absolute. When estates in property are described, the presumption is that ownership is of the complete bundle of sticks that make up fee simple absolute ownership. This would include the land, mineral, and air rights. This would also include the full freehold estate not subject to any life estates or defeasible conditions. In other words, the title unites all present and future interests.

From the question as given, it appears that Dufus owns all the property in his name alone. As much as he would like to tell that to the mysterious Margaux, Dufus cannot readily make that sort of claim. A number of the properties in his inventory are situated in community property states: the homes in California, New Mexico, and the plant in Louisiana. At the very minimum, Loquacia would be able to claim a community property interest in those parcels. Under the general rules of community property, equal coownership is attributed to all property earned during the marriage regardless of who earned it. Separate property status is generally allowed for property brought into the marriage and gifts of inheritance acquired during the marriage. Even in those instances, it must be shown that the property was not commingled.

In addition to the community claims that Loquacia may make, most noncommunity property states have effectively adopted community property-like results in divorce proceedings. The basic and equitable premise is that a marriage is more than just a spiritual and sexual union. It is also an economic partnership that provides mutual support and opportunity for each other's enrichment. Here Dufus did well in business because he had the support system of having his wife care for his personal and family needs. It is only fair to recognize that work in economic terms if the property has to be divided due to divorce.

It appears from the facts given that there are no claims to the real property owned by Dufus and Loquacia that are based on easement or license.

Even though it appears that Dufus sought to avoid any transfer of these assets to his wife's name by structuring the Big T. Brand Holding Company in his name, any number of transfer methods would have been available to him had he been more considerate of his wife's interests in the marital property. He could have made gift transfers and the like. Because he has failed to do so, the most likely result will be that all the properties, including the Big T. Brand Holding Company, will be ordered equally divided. Dufus can then go back to Myorca, and Loquacia can get on with her life.

Chapter 38
INSURANCE AND ESTATES

Education never ends... It is a series of lessons with the greatest for the last.
Sir Arthur Conan Doyle

I. Overview

Insurance

Insurance is defined as a contractual arrangement whereby one undertakes to indemnify another against loss, damage, or liability. Even though the law of insurance is derived from contract, the need for insurance is rooted in reality. Few activities in today's overly litigious world are not in one way or another involved with insurance. Contracts of insurance may concern us from our moment of birth to our moment of death and every time in between. Insurance is more important than ever because it appears that new risks never even fathomed by our forefathers emerge every day. One very real and unfortunate truth of conducting one's personal and business affairs is that financial safety nets are more necessary than ever. How many of us can really afford to be self-insured? What are some of the factors that have led to this state of affairs?

One large element in this picture is found in the results of having a legal system that is willing to attach responsibility for many heretofore-unforeseen situations. Modern technology has raised expectations to the point where every product must be perfect, every operation a success, and every accident avoidable. Twenty/twenty hindsight seems to be a particularly costly form of myopia to the insurance industry.

Another element is found in the insurance industries own backyard. State regulations, and policy owner political initiatives, such as those seen in California recently are in many ways an outgrowth of perceived mismanagement and greed in the halls of insurance companies. The legal environment has spawned all sorts of new insurance sales opportunities for insurance companies such as Medicare gap coverage. When these coverages fail to deliver, resentment grows.

Yet another problem is found in the legal departments of many insurance companies. Attempts to obfuscate the real coverage by convoluted technical legal language and words of art has led to much resentment over what insurance companies really can or will do in case of need. This problem is exacerbated by perceived bad faith efforts to avoid coverage for large scale costs arising out of ailments such as AIDS. Expensive experimental treatments have also severely tested insurance companies willingness to pay. Large-scale disasters like Hurricane Andrew have led to many examples of insurance company flight from high risk areas.

In spite of all these problems, the role of insurance is more vital than ever. Witness the recent calamitous fire in the hills of Oakland and Berkeley, California costing many lives and over a $1 billion in property damage. Not to have insurance would surely be pouring salt on an already grievous wound. Insurance law is ripe with controversy and litigation. Yet no one really wants or can imagine a legal, political, or social environment without it.

Estates

You can't take it with you, but you can control where it goes. The laws of wills and trust are about control. This control is not absolute now and never has been. Yet with intelligent foresight and timely estate planning, the extent and breadth of possible control would surprise most people. It is no coincidence that the money of large family fortunes first accumulated in the early industrial age remain essentially intact today with the help of wills, trusts, and proper estate planning.

The origins of the laws of wills, trusts, and estates go back to the *canon* laws of the medieval church. The church had much control not only over the spiritual life of its members, but also over the more temporal affairs such as the disposition of worldly goods. The early church-based patterns of disposition of property based on family relationships are still reflected in more modern day laws such as the Uniform Probate Code. These traditions are strongly evidenced in civil law countries where the notion of forced heirship is still alive and well. England's break from the church under Henry VIII started a more tolerant tradition within the common law. In other parts of the world, such as Muslim countries, religion remains the cornerstone of the laws of devolution.

Transfers of property can be generally categorized into two main classifications; lifetime or *inter vivos* and death or testamentary transfers. With lifetime transfers, the main motivation elements are taxes, control, and present needs. The federal wealth transfer tax structure is extensive and very complicated. The three main components of that tax structure are the gift, estate, and generation skipping taxes. In addition, states have also developed their own property transfer tax structures based on inheritance taxes, estate pick-up taxes, or some combination thereof. All these taxes are premised on the notion that wealth accumulation was done with the blessings of the law structure, and transfers of that wealth afford an opportunity for the government to share in those riches.

The control element also goes to the issue of how much the owner is actually willing to give up before he or she departs this life. Tax laws are specifically designed to tax the substance of control rather than just the appearance of having given up control. Various code provisions with regard to life estates, incomplete transfers, retained powers over trusts, and possibilities of reverter are all patterned on that premise. In addition, simple human nature is such that it can be most difficult to give up control over assets that took the better part of a lifetime to accumulate. This problem is further exacerbated if the intended donee is too young, too immature, or an out and out spendthrift.

Lifetime needs make *inter vivos* transfers ever more difficult. In an age of being exposed to possible catastrophic health care costs late in life, many people simply cannot afford to make lifetime transfers that would otherwise make for good estate planning. The burgeoning growth of living trusts and similar financial planning devices are testaments to efforts to try to resolve this dilemma.

Once the end has come, the next classification of transfers can be subcategorized into testate and intestate death transfers. Testate means that the person has spoken by way of a legally recognized will or will substitute. The law of wills is much like the laws of death taxes--very complicated and formalized. The general public often looks at the cases on admission of wills to the probate process and sees them as being overly harsh or doctrinaire. The basic point that must be remembered is that the decedent is not there to challenge any arguments against his or her purported intent. There is a rhyme and reason to it all in spite of the sometimes arbitrary nature of such proceedings.

If a person has not written a will, or the will is not admitted to probate, he or she is deemed to have died intestate. Every state has stepped into this void and written a will by way of an intestate statute. Under these laws, the devolution follows the pattern originally created by the medieval church. This pattern is basically designed to convey the property on to the person's perceived natural bounty. The problem is that in an age of a 50 percent national divorce rate, the possibility of multiple families claims makes having a will more necessary than ever. In all events, if you want to control the disposition of your property at your death, have an up to date will at the ready.

Once a probate proceeding has begun, the person appointed by the court to administer the estate will be charged with three major duties:

1. To garner the assets of the estate.
2. To net out those assets by paying all proper claims, taxes, and costs of administration.
3. To distribute the net proceeds of the estate per the will or the intestate statute.

It is all very complex and emotionally draining for the parties directly involved, but death has always been the most inescapable fact of life.

II. Hypothetical Multi-Issue Essay Question

Autocross is an automobile-related sport wherein the driver tries to navigate his or her car through a set of difficult turns and roadways without going off course or hitting pylon markers. The driver who gets through this obstacle course in the fastest time wins. Because autocrossing does not involve any head-to-head racing, it affords a low cost opportunity for many of the Walter Mittys of the world to try their hand at auto sports and dream of moving up to Formula One someday.

One such dreamer was Enzo Autre Droit. All his life he wanted to race at Ring, LeMans, Indianapolis, and other such famous venues. Alas, his talents and his pocketbook confined his racing exploits to the local autocross held on weekends at the County Fairgrounds. Boy Racers Autoclub sponsored the events.

Boy Racers obtained event insurance for its autocross events. This insurance required that all participants sign an exculpatory clause releasing Boy Racers from any harm or liability arising out of the event. In addition, Enzo's own auto insurance company, Airbag Insurance, excluded coverage for "motorsports" events.

Last week, Enzo lost it on a hard left banking turn and went into a trailing throttle oversteer. His mean, green, driving machine went into a roll and was badly damaged. Enzo and members of the Boy Racers Autoclub pushed the car into a nearby ditch. Enzo now claims it was all a normal accident due to bad road conditions. What results?

III. Outline

Insurance Contracts
- Insured
  - The party who pays a premium to a particular insurance company for insurance coverage
- Insurer
  - The insurance company that underwrites the insurance coverage
- Policy
  - The insurance contract
- Premium
  - The money paid to the insurance company

Deductions and Exclusions
- Deductible clause
  - Clause in an insurance policy that provides that insurance proceeds are payable only after the insured has paid a certain amount of the loss

- Exclusion
  - Part of an insurance policy that says what is not covered under certain circumstances

Parties to a Life Insurance Contract
- Insurance company that issues the policy
- Owner of the policy who contracts with the insurance company
- Insured
  - The person whose life is insured
- Beneficiary
  - Person who is to receive insurance proceeds when the insured dies

Types of Health Insurance
- Health
- Disability

Fire and Homeowners Insurance
- Standard fire insurance
- Homeowners
- Renters

Automobile Insurance
- Collision
- Comprehensive
- Liability
- Medical payment
- Uninsured motorist

Business Insurance
- Business interruption
- Product liability
- Directors' and officers' liability
- Workers' compensation
- Professional malpractice
- Fidelity
- Credit

Other Types of Insurance
- Title
- Marine
- Umbrella
- Group

Wills
- Requirements for making a will
  - Testamentary capacity
  - Writing
  - Testator's signature
- Wills must be attested to by mentally competent witnesses
- Wills can only be legally changed by codicils
- A will may be revoked by acts of the testator or by operation of law

Special Types of Wills
- Holographic
  - Will that is entirely handwritten and signed by the testator
- Noncupative
  - Oral wills made before witnesses

Types of Testamentary Gifts
- Devise
  - A gift of real estate by will
- Bequest
  - A gift of personal property by will
- Specific gift
  - A gift of a specifically named piece of property
- General gift
  - A gift that does not identify the specific property from which the gift is to be made
- Residuary gift
  - A gift of the estate left after debts, taxes, and specific and general gifts have been paid

Per Stirpes and Per Capita Distribution
- Per stirpes
  - Lineal descendants inherit by representation of their parent, i.e. they split what their deceased parent would have received
- Per capita
  - Lineal descendants equally share the property of the estate regardless of their degree of relationship to the testator

Undue Influence
- Occurs where one person takes advantage of another person's mental, emotional, or physical weakness and unduly persuades that person to make a will
- The persuasion by the wrongdoer must overcome the free will of the testator

Comparison of Dying with and without a Valid Will
- Deceased dies with a valid will
  - Beneficiaries are named in the will
- Deceased died without a valid will
  - Heirs set forth in the applicable intestacy statute
  - If there are no heirs, the deceased's property escheats to the state

Parties to a Trust

Types of Trusts
- Express
  - Voluntarily created by the settlor
    - Inter vivos or living trusts
    - Testamentary
- Implied
  - Imposed by law
- Constructive
  - Equitable trusts implied by law to avoid fraud
- Resulting
  - Created by conduct of the parties

Special
    Charitable
    Spendthrift
    Totten

IV. Objective Questions

Terms:

1. Insurance is defined as a contract whereby one undertakes to _______________ another against loss, damage, or liability arising from a contingent or unknown event.

2. This act, passed in 1945, gave the regulation of insurance to the states. It is called the _______________ _______________ Act.

3. Insurance that features both term and whole life insurance is generally called _______________ _______________.

4. An employer can purchase _______________ insurance to protect against dishonesty and defalcation of employees.

5. When an insurance company pays a claim, it succeeds to the right of the insured to recover from a third party. This right is called _______________.

6. A person who dies without having a will or whose will is not admitted to probate dies _______________.

7. Nieces, nephews, aunts, and uncles are _______________ heirs.

8. When a specific devise of property to a beneficiary is no longer in existence when the testator dies, _______________ occurs.

9. A trust where creditors of the beneficiary cannot recover the trust's assets to satisfy personal debts owed them by the beneficiary is a _______________ trust.

10. A trust which is created when a person deposits money in a bank account in his or her own name and holds it for the benefit of another person is a _______________ trust.

True/False:

1. ____ An insurance broker works exclusively for one insurance company.

2. ____ Insurance companies are regulated by the federal antitrust (Sherman and Clayton Acts) laws.

3. ____ There are two parties to a life insurance contract.

4. ____ A fire insurance policy is assignable without consent of the insurance company.

5. ____ When an insurance company reinsures, it sells a portion of the policy's risk and right to receive premiums to another insurance company.

6. ____ The trustee is the beneficial owner of the trust corpus.

7. ____ Nuncupative wills are oral wills made before witnesses.

8. ____ The Uniform Probate Code does not require any attestation of a will by witnesses.

9. ____ A joint will is where two or more testators execute the same instrument as their will.

10. ____ *Inter vivos* trusts are created by will.

Multiple Choice:

1. Which of the following does not represent a recognized insurable interest?
   A. Life of a partner.
   B. Life of a friend.
   C. Life of a parent.
   D. Life of a debtor.

2. If Fred pays $100 a month for insurance all his life, and it now has a cash value of $10,000, he most likely has a:
   A. Term life policy.
   B. A comprehensive policy.
   C. An umbrella policy.
   D. Straight life policy.

3. Mr. Big owns Mega-Widgets Corporation. To insure against the financial loss of Mr. Big's death, Mega-Widgets can take out what kind of policy on his life?
   A. Job title insurance.
   B. Job disability insurance.
   C. Key person life insurance.
   D. Standard New York fire policy.

4. Which of the following events would not be covered by a homeowner's comprehensive personal liability insurance policy?
   A. Professional malpractice.
   B. Injuries sustained from a slip and fall on the insured's property.
   C. Property damaged by the insured.
   D. Injuries inflicted by members of the insured's family.

5. Which of the following are not defenses of the insurer from having to pay on a covered claim?
   A. Misrepresentation of a material fact by insured.
   B. Breach of warranty by insured.
   C. Commission of suicide after period of incontestability has expired on a life insurance policy.
   D. Concealment of material fact.

6. To have testamentary capacity, which of the following is not required?
   A. Extreme wealth.
   B. A comprehension of what property is owned.
   C. A comprehension of who is intended to benefit under the will.
   D. Legal age.

7. Assuming Massazona has a per capita statute, if Gramps dies, leaving no spouse, two living sons, and two grandchildren of a predeceased daughter, each grandchild will get what percentage of the estate?
   A. 25 percent?
   B. 33 percent?
   C. 16 percent?
   D. 10 percent?

8. Under the Uniform Probate Code, if a decedent leaves a surviving spouse, no children, and one surviving parent, the surviving spouse gets:
   A. All of the estate?
   B. None of the estate?
   C. Half the estate?
   D. Half the estate plus $50,000?

9. Fred dies leaving specific devise to Ethel of his house at 123 Spruce Street. At the time of his death he had already sold his house on Spruce and moved to Main Street. His devise has:
   A. Abated?
   B. Adeemed?
   C. Been inherited?
   D. Been gifted?

10. Which of the following does not act as a revocation of a will?
    A. Accidental destruction of the will?
    B. Intentional burning of the will?
    C. Intention obliteration of the will?
    D. Directing another person to tear up the will?

V. Answers to Objective Questions

Terms:

1. *Indemnity*. This means to restore the victim for a loss, in whole or in part, by payment, repair, or replacement. It is a means of transferring and distributing risk of loss.
2. *McCarran-Ferguson*. The federal government enacted this act with two objectives in mind: to allow the states to take primary control over the regulation of the insurance service industry and to exempt insurance companies from federal antitrust laws.
3. *Universal life*. This type of policy features elements of both term and whole life insurance. Most of these policies provide for a variable interest rate on the savings element of the contract.
4. *Fidelity*. Often these policies will call for certain procedures to be carried out by employers before the insurance is issued. For example, background and credit checks are commonly conducted on potential employees who are expected to be entrusted with the employer's property.
5. *Subrogation*. If an insurance company pays a claim to an insured for liability or property damage caused by a third party, the insurer succeeds to the right of the insured to recover from the third party.

6. *Intestate.* This term literally means "not spoken." Under such circumstances, the state steps in and prescribes who will receive the decedent's property under the terms of its intestate succession statute.
7. *Collateral.* This is a person who is not of the direct line of the deceased but rather from a collateral line such as a brother or sister.
8. *Ademption.* The original devise may have been withdrawn, given in life, or revoked. In any event, because it is no longer in existence at the time of death, the doctrine of ademption usually applies.
9. *Spendthrift.* These trusts are designed to protect the improvident against their wasteful actions. Most states allow these trusts in spite of the possible harm that may come to creditors who rely on the beneficiary's appearance of attachable wealth. England has curtailed the use of these sorts of trusts.
10. *Totten.* It is a tentative trust created by a bank account and is revocable at will until the trustor dies or makes a completed gift during life

True/False:

1. False. A broker is an independent contractor who can represent a number of companies.
2. False. The McCarran-Ferguson Act of 1945 exempts this from federal antitrust laws.
3. False. There are four parties: the insurance company, the owner of the policy, the person whose life is insured, and the beneficiary.
4. False. Because the insurer looks at the character of the insured in making its decision to issue the policy, the insurance company must approve assignments.
5. True. This practice is especially prevalent in situations that involve very high dollar loss exposure. Examples would include the insurance for a supertanker or satellite.
6. True. This is true, although more and more people are using videotaped and electronic wills to supplement written wills.
7. False. This is effective when the applicant receives the receipt from the machine.
8. True. These rules are generally designed to have the insured time to obtain alternative coverage in case of cancellation.
9. True. "Drive-other clauses" cover the insured while he or she is driving automobiles other than his or her own.
10. False. This protects against defects in title and liens or encumbrances that are not disclosed on the title insurance policy.

Multiple Choice:

1. B. Mere friendship does not constitute a recognized monetary or family based insurable interest.
2. D. A straight life policy provides coverage for the entire life of the insured and usually involves an element of savings.
3. C. Key person insurance may be taken out by a company to protect its pecuniary interest in a key executive's life. The other policies listed are not life insurance policies.
4. A. Professional malpractice is not normally covered by homeowners policies.
5. C. A, B, and D are all grounds for defense by the insurance company. The period of incontestability is a period after which the insurance company cannot raise certain defenses. Suicide after a certain period is a common example.
6. A. B, C, and D, are components of testamentary capacity that go to "sound mind" and legal age. Wealth is not a prerequisite for being able to execute a will.
7. A. 25 percent. Per capita gives surviving grandchildren the same standing as children of the deceased. Here the two surviving children and the two surviving grandchildren would all take equally at 25 percent each of the estate.
8. C. The surviving spouse gets half the estate.

9. The house on Spruce Street has been adeemed. It is not part of the estate at death, and the specific devise fails.
10. A. The accidental destruction of a will does not revoke it. The other methods listed all constitute a revocation.

VI. Answers to Essay Question

Both Enzo and Boy Racers Autoclub had the legal capacity to enter into insurance contracts. There is no evidence of any age or mental limitation basis for limiting their capacity to contract.

They each had the requisite insurance interest to purchase insurance in their respective contracts. Enzo had an economic property interest in his mean, green, driving machine (his auto). Boy Racers had an economic interest in insuring against liabilities that could arise out of staging such an event. In both cases, the insurance companies imposed definite key provisions with regard to certain conditions of coverage.

With regard to the auto insurance contract signed by Enzo, his actions were clearly in violation of the provisions of the policy. He is certainly free to engage in such events, but he could not expect to be covered in case of any mishaps traceable to engaging in auto sports. The accident as it occurred was not covered by the policy. To roll the car into a ditch and then claim coverage is even worse. This is insurance fraud and is punishable as a crime in addition to any civil consequences.

With regard to Boy Racers, their liability for aiding and assisting Enzo to commit a fraud on his insurance company would not likely be covered by their event liability policy. Most such policies do not cover illegal acts. In addition, they would also be criminally liable along with Enzo for insurance fraud.

Maybe they should all stick to simply dreaming about racing in Formula One. They will have plenty of time to dream if convicted.